PEARSON EDEXCEL INTERNATIONAL A LEVEL
FURTHER PURE MATHEMATICS 3
Student Book

Series Editors: Joe Skrakowski and Harry Smith

Authors: Greg Attwood, Jack Barraclough, Tom Begley, Ian Bettison, Lee Cope, Alistair Macpherson, Bronwen Moran, Johnny Nicholson, Laurence Pateman, Joe Petran, Keith Pledger, Joe Skrakowski, Harry Smith, Geoff Staley, Dave Wilkins

Published by Pearson Education Limited, 80 Strand, London, WC2R 0RL.

www.pearsonglobalschools.com

Copies of official specifications for all Pearson qualifications may be found on the website: https://qualifications.pearson.com

Text © Pearson Education Limited 2020
Edited by Linnet Bruce
Typeset by Tech-Set Ltd, Gateshead, UK
Original illustrations © Pearson Education Limited 2020
Illustrated by © Tech-Set Ltd, Gateshead, UK
Cover design by © Pearson Education Limited 2020

The rights of Greg Attwood, Jack Barraclough, Tom Begley, Ian Bettison, Lee Cope, Charles Garnet Cox, Alistair Macpherson, Bronwen Moran, Johnny Nicholson, Laurence Pateman, Joe Petran, Keith Pledger, Joe Skrakowski, Harry Smith, Geoff Staley and Dave Wilkins to be identified as the authors of this work have been asserted by them in accordance with the Copyright, Designs and Patents Act 1988.

First published 2020

23 22 21 20
10 9 8 7 6 5 4 3 2 1

British Library Cataloguing in Publication Data
A catalogue record for this book is available from the British Library

ISBN 978 1 292244 66 2

Copyright notice
All rights reserved. No part of this may be reproduced in any form or by any means (including photocopying or storing it in any medium by electronic means and whether or not transiently or incidentally to some other use of this publication) without the written permission of the copyright owner, except in accordance with the provisions of the Copyright, Designs and Patents Act 1988 or under the terms of a licence issued by the Copyright Licensing Agency, 5th Floor, Shackleton House, 4 Battlebridge Lane, London, SE1 2HX (www.cla.co.uk). Applications for the copyright owner's written permission should be addressed to the publisher.

Picture Credits
The authors and publisher would like to thank the following individuals and organisations for permission to reproduce photographs:

Shutterstock.com: spacedrone808 46; **Getty Images:** MediaProduction 1, Westend61 54, Science Photo Library 100, Abstract Aerial Art 137; **123rf.com:** destinacigdem 17

Cover images: *Front*: **Getty Images:** Werner Van Steen
Inside front cover: **Shutterstock.com:** Dmitry Lobanov

All other images © Pearson Education Limited 2020
All artwork © Pearson Education Limited 2020

Endorsement Statement
In order to ensure that this resource offers high-quality support for the associated Pearson qualification, it has been through a review process by the awarding body. This process confirms that this resource fully covers the teaching and learning content of the specification or part of a specification at which it is aimed. It also confirms that it demonstrates an appropriate balance between the development of subject skills, knowledge and understanding, in addition to preparation for assessment.

Endorsement does not cover any guidance on assessment activities or processes (e.g. practice questions or advice on how to answer assessment questions) included in the resource, nor does it prescribe any particular approach to the teaching or delivery of a related course.

While the publishers have made every attempt to ensure that advice on the qualification and its assessment is accurate, the official specification and associated assessment guidance materials are the only authoritative source of information and should always be referred to for definitive guidance.

Pearson examiners have not contributed to any sections in this resource relevant to examination papers for which they have responsibility.

Examiners will not use endorsed resources as a source of material for any assessment set by Pearson. Endorsement of a resource does not mean that the resource is required to achieve this Pearson qualification, nor does it mean that it is the only suitable material available to support the qualification, and any resource lists produced by the awarding body shall include this and other appropriate resources.

CONTENTS

COURSE STRUCTURE	**iv**
ABOUT THIS BOOK	**vi**
QUALIFICATION AND ASSESSMENT OVERVIEW	**viii**
EXTRA ONLINE CONTENT	**x**
1 HYPERBOLIC FUNCTIONS	**1**
2 FURTHER COORDINATE SYSTEMS	**17**
3 DIFFERENTIATION	**46**
4 INTEGRATION	**54**
REVIEW EXERCISE 1	**93**
5 VECTORS	**100**
6 FURTHER MATRIX ALGEBRA	**137**
REVIEW EXERCISE 2	**191**
EXAM PRACTICE	**199**
GLOSSARY	**201**
ANSWERS	**204**
INDEX	**244**

CHAPTER 1 HYPERBOLIC FUNCTIONS — 1
1.1 INTRODUCTION TO HYPERBOLIC FUNCTIONS — 2
1.2 SKETCHING GRAPHS OF HYPERBOLIC FUNCTIONS — 4
1.3 INVERSE HYPERBOLIC FUNCTIONS — 7
1.4 IDENTITIES AND EQUATIONS — 10
CHAPTER REVIEW 1 — 14

CHAPTER 2 FURTHER COORDINATE SYSTEMS — 17
2.1 ELLIPSES — 18
2.2 HYPERBOLAS — 20
2.3 ECCENTRICITY — 22
2.4 TANGENTS AND NORMALS TO AN ELLIPSE — 29
2.5 TANGENTS AND NORMALS TO A HYPERBOLA — 33
2.6 LOCI — 38
CHAPTER REVIEW 2 — 42

CHAPTER 3 DIFFERENTIATION — 46
3.1 DIFFERENTIATING HYPERBOLIC FUNCTIONS — 47
3.2 DIFFERENTIATING INVERSE HYPERBOLIC FUNCTIONS — 49
3.3 DIFFERENTIATING INVERSE TRIGONOMETRIC FUNCTIONS — 50
CHAPTER REVIEW 3 — 52

CHAPTER 4 INTEGRATION — 54
4.1 STANDARD INTEGRALS — 55
4.2 INTEGRATION — 58
4.3 TRIGONOMETRIC AND HYPERBOLIC SUBSTITUTIONS — 61
4.4 INTEGRATING EXPRESSIONS — 67
4.5 INTEGRATING INVERSE TRIGONOMETRIC AND HYPERBOLIC FUNCTIONS — 71
4.6 DERIVING AND USING REDUCTION FORMULAE — 73
4.7 FINDING THE LENGTH OF AN ARC OF A CURVE — 79
4.8 FINDING THE AREA OF A SURFACE OF REVOLUTION — 82
CHAPTER REVIEW 4 — 87

REVIEW EXERCISE 1 — 93

CHAPTER 5 VECTORS — 100
5.1 VECTOR PRODUCT — 101
5.2 FINDING AREAS — 106
5.3 SCALAR TRIPLE PRODUCT — 110
5.4 STRAIGHT LINES — 115
5.5 VECTOR PLANES — 117
5.6 SOLVING GEOMETRIC PROBLEMS — 121
CHAPTER REVIEW 5 — 130

CHAPTER 6 FURTHER MATRIX ALGEBRA 137

- 6.1 TRANSPOSING A MATRIX 138
- 6.2 THE DETERMINANT OF A 3 × 3 MATRIX 142
- 6.3 THE INVERSE OF A 3 × 3 MATRIX WHERE IT EXISTS 146
- 6.4 USING MATRICES TO REPRESENT LINEAR TRANSFORMATIONS IN 3 DIMENSIONS 152
- 6.5 USING INVERSE MATRICES TO REVERSE THE EFFECT OF A LINEAR TRANSFORMATION 160
- 6.6 THE EIGENVALUES AND EIGENVECTORS OF 2 × 2 AND 3 × 3 MATRICES 165
- 6.7 REDUCING A SYMMETRIC MATRIX TO DIAGONAL FORM 175

CHAPTER REVIEW 6 185

REVIEW EXERCISE 2 191

EXAM PRACTICE 199

GLOSSARY 201

ANSWERS 204

INDEX 244

ABOUT THIS BOOK

The following three themes have been fully integrated throughout the Pearson Edexcel International Advanced Level in Mathematics series, so they can be applied alongside your learning.

1. Mathematical argument, language and proof
- Rigorous and consistent approach throughout
- Notation boxes explain key mathematical language and symbols

2. Mathematical problem-solving
- Hundreds of problem-solving questions, fully integrated into the main exercises
- Problem-solving boxes provide tips and strategies
- Challenge questions provide extra stretch

The Mathematical Problem-Solving Cycle

specify the problem → collect information → process and represent information → interpret results → (cycle)

3. Transferable skills
- Transferable skills are embedded throughout this book, in the exercises and in some examples
- These skills are signposted to show students which skills they are using and developing

Finding your way around the book

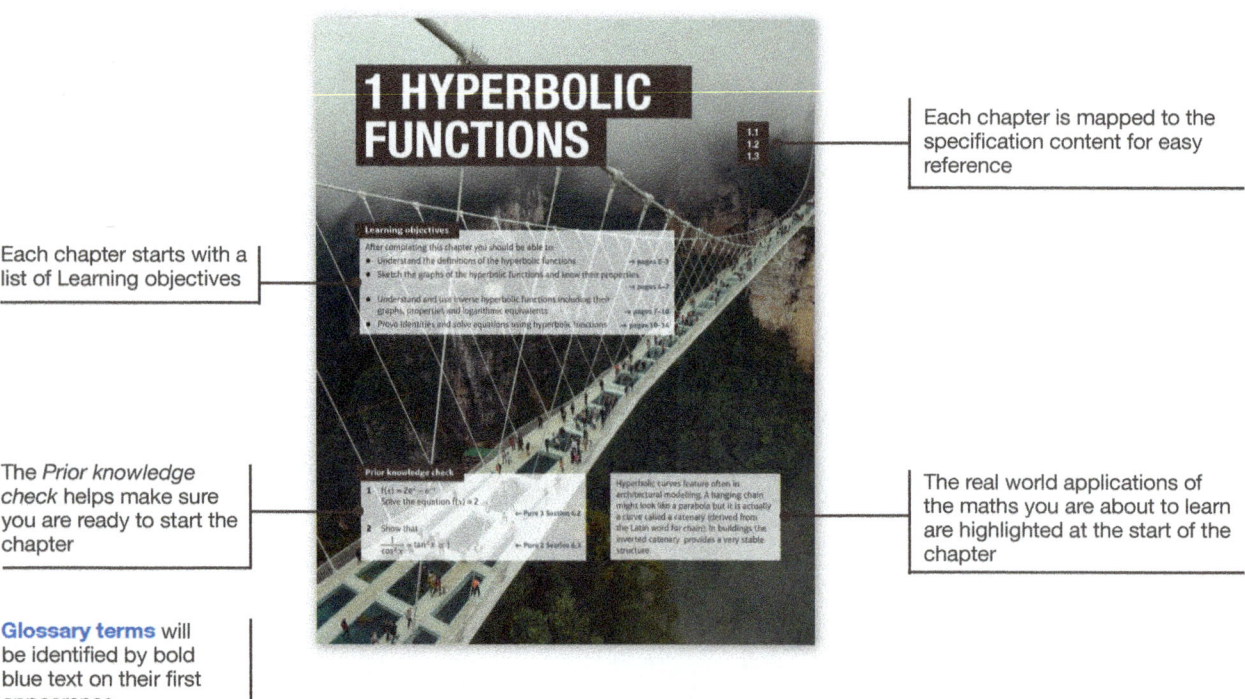

Each chapter is mapped to the specification content for easy reference

Each chapter starts with a list of Learning objectives

The *Prior knowledge check* helps make sure you are ready to start the chapter

The real world applications of the maths you are about to learn are highlighted at the start of the chapter

Glossary terms will be identified by bold blue text on their first appearance

ABOUT THIS BOOK

Step-by-step worked examples focus on the key types of questions you'll need to tackle

Exercise questions are carefully graded so they increase in difficulty and gradually bring you up to exam standard

Exercises are packed with exam-style questions to ensure you are ready for the exams

Exam-style questions are flagged with Ⓔ

Problem-solving questions are flagged with Ⓟ

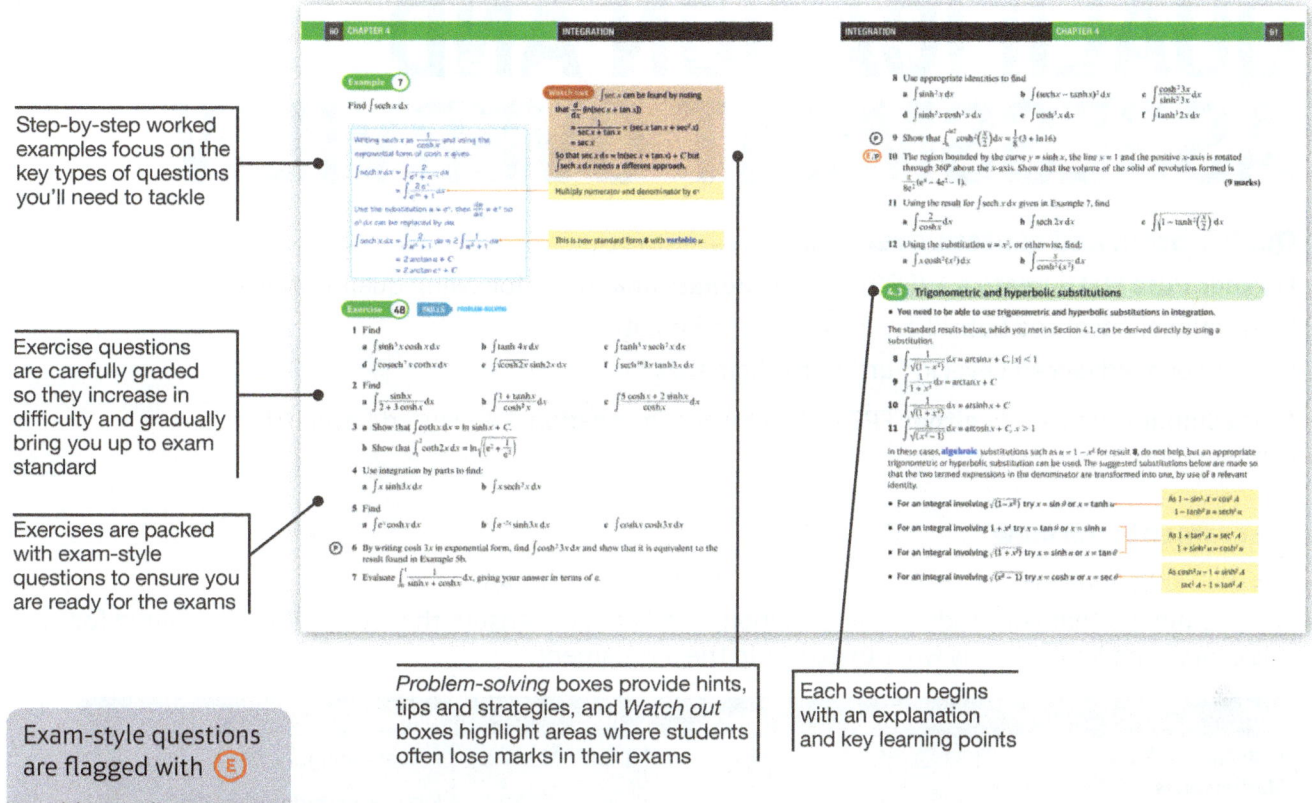

Problem-solving boxes provide hints, tips and strategies, and Watch out boxes highlight areas where students often lose marks in their exams

Each section begins with an explanation and key learning points

Each chapter ends with a Chapter review and a Summary of key points

After every few chapters, a Review exercise helps you consolidate your learning with lots of exam-style questions

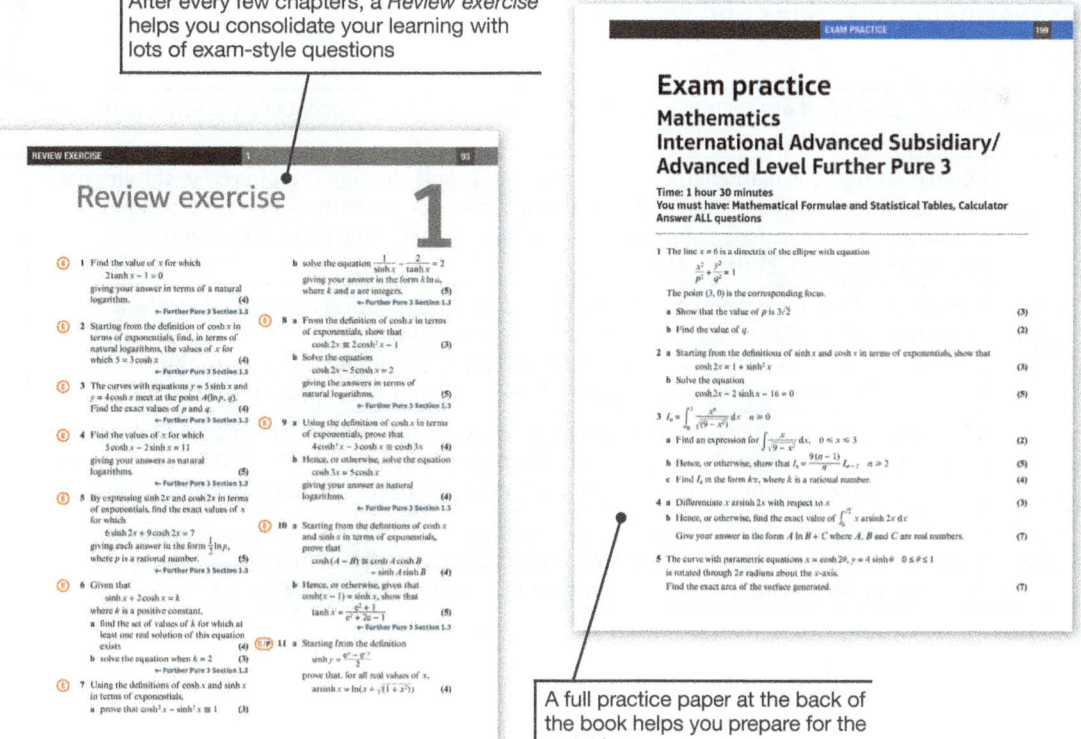

A full practice paper at the back of the book helps you prepare for the real thing

QUALIFICATION AND ASSESSMENT OVERVIEW

Qualification and content overview

Further Pure Mathematics 3 (FP3) is an **optional*** unit in the following qualifications:

International Advanced Subsidiary in Further Mathematics

International Advanced Level in Further Mathematics

*It is compulsory to study **either** FP2 **or** FP3 for the International Advanced Level in Further Mathematics.

Assessment overview

The following table gives an overview of the assessment for this unit.

We recommend that you study this information closely to help ensure that you are fully prepared for this course and know exactly what to expect in the assessment.

Unit	Percentage	Mark	Time	Availability
FP3: Further Pure Mathematics 3 Paper code WFM03/01	$33\frac{1}{3}$ % of IAS $16\frac{2}{3}$ % of IAL	75	1 hour 30 mins	January and June First assessment June 2020

IAS: International Advanced Subsidiary, IAL: International Advanced A Level.

Assessment objectives and weightings

		Minimum weighting in IAS and IAL
AO1	Recall, select and use their knowledge of mathematical facts, concepts and techniques in a variety of contexts.	30%
AO2	Construct rigorous mathematical arguments and proofs through use of precise statements, logical deduction and inference and by the manipulation of mathematical expressions, including the construction of extended arguments for handling substantial problems presented in unstructured form.	30%
AO3	Recall, select and use their knowledge of standard mathematical models to represent situations in the real world; recognise and understand given representations involving standard models; present and interpret results from such models in terms of the original situation, including discussion of the assumptions made and refinement of such models.	10%
AO4	Comprehend translations of common realistic contexts into mathematics; use the results of calculations to make predictions, or comment on the context; and, where appropriate, read critically and comprehend longer mathematical arguments or examples of applications.	5%
AO5	Use contemporary calculator technology and other permitted resources (such as formulae booklets or statistical tables) accurately and efficiently; understand when not to use such technology, and its limitations. Give answers to appropriate accuracy.	5%

Relationship of assessment objectives to units

FP3	Assessment objective				
	AO1	AO2	AO3	AO4	AO5
Marks out of 75	25–30	25–30	0–5	7–12	5–10
%	$33\frac{1}{3}$–40	$33\frac{1}{3}$–40	$6\frac{2}{3}$	$9\frac{1}{3}$–13	$6\frac{2}{3}$–$13\frac{1}{3}$

Calculators

Students may use a calculator in assessments for these qualifications. Centres are responsible for making sure that calculators used by their students meet the requirements given in the table below.

Students are expected to have available a calculator with at least the following keys: $+, -, \times, \div, \pi, x^2$, $\sqrt{x}, \frac{1}{x}, x^y, \ln x, e^x, x!$, sine, cosine and tangent and their inverses in degrees and decimals of a degree, and in radians; memory.

Prohibitions

Calculators with any of the following facilities are prohibited in all examinations:
- databanks
- retrieval of text or formulae
- built-in symbolic algebra manipulations
- symbolic differentiation and/or integration
- language translators
- communication with other machines or the internet

EXTRA ONLINE CONTENT

Extra online content

Whenever you see an *Online* box, it means that there is extra online content available to support you.

SolutionBank
SolutionBank provides worked solutions for questions in the book. Download the solutions as a PDF or quickly find the solution you need online.

Use of technology
Explore topics in more detail, visualise problems and consolidate your understanding. Use pre-made GeoGebra activities or Casio resources for a graphic calculator.

Online Find the point of intersection graphically using technology.

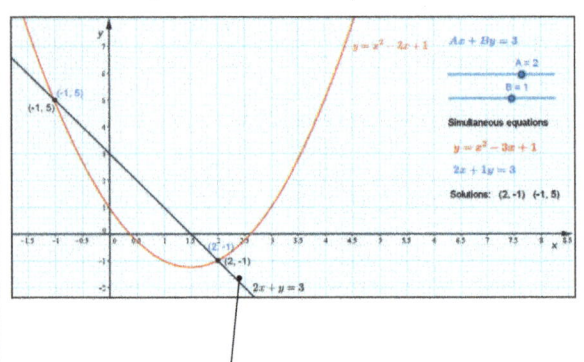

GeoGebra-powered interactives

Interact with the maths you are learning using GeoGebra's easy-to-use tools

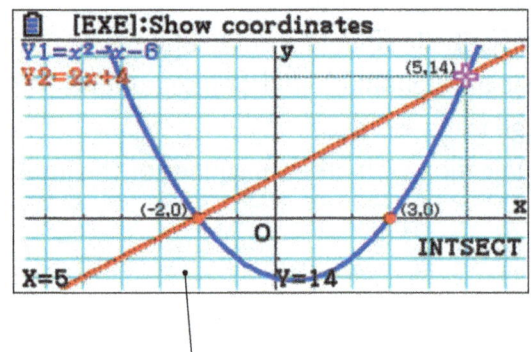

Graphic calculator interactives

Explore the maths you are learning and gain confidence in using a graphic calculator

Calculator tutorials
Our helpful video tutorials will guide you through how to use your calculator in the exams. They cover both Casio's scientific and colour graphic calculators.

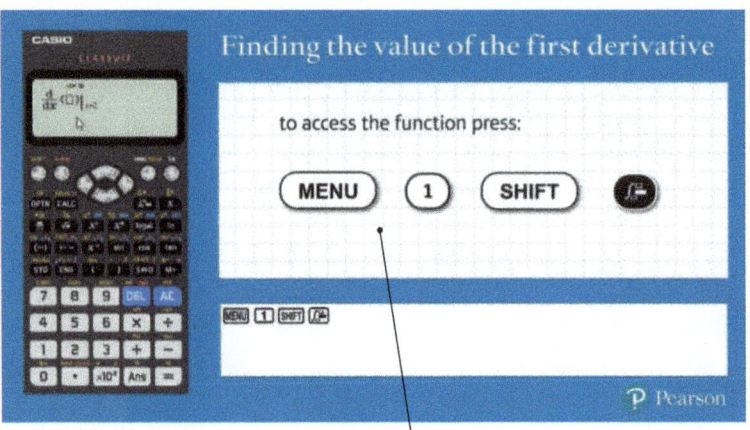

Online Work out each coefficient quickly using the ^nC_r and power functions on your calculator.

Step-by-step guide with audio instructions on exactly which buttons to press and what should appear on your calculator's screen

1 HYPERBOLIC FUNCTIONS

1.1
1.2
1.3

Learning objectives

After completing this chapter you should be able to:
- Understand the definitions of the hyperbolic functions → pages 2–3
- Sketch the graphs of the hyperbolic functions and know their properties → pages 4–7
- Understand and use inverse hyperbolic functions including their graphs, properties and logarithmic equivalents → pages 7–10
- Prove identities and solve equations using hyperbolic functions → pages 10–14

Prior knowledge check

1 $f(x) = 2e^x - e^{-x}$
   Solve the equation $f(x) = 2$
   ← Pure 3 Section 4.2

2 Show that
   $\dfrac{1}{\cos^2 x} - \tan^2 x \equiv 1$
   ← Pure 2 Section 6.3

Hyperbolic curves feature often in architectural modelling. A hanging chain might look like a parabola but it is actually a curve called a catenary (derived from the Latin word for chain). In buildings the inverted catenary provides a very stable structure.

1.1 Introduction to hyperbolic functions

Hyperbolic functions have several properties in common with trigonometric **functions**, but they are defined in terms of **exponential** functions.

> **Notation** x belongs to the set of real numbers, using the correct mathematical notation.

- Hyperbolic **sine** (or sinh) is defined as $\sinh x \equiv \dfrac{e^x - e^{-x}}{2}$

> **Notation** Often pronounced 'shine'.

- Hyperbolic **cosine** (or cosh) is defined as $\cosh x \equiv \dfrac{e^x + e^{-x}}{2}$

> **Notation** Often pronounced 'cosh'.

- Hyperbolic **tangent** (or tanh) is defined as $\tanh x \equiv \dfrac{\sinh x}{\cosh x}$

> **Notation** Often pronounced 'tanch' or 'than'.

You can use the definitions of $\sinh x$ and $\cosh x$ to write $\tanh x$ in exponential form.

$$\tanh x \equiv \frac{\sinh x}{\cosh x} \equiv \frac{e^x - e^{-x}}{2} \times \frac{2}{e^x + e^{-x}} \equiv \frac{e^x - e^{-x}}{e^x + e^{-x}}$$

Multiplying the numerator and denominator of the final expression through by e^x gives:

- $\tanh x \equiv \dfrac{e^{2x} - 1}{e^{2x} + 1}$

There are also hyperbolic functions corresponding to (i.e. connected to) the **reciprocal** trigonometric functions:

- Hyperbolic **cosecant** (or cosech) is defined as $\operatorname{cosech} x \equiv \dfrac{1}{\sinh x} \equiv \dfrac{2}{e^x - e^{-x}}$

> **Notation** Often pronounced 'cosech' or 'cosheck'.

- Hyperbolic **secant** (or sech) is defined as $\operatorname{sech} x \equiv \dfrac{1}{\cosh x} \equiv \dfrac{2}{e^x + e^{-x}}$

> **Notation** Often pronounced 'sheck' or 'setch'.

- Hyperbolic **cotangent** (or coth) is defined as $\coth x \equiv \dfrac{1}{\tanh x} \equiv \dfrac{e^{2x} + 1}{e^{2x} - 1}$

> **Notation** Often pronounced 'coth'.

Example 1 SKILLS ANALYSIS

Find, to 2 decimal places, the values of

a $\sinh 3$ **b** $\cosh 1$ **c** $\tanh 0.8$

a $\sinh 3 = \dfrac{e^3 - e^{-3}}{2} = 10.02$ (2 d.p.)

b $\cosh 1 = \dfrac{e^1 + e^{-1}}{2} = 1.54$ (2 d.p.)

c $\tanh 0.8 = \dfrac{e^{1.6} - 1}{e^{1.6} + 1} = 0.66$ (2 d.p.)

HYPERBOLIC FUNCTIONS CHAPTER 1

Example 2

Find the exact value of $\tanh(\ln 4)$.

$$\tanh(\ln 4) = \frac{e^{2\ln 4} - 1}{e^{2\ln 4} + 1} = \frac{e^{\ln 4^2} - 1}{e^{\ln 4^2} + 1} = \frac{e^{\ln 16} - 1}{e^{\ln 16} + 1}$$

$$= \frac{16 - 1}{16 + 1} = \frac{15}{17}$$

— Use $e^{\ln k} = k$.

Example 3

Use the definition of $\sinh x$ to find, to 2 decimal places, the value of x for which $\sinh x = 5$

$$\frac{e^x - e^{-x}}{2} = 5 \Rightarrow e^x - e^{-x} = 10$$

$e^{2x} - 1 = 10e^x$ — Multiply both sides by e^x.

$e^{2x} - 10e^x - 1 = 0$ — The substitution $u = e^x$ turns this into the quadratic equation $u^2 - 10u - 1 = 0$

$e^x = 5 \pm \sqrt{26}$

$\Rightarrow e^x = 5 + \sqrt{26}$ — e^x cannot be negative.

So $x = \ln(5 + \sqrt{26}) = 2.31$ (2 d.p.)

Exercise 1A SKILLS ANALYSIS

1 Use your calculator to find, to 2 decimal places, the value of:

   a $\sinh 4$ **b** $\cosh\left(\frac{1}{2}\right)$ **c** $\tanh(-2)$ **d** $\text{sech}\, 5$

2 Write, in terms of e:

   a $\sinh 1$ **b** $\cosh 4$ **c** $\tanh 0.5$ **d** $\text{sech}(-1)$

3 Find the exact values of:

   a $\sinh(\ln 2)$ **b** $\cosh(\ln 3)$ **c** $\tanh(\ln 2)$ **d** $\text{cosech}(\ln \pi)$

In questions **4** to **8**, use the definitions of the hyperbolic functions (in terms of exponentials) to find each answer, then check your answers using an inverse hyperbolic function on your calculator.

4 Find, to 2 decimal places, the values of x for which $\cosh x = 2$

5 Find, to 2 decimal places, the values of x for which $\sinh x = 1$

6 Find, to 2 decimal places, the values of x for which $\tanh x = -\frac{1}{2}$

7 Find, to 2 decimal places, the values of x for which $\coth x = 10$

8 Find, to 2 decimal places, the values of x for which $\text{sech}\, x = \frac{1}{8}$

1.2 Sketching graphs of hyperbolic functions

You can sketch the graphs of the hyperbolic functions by considering the graphs of $y = e^x$ and $y = e^{-x}$

$$\sinh x = \frac{e^x - e^{-x}}{2} = \frac{e^x + (-e^{-x})}{2}$$

so the graph of $y = \sinh x$ is the 'average' of the graphs of $y = e^x$ and $y = -e^{-x}$

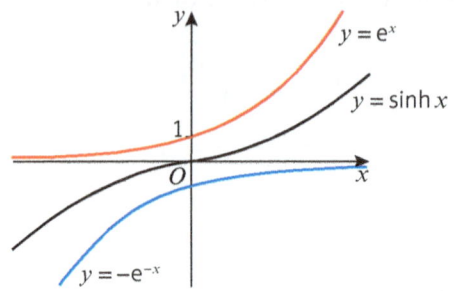

For the graph of $y = \sinh x$

- when x is large and positive, e^{-x} is small, so $\sinh x \approx \frac{1}{2}e^x$
- when x is large and negative, e^x is small, so $\sinh x \approx -\frac{1}{2}e^{-x}$
- **For any value a, $\sinh(-a) = -\sinh a$**

Notation $f(x) = \sinh x$ is an **odd** function since $f(-x) = -f(x)$

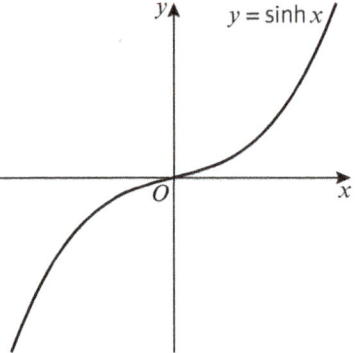

Consider the graphs of $y = e^x$ and $y = e^{-x}$

$$\cosh x = \frac{e^x + e^{-x}}{2}$$

so the graph of $y = \cosh x$ is the 'average' of the graphs of $y = e^x$ and $y = e^{-x}$

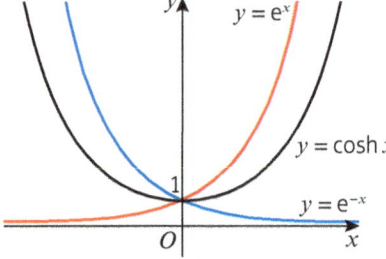

For the graph of $y = \cosh x$

- when x is large and positive, e^{-x} is small, so $\cosh x \approx \frac{1}{2}e^x$
- when x is large and negative, e^x is small, so $\cosh x \approx \frac{1}{2}e^{-x}$
- **For any value a, $\cosh(-a) = \cosh a$**

Notation $f(x) = \cosh x$ is an **even** function because $f(-x) = f(x)$

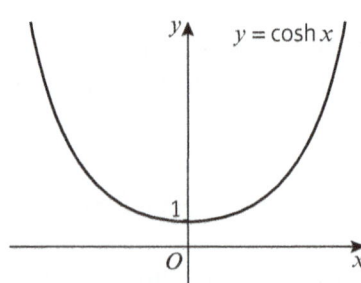

Example 4 — SKILLS: CRITICAL THINKING

Sketch the graph of $y = \tanh x$

$\tanh x = \dfrac{\sinh x}{\cosh x}$

When $x = 0$, $\tanh x = \dfrac{0}{1} = 0$

When x is large and positive, $\sinh x \approx \dfrac{1}{2}e^x$ and $\cosh x \approx \dfrac{1}{2}e^x$, so $\tanh x \approx 1$.

When x is large and negative, $\sinh x \approx -\dfrac{1}{2}e^{-x}$ and $\cosh x \approx \dfrac{1}{2}e^{-x}$, so $\tanh x \approx -1$.

As $x \to \infty$, $\tanh x \to 1$ and as $x \to -\infty$, $\tanh x \to -1$

For $f(x) = \tanh x$, $x \in \mathbb{R}$, the range of f is $-1 < f(x) < 1$

$y = -1$ and $y = 1$ are asymptotes to the curve.

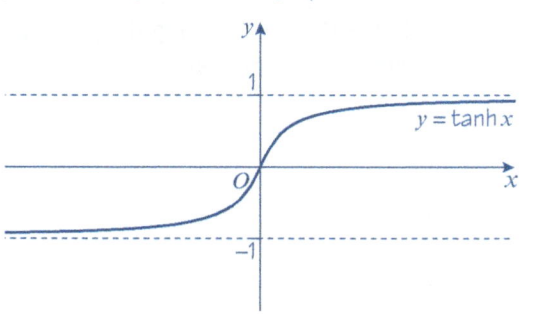

Online Explore graphs of hyberbolic functions using GeoGebra.

Consider the graphs of $y = \sinh x$ and $y = \cosh x$ to work out the behaviour of $y = \tanh x$ as $x \to \infty$ and $x \to -\infty$

You should always include any **asymptotes** on a sketch graph.

Example 5 — SKILLS: CRITICAL THINKING

Sketch the graph of $y = \operatorname{sech} x$

Using $\operatorname{sech} x = \dfrac{1}{\cosh x}$

When $x = 0$, $\operatorname{sech} x = \dfrac{1}{1} = 1$

As $x \to \infty$, $\cosh x \to \infty$, so $\operatorname{sech} x \to 0$

As $x \to -\infty$, $\cosh x \to \infty$, so $\operatorname{sech} x \to 0$

The x-axis is an asymptote to the curve.

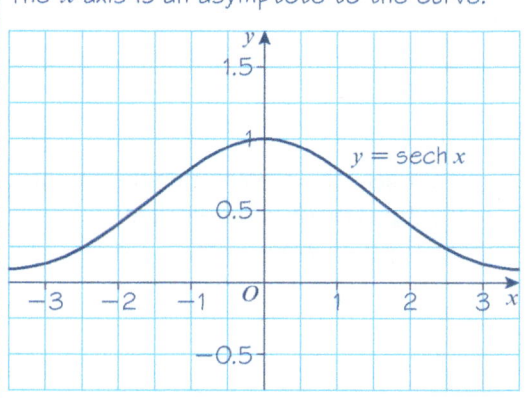

Check this sketch using the graphic function on your calculator.

Example 6 — SKILLS: CRITICAL THINKING

Sketch the graph of $y = \operatorname{cosech} x$, $x \neq 0$

Using $\operatorname{cosech} x = \dfrac{1}{\sinh x}$

For positive x, as $x \to 0$, $\operatorname{cosech} x \to \infty$

For negative x, as $x \to 0$, $\operatorname{cosech} x \to -\infty$

As $x \to \infty$, $\sinh x \to \infty$, so $\operatorname{cosech} x \to 0$

As $x \to -\infty$, $\sinh x \to -\infty$, so $\operatorname{cosech} x \to 0$

The x- and y-axes are asymptotes to the curve.

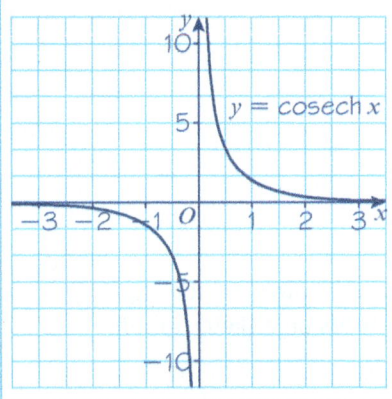

Check this sketch using the graphic function on your calculator.

Example 7 — SKILLS: CRITICAL THINKING

Sketch the graph of $y = \coth x$, $x \neq 0$.

Using $\coth x = \dfrac{1}{\tanh x}$

For positive x, as $x \to 0$, $\coth x \to \infty$

For negative x, as $x \to 0$, $\coth x \to -\infty$

As $x \to \infty$, $\tanh x \to 1$, so $\coth x \to 1$

As $x \to -\infty$, $\tanh x \to -1$, so $\coth x \to -1$

The y-axis is an asymptote to the curve.

$y = -1$ and $y = 1$ are asymptotes to the curve.

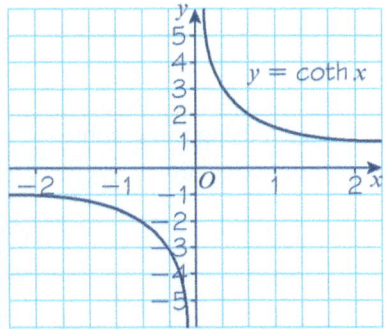

Check this sketch using the graphic function on your calculator.

Exercise 1B

1. On the same diagram, sketch the graphs of $y = \cosh 2x$ and $y = 2\cosh x$

2. **a** On the same diagram, sketch the graphs of $y = \operatorname{sech} x$ and $y = \sinh x$
   b Show that, at the point of intersection of the graphs, $x = \frac{1}{2}\ln(2 = \sqrt{5})$

3. Find the range of each hyberbolic function.
   a $f(x) = \sinh x, x \in \mathbb{R}$
   b $f(x) = \cosh x, x \in \mathbb{R}$
   c $f(x) = \tanh x, x \in \mathbb{R}$
   d $f(x) = \operatorname{sech} x, x \in \mathbb{R}$
   e $f(x) = \operatorname{cosech} x, x \in \mathbb{R}, x \neq 0$
   f $f(x) = \coth x, x \in \mathbb{R}, x \neq 0$

4. **a** Sketch the graph of $y = 1 + \coth x, x \in \mathbb{R}, x \neq 0$
   b Write down the equations of the asymptotes to this curve.

5. **a** Sketch the graph of $y = 3\tanh x, x \in \mathbb{R}$
   b Write down the equations of the asymptotes to this curve.

Challenge Sketch the graph of $y = \sinh x + \cosh x$

1.3 Inverse hyperbolic functions

You can define and use the inverses of the hyperbolic functions.

If $f(x) = \sinh x$, the inverse function f^{-1} is called $\operatorname{arsinh} x$.

The graph of $y = \operatorname{arsinh} x$ is the reflection of the graph of $y = \sinh x$ in the line $y = x$.

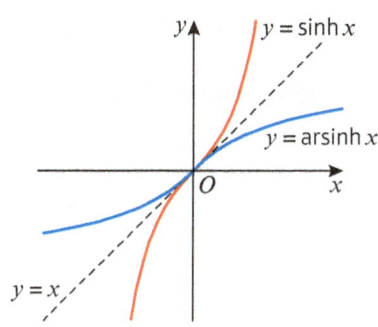

The inverse of a function is defined only if the function is one-to-one, so for $\cosh x$ the **domain** must be restricted in order to define an inverse.

For $f(x) = \cosh x, x \geq 0$, $f^{-1}(x) = \operatorname{arcosh} x, x \geq 1$

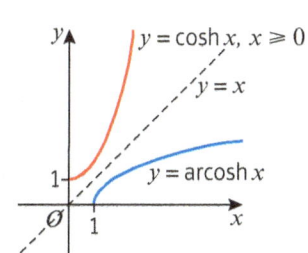

- The following table shows the inverse hyperbolic functions, with domains restricted where necessary.

Hyperbolic function	Inverse hyperbolic function
$y = \sinh x$	$y = \text{arsinh } x$
$y = \cosh x, x \geq 0$	$y = \text{arcosh } x, x \geq 1$
$y = \tanh x$	$y = \text{artanh } x, |x| < 1$
$y = \text{sech } x, x \geq 0$	$y = \text{arsech } x, 1 < x \leq 1$
$y = \text{cosech } x, x \neq 0$	$y = \text{arcosech } x, x \neq 0$
$y = \coth x, x \neq 0$	$y = \text{arcoth } x, |x| > 1$

Notation arsinh, arcosh and artanh are sometimes written as $\sinh^{-1}$, $\cosh^{-1}$ and $\tanh^{-1}$.

Example SKILLS REASONING/CRITICAL THINKING

Sketch the graph of $y = \text{arcoth } x, |x| > 1$

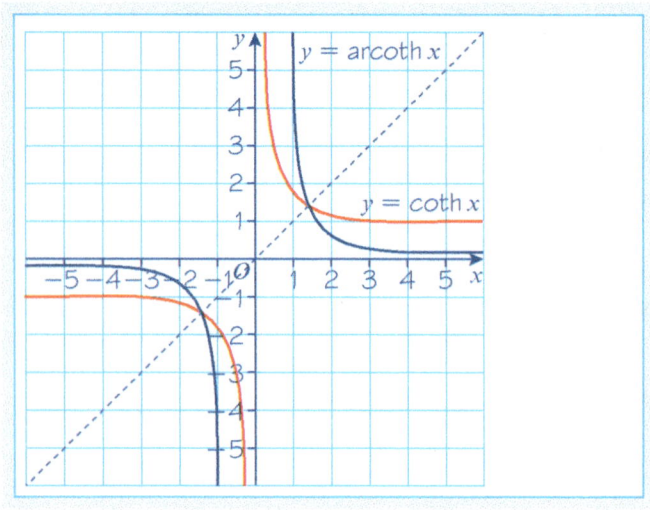

Reflect the graph of $y = \coth x$ in the line $y = x$

You can express the inverse hyperbolic functions in terms of **natural logarithms**.

Example SKILLS ANALYSIS

Show that $\text{arsinh } x = \ln\left(x + \sqrt{x^2 + 1}\right)$

Let $y = \text{arsinh } x$
$x = \sinh y$
$x = \dfrac{e^y - e^{-y}}{2}$ — Use the definition of sinh.
$e^y - e^{-y} = 2x$
$e^{2y} - 1 = 2xe^y$ — Multiply by e^y
$e^{2y} - 2xe^y - 1 = 0$
$(e^y - x)^2 - x^2 - 1 = 0$
$e^y = x \pm \sqrt{x^2 + 1}$
$e^y = x - \sqrt{x^2 + 1}$ can be ignored since $\sqrt{x^2 + 1} > x$, and would give a negative value of e^y, which is not possible.
So $e^y = x + \sqrt{x^2 + 1}$
$y = \ln(x + \sqrt{x^2 + 1}) \Rightarrow \text{arsinh } x = \ln(x + \sqrt{x^2 + 1})$

Problem-solving

$e^{2y} - 2xe^y - 1 = 0$ is a quadratic in e^y. You can write it as $(e^y)^2 - 2xe^y - 1 = 0$ and then complete the square.

Example 10

Show that $\operatorname{arcosh} x = \ln(x + \sqrt{x^2 - 1})$, $x \geqslant 1$

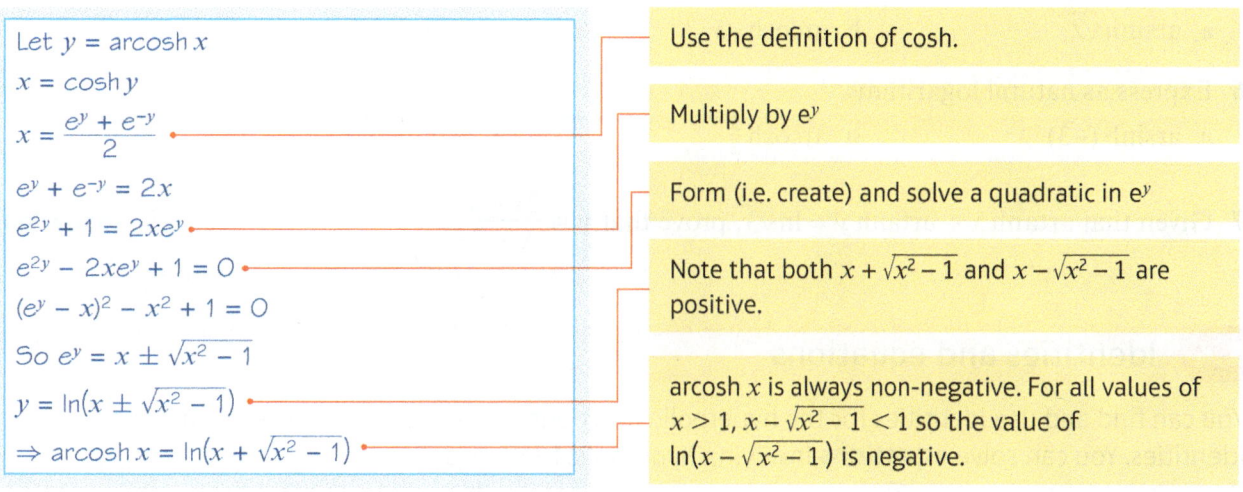

- Use the definition of cosh.
- Multiply by e^y
- Form (i.e. create) and solve a quadratic in e^y
- Note that both $x + \sqrt{x^2 - 1}$ and $x - \sqrt{x^2 - 1}$ are positive.
- arcosh x is always non-negative. For all values of $x > 1$, $x - \sqrt{x^2 - 1} < 1$ so the value of $\ln(x - \sqrt{x^2 - 1})$ is negative.

Let $y = \operatorname{arcosh} x$
$x = \cosh y$
$x = \dfrac{e^y + e^{-y}}{2}$
$e^y + e^{-y} = 2x$
$e^{2y} + 1 = 2xe^y$
$e^{2y} - 2xe^y + 1 = 0$
$(e^y - x)^2 - x^2 + 1 = 0$
So $e^y = x \pm \sqrt{x^2 - 1}$
$y = \ln(x \pm \sqrt{x^2 - 1})$
$\Rightarrow \operatorname{arcosh} x = \ln(x + \sqrt{x^2 - 1})$

You can use a similar method to express artanh x in terms of natural logarithms.

The following formulae are provided in the formula booklet and can be used directly unless you are asked to prove them.

- $\operatorname{arsinh} x = \ln(x + \sqrt{x^2 + 1})$
- $\operatorname{arcosh} x = \ln(x + \sqrt{x^2 - 1})$, $x \geqslant 1$
- $\operatorname{artanh} x = \dfrac{1}{2}\ln\left(\dfrac{1 + x}{1 - x}\right)$, $|x| < 1$

Example 11

Express as natural logarithms

a arsinh 1 **b** arcosh 2 **c** artanh $\dfrac{1}{3}$

a $\operatorname{arsinh} 1 = \ln(1 + \sqrt{1^2 + 1}) = \ln(1 + \sqrt{2})$

b $\operatorname{arcosh} 2 = \ln(2 + \sqrt{2^2 - 1}) = \ln(2 + \sqrt{3})$

c $\operatorname{artanh} \dfrac{1}{3} = \dfrac{1}{2}\ln\left(\dfrac{1 + \frac{1}{3}}{1 - \frac{1}{3}}\right) = \dfrac{1}{2}\ln 2 = \ln\sqrt{2}$ — Use $a \ln x = \ln x^a$

Exercise 1C   SKILLS   ANALYSIS

1 Sketch the graph of $y = \operatorname{artanh} x$, $|x| < 1$

(P) 2 Sketch the graph of $y = \operatorname{arcosech} x$, $0 < x < 1$

(E/P) 3 Prove that $\operatorname{artanh} x = \dfrac{1}{2}\ln\left(\dfrac{1 + x}{1 - x}\right)$, $|x| < 1$ **(5 marks)**

4 Express as natural logarithms

  a arsinh 2   **b** arcosh 3   **c** artanh $\frac{1}{2}$

5 Express as natural logarithms

  a arsinh $\sqrt{2}$ **b** arcosh $\sqrt{5}$ **c** artanh 0.1

6 Express as natural logarithms

  a arsinh (−3)   **b** arcosh $\frac{3}{2}$ **c** artanh $\frac{1}{\sqrt{3}}$

 7 Given that artanh x + artanh $y = \ln\sqrt{3}$, prove that $y = \dfrac{2x - 1}{x - 2}$ **(6 marks)**

1.4 Identities and equations

You can find and use identities for the hyperbolic functions that are similar to the trigonometric identities. You can solve **equations** involving hyperbolic functions.

Example 12 SKILLS REASONING/ARGUMENTATION

Prove that $\cosh^2 A - \sinh^2 A \equiv 1$

$$\text{LHS} \equiv \cosh^2 A - \sinh^2 A \equiv \left(\frac{e^A + e^{-A}}{2}\right)^2 - \left(\frac{e^A - e^{-A}}{2}\right)^2$$

$$\equiv \left(\frac{e^{2A} + 2 + e^{-2A}}{4}\right) - \left(\frac{e^{2A} - 2 + e^{-2A}}{4}\right)$$

$$\equiv \frac{4}{4} \equiv 1 \equiv \text{RHS}$$

Use the definitions of cosh and sinh.

$e^A \times e^{-A} = \dfrac{e^A}{e^A} = 1$

- $\cosh^2 A - \sinh^2 A \equiv 1$ (this formula is given to you in the formula booklet)

Example 13

Prove that $\sinh(A + B) \equiv \sinh A \cosh B + \cosh A \sinh B$

$$\text{RHS} \equiv \sinh A \cosh B + \cosh A \sinh B$$

$$\equiv \left(\frac{e^A - e^{-A}}{2}\right)\left(\frac{e^B + e^{-B}}{2}\right) + \left(\frac{e^A + e^{-A}}{2}\right)\left(\frac{e^B - e^{-B}}{2}\right)$$

$$\equiv \left(\frac{e^{A+B} + e^{A-B} - e^{-A+B} - e^{-A-B}}{4}\right) + \left(\frac{e^{A+B} - e^{A-B} + e^{-A+B} - e^{-A-B}}{4}\right)$$

$$\equiv \frac{2e^{A+B} - 2e^{-A-B}}{4} \equiv \frac{e^{A+B} - e^{-(A+B)}}{2} \equiv \sinh(A + B) \equiv \text{LHS}$$

You can prove other sinh and cosh addition formulae similarly, giving:

- $\sinh(A \pm B) \equiv \sinh A \cosh B \pm \cosh A \sinh B$
- $\cosh(A \pm B) \equiv \cosh A \cosh B \pm \sinh A \sinh B$

Example 14

Prove that $\cosh 2A \equiv 1 + 2\sinh^2 A$

$$\text{RHS} \equiv 1 + 2\sinh^2 A$$
$$\equiv 1 + 2\left(\frac{e^A - e^{-A}}{2}\right)\left(\frac{e^A - e^{-A}}{2}\right)$$
$$\equiv 1 + 2\left(\frac{e^{2A} - 2 + e^{-2A}}{4}\right) \equiv 1 - 1 + \left(\frac{e^{2A} + e^{-2A}}{2}\right)$$
$$\equiv \cosh 2A \equiv \text{LHS}$$

Use the definition of sinh.

Given a trigonometric **identity**, it is generally possible to write down the corresponding hyperbolic identity using what is known as **Osborn's rule**:

- Replace cos by cosh: $\cos A \rightarrow \cosh A$
- Replace sin by sinh: $\sin A \rightarrow \sinh A$

However …

- replace any product (or **implied** product) of two sin terms by **minus** the product of two sinh terms:

  e.g. $\sin A \sin B \rightarrow -\sinh A \sinh B$

  $\tan^2 A \rightarrow -\tanh^2 A$

This is the implied product of two sin terms because $\tan^2 A \equiv \dfrac{\sin^2 A}{\cos^2 A}$

Example 15 SKILLS ANALYSIS

Write down the hyperbolic identity corresponding to

a $\cos 2A \equiv 2\cos^2 A - 1$ **b** $\tan(A - B) \equiv \dfrac{\tan A - \tan B}{1 + \tan A \tan B}$

a $\cosh 2A \equiv 2\cosh^2 A - 1$

b $\tanh(A - B) \equiv \dfrac{\tanh A - \tanh B}{1 - \tanh A \tanh B}$

Implied product of two sin terms because $\tan A \tan B \equiv \dfrac{\sin A \sin B}{\cos A \cos B}$

Example 16

Given that $\sinh x = \frac{3}{4}$, find the exact value of

a $\cosh x$ **b** $\tanh x$ **c** $\sinh 2x$

a Using $\cosh^2 x - \sinh^2 x \equiv 1$,

$\cosh^2 x - \dfrac{9}{16} = 1 \Rightarrow \cosh^2 x = \dfrac{25}{16}$

$\Rightarrow \cosh x = \dfrac{5}{4}$

b Using $\tanh x \equiv \dfrac{\sinh x}{\cosh x}$

$\tanh x = \dfrac{3}{4} \div \dfrac{5}{4} = \dfrac{3}{5}$

c Using $\sinh 2x \equiv 2\sinh x \cosh x$

$\sinh 2x = 2 \times \dfrac{3}{4} \times \dfrac{5}{4} = \dfrac{15}{8}$

$\cosh x \geq 1$, so $\cosh x = -\dfrac{5}{4}$ is not possible.

Example 17

Solve $6\sinh x - 2\cosh x = 7$ for real values of x.

$6\left(\dfrac{e^x - e^{-x}}{2}\right) - 2\left(\dfrac{e^x + e^{-x}}{2}\right) = 7$

$3e^x - 3e^{-x} - e^x - e^{-x} = 7$

$2e^x - 7 - 4e^{-x} = 0$

$2e^{2x} - 7e^x - 4 = 0$

$(2e^x + 1)(e^x - 4) = 0$

$e^x = -\dfrac{1}{2},\ e^x = 4$

$e^x = 4$

$x = \ln 4$

There is no hyperbolic identity that will easily transform the equation into an equation in just one hyperbolic function, so use the basic definitions.

There are no real values of x for which $e^x = -\dfrac{1}{2}$.

Example 18

Solve $2\cosh^2 x - 5\sinh x = 5$, giving your answers as natural logarithms.

Using $\cosh^2 x - \sinh^2 x \equiv 1$,

$2(1 + \sinh^2 x) - 5\sinh x = 5$

$2\sinh^2 x - 5\sinh x - 3 = 0$

$(2\sinh x + 1)(\sinh x - 3) = 0$

So $\sinh x = -\dfrac{1}{2}$ or $\sinh x = 3$

Then,

$x = \operatorname{arsinh}\left(-\dfrac{1}{2}\right)$ or $x = \operatorname{arsinh} 3$

$\Rightarrow x = \ln\left(-\dfrac{1}{2} + \sqrt{\dfrac{1}{4} + 1}\right)$ or $x = \ln(3 + \sqrt{9+1})$

$\Rightarrow x = \ln\left(-\dfrac{1}{2} + \dfrac{\sqrt{5}}{2}\right)$ or $x = \ln(3 + \sqrt{10})$

Use this identity to transform the equation into an equation in just one hyperbolic function.

Use $\operatorname{arsinh} x = \ln(x + \sqrt{x^2 + 1})$.

Example 19

Solve $\cosh 2x - 5\cosh x + 4 = 0$, giving your answers as natural logarithms where appropriate.

Using $\cosh 2x \equiv 2\cosh^2 x - 1$,

$2\cosh^2 x - 1 - 5\cosh x + 4 = 0$

$2\cosh^2 x - 5\cosh x + 3 = 0$

$(2\cosh x - 3)(\cosh x - 1) = 0$

So $\cosh x = \dfrac{3}{2}$ or $\cosh x = 1$

$\Rightarrow x = \ln\left(\dfrac{3}{2} \pm \sqrt{\dfrac{9}{4} - 1}\right)$ or $x = 0$

$\Rightarrow x = \ln\left(\dfrac{3}{2} \pm \dfrac{\sqrt{5}}{2}\right)$ or $x = 0$

Use this identity to transform the equation into an equation in just one hyperbolic function.

You can use $\operatorname{arcosh} x = \ln(x + \sqrt{x^2 - 1})$, but remember that both $\ln(x + \sqrt{x^2 - 1})$ and $\ln(x - \sqrt{x^2 - 1})$ are possible.

For any value of k greater than 1, $\cosh x = k$ will give two values of x, one positive and one negative.

HYPERBOLIC FUNCTIONS   CHAPTER 1

Exercise 1D   **SKILLS**   REASONING/ARGUMENTATION

1 Prove the following identities, using the definitions of $\sinh x$ and $\cosh x$.

   a $\sinh 2A \equiv 2 \sinh A \cosh A$

   b $\cosh(A - B) \equiv \cosh A \cosh B - \sinh A \sinh B$

   c $\cosh 3A \equiv 4\cosh^3 A - 3 \cosh A$

   d $\sinh A - \sinh B \equiv 2 \sinh\left(\dfrac{A - B}{2}\right) \cosh\left(\dfrac{A + B}{2}\right)$

2 Use Osborn's rule to write down the hyperbolic identities corresponding to the following trigonometric identities.

   a $\sin(A - B) \equiv \sin A \cos B - \cos A \sin B$

   b $\sin 3A \equiv 3 \sin A - 4 \sin^3 A$

   c $\cos A + \cos B \equiv 2 \cos\left(\dfrac{A + B}{2}\right) \cos\left(\dfrac{A - B}{2}\right)$

   d $\cos 2A \equiv \dfrac{1 - \tan^2 A}{1 + \tan^2 A}$

   e $\cos 2A \equiv \cos^4 A - \sin^4 A$

3 Given that $\cosh x = 2$, find the exact values of

   a $\sinh x$ **b** $\tanh x$ **c** $\cosh 2x$

4 Given that $\sinh x = -1$, find the exact values of

   a $\cosh x$ **b** $\sinh 2x$ **c** $\tanh 2x$

5 Solve the following equations, giving your answers as natural logarithms.

   a $3 \sinh x + 4 \cosh x = 4$

   b $7 \sinh x - 5 \cosh x = 1$

   c $30 \cosh x = 15 + 26 \sinh x$

   d $13 \sinh x - 7 \cosh x + 1 = 0$

   e $\cosh 2x - 5 \sinh x = 13$

   f $3 \sinh^2 x - 13 \cosh x + 7 = 0$

   g $\sinh 2x - 7 \sinh x = 0$

   h $4 \cosh x + 13e^{-x} = 11$

   i $2 \tanh x = \cosh x$

(E) 6 **a** Starting from the definitions of $\sinh x$ and $\cosh x$ in terms of exponentials, prove that

   $\cosh 2x \equiv 2 \cosh^2 x - 1$ **(3 marks)**

   b Solve the equation

   $\cosh 2x - 3 \cosh x = 8$

   giving your answers as exact logarithms.   **(5 marks)**

(E) 7 Solve the equation

   $2 \sinh^2 x - 5 \cosh x = 5$

   giving your answer in terms of natural logarithms in simplest form.   **(6 marks)**

8 Joshua is asked to prove the following identity:

$$\frac{1 + \tanh^2 x}{1 - \tanh^2 x} \equiv 2\cosh^2 x - 1$$

His answer is below.

> $\frac{1 + \tanh^2 x}{1 - \tanh^2 x} \equiv \frac{\text{sech}^2 x}{2 - \text{sech}^2 x}$ (using $\text{sech}^2 x \equiv 1 + \tanh^2 x$: same identity as the trig one)
>
> $\equiv \frac{\text{sech}^2 x}{2} - 1$ (splitting the fraction up and cancelling)
>
> $\equiv \frac{2}{\text{sech}^2 x} - 1$ (taking the reciprocal of both terms)
>
> $\equiv 2\cosh^2 x - 1$

Joshua has made three errors. Explain the errors and provide a correct proof.

9 a Express $10\cosh x + 6\sinh x$ in the form $R\cosh(x + a)$ where $R > 0$. Give the value of a correct to 3 decimal places. **(4 marks)**

Hint Use the identity for $\cosh(A + B)$

b Write down the minimum value of $10\cosh x + 6\sinh x$ **(1 mark)**

c Use your answer to part **a** to solve the equation $10\cosh x + 6\sinh x = 11$ Give your answers to 3 decimal places. **(4 marks)**

Chapter review 1

SKILLS PROBLEM-SOLVING, REASONING/ARGUMENTATION

1 Find the exact value of **a** $\sinh(\ln 3)$ **b** $\cosh(\ln 5)$ **c** $\tanh\left(\ln \frac{1}{4}\right)$

2 Given that $\operatorname{artanh} x - \operatorname{artanh} y = \ln 5$, find y in terms of x.

3 Using the definitions of $\sinh x$ and $\cosh x$, prove that

$\sinh(A - B) \equiv \sinh A \cosh B - \cosh A \sinh B$ **(5 marks)**

4 Using definitions in terms of exponentials, prove that

$\sinh x \equiv \dfrac{2\tanh \frac{1}{2}x}{1 - \tanh^2 \frac{1}{2}x}$ **(5 marks)**

5 Solve, giving your answers as natural logarithms

$9\cosh x - 5\sinh x = 15$ **(6 marks)**

6 Solve, giving your answers as natural logarithms

$23\sinh x - 17\cosh x + 7 = 0$ **(6 marks)**

7 Solve, giving your answers as natural logarithms

$3\cosh^2 x + 11\sinh x = 17$ **(6 marks)**

8 a On the same diagram, sketch the graphs of $y = 6 + \sinh x$ and $y = \sinh 3x$ (2 marks)
   b Using the identity $\sinh 3x \equiv 3\sinh x + 4\sinh^3 x$, show that the graphs **intersect** where $\sinh x = 1$ and hence find the exact **coordinates** of the point of **intersection**. (5 marks)

9 a Given that $13\cosh x + 5\sinh x = R\cosh(x + \alpha)$, $R > 0$, use the identity $\cosh(A + B) \equiv \cosh A \cosh B + \sinh A \sinh B$ to find the values of R and α, giving the value of α to 3 decimal places. (4 marks)
   b Write down the minimum value of $13\cosh x + 5\sinh x$ (1 mark)

10 a Express $3\cosh x + 5\sinh x$ in the form $R\sinh(x + \alpha)$, where $R > 0$ Give α to 3 decimal places. (4 marks)
   b Use the answer to part **a** to solve the equation $3\cosh x + 5\sinh x = 8$, giving your answer to 2 decimal places. (3 marks)
   c Solve $3\cosh x + 5\sinh x = 8$ by using the definitions of $\cosh x$ and $\sinh x$. (4 marks)

Challenge

SKILLS
CREATIVITY

Sketch the graph of $y = (\operatorname{arsinh} x)^2$

Summary of key points

1. • Hyperbolic sine (or **sinh**) is defined as $\sinh x \equiv \dfrac{e^x - e^{-x}}{2}$, $x \in \mathbb{R}$

   • Hyperbolic cosine (or **cosh**) is defined as $\cosh x \equiv \dfrac{e^x + e^{-x}}{2}$, $x \in \mathbb{R}$

   • Hyperbolic tangent (or **tanh**) is defined as $\tanh x \equiv \dfrac{\sinh x}{\cosh x} \equiv \dfrac{e^{2x} - 1}{e^{2x} + 1}$, $x \in \mathbb{R}$

   • Hyperbolic cosecant (or **cosech**) is defined as $\operatorname{cosech} x \equiv \dfrac{2}{e^x - e^{-x}}$, $x \in \mathbb{R}$

   • Hyperbolic secant (or **sech**) is defined as $\operatorname{sech} x \equiv \dfrac{2}{e^x + e^{-x}}$, $x \in \mathbb{R}$

   • Hyperbolic cotangent (or **coth**) is defined as $\coth x \equiv \dfrac{e^{2x} + 1}{e^{2x} - 1}$, $x \in \mathbb{R}$

2. • The graph of $y = \sinh x$:

   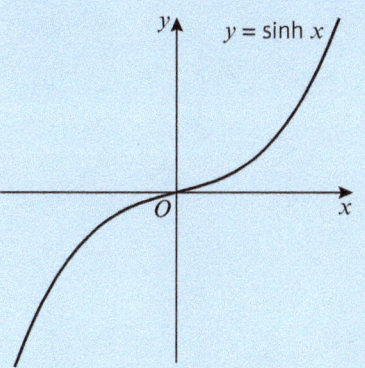

   For any value a, $\sinh(-a) = -\sinh a$.

   • The graph of $y = \cosh x$:

   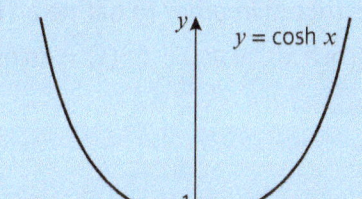

   For any value a, $\cosh(-a) = \cosh a$.

3 The table shows the inverse hyperbolic functions, with domains restricted where necessary.

Hyberbolic function	Inverse hyperbolic function
$y = \sinh x$	$y = \operatorname{arsinh} x$
$y = \cosh x, x \geq 0$	$y = \operatorname{arcosh} x, x \geq 1$
$y = \tanh x$	$y = \operatorname{artanh} x, |x| < 1$
$y = \operatorname{sech} x, x \geq 0$	$y = \operatorname{arsech} x, 0 < x \leq 1$
$y = \operatorname{cosech} x, x \neq 0$	$y = \operatorname{arcosech} x, x \neq 0$
$y = \coth x, x \neq 0$	$y = \operatorname{arcoth} x, |x| > 0$

4 The following formulae are given to you in the formula booklet in the examination.
- $\cosh^2 x - \sinh^2 x = 1$
- $\sinh 2x = 2 \sinh x \cosh x$
- $\cosh 2x = \cosh^2 x + \sinh^2 x$
- $\operatorname{arcosh} x = \ln\{x + \sqrt{x^2 - 1}\} \quad x \geq 1$
- $\operatorname{arsinh} x = \ln\{x + \sqrt{x^2 + 1}\}$
- $\operatorname{artanh} x = \frac{1}{2} \ln\left(\frac{1+x}{1-x}\right) \quad (|x| < 1)$

5 $\operatorname{sech}^2 A = 1 - \tanh^2 A \qquad \operatorname{cosech}^2 A = \coth^2 A - 1$

6
- $\sinh(A \pm B) \equiv \sinh A \cosh B \pm \cosh A \sinh B$
- $\cosh(A \pm B) \equiv \cosh A \cosh B \pm \sinh A \sinh B$

7 If $f(x) = \sinh x$, then the inverse function f^{-1} is called $\operatorname{arsinh} x$ (sometimes written as $\sinh^{-1} x$).

8 If $y = \operatorname{arsinh} x$ then $x = \sinh y$

9 The graph of $y = \operatorname{arsinh} x$ is the reflection of the graph $y = \sinh x$ in the line $y = x$.

10 The inverse of a function is defined only if the function is one-to-one, so for $\cosh x$ the domain must be restricted in order to define an inverse.
For $f(x) = \cosh x \quad x \geq 0, \quad f^{-1}(x) = \operatorname{arcosh} x \quad (x \geq 1)$

2 FURTHER COORDINATE SYSTEMS

2.1
2.2
2.3
2.4

Learning objectives

After completing this chapter you should be able to:
- Identify an ellipse or a hyperbola from its Cartesian or parametric equations → pages 18–22
- Find the foci, directrices, and eccentricity for an ellipse or a hyperbola → pages 22–29
- Find tangents and normals to these curves → pages 29–38
- Solve simple loci questions → pages 38–42

Prior knowledge check

1. The curve C has equation $x^2 - 9y^2 = 20$. Find the gradient of C at the point $\left(6, \frac{4}{3}\right)$.
   ← Pure 4 Section 5.2

2. Find the x-coordinates of the points of intersection of the circle with equation $x^2 + y^2 = a^2$ and the line $y = kx$, giving your answer in terms of a and k.
   ← Pure 1 Section 3.2

3. A curve has parametric equations $x = at^2$, $y = 2at$, $t \in \mathbb{R}$ where a is a positive constant. Find the Cartesian equation of the curve.
   ← Further Pure 1 Section 4.2

The Earth's motion around the Sun can be modelled as following an elliptical path, where the Sun is located at one focus of the ellipse.

2.1 Ellipses

In Further Pure 1 you encountered the **parabola** and the **rectangular hyperbola**, which are both examples of **conic sections**.

If you slice a cone in such a way as to produce a **closed** curve, the resulting curve is called an **ellipse**.

A circle is a special case of an ellipse

Ellipse

Online Explore conic sections using GeoGebra.

- A standard ellipse has the **Cartesian** equation
$$\frac{x^2}{a^2} + \frac{y^2}{b^2} = 1$$

When $x = 0$, $\frac{y^2}{b^2} = 1$ and so $y = \pm b$

When $y = 0$, $\frac{x^2}{a^2} = 1$ and so $x = \pm a$

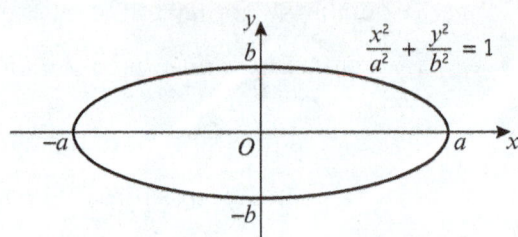

You can define a **general point** P on the ellipse in terms of a parameter, t.

- The standard ellipse has **parametric equations**
$$x = a\cos t, y = b\sin t, 0 \leq t < 2\pi$$

- A general point P on an ellipse has coordinates $(a\cos t, b\sin t)$.

Notation Substituting $x = a\cos t$ and $y = b\sin t$ into $\frac{x^2}{a^2} + \frac{y^2}{b^2}$ produces $\cos^2 t + \sin^2 t$ which is equal to 1.

← Pure 2 Section 6.3

Example 1   SKILLS   PROBLEM-SOLVING

The ellipse E has equation $4x^2 + 9y^2 = 36$
a Sketch E.
b Write down parametric equations for E.

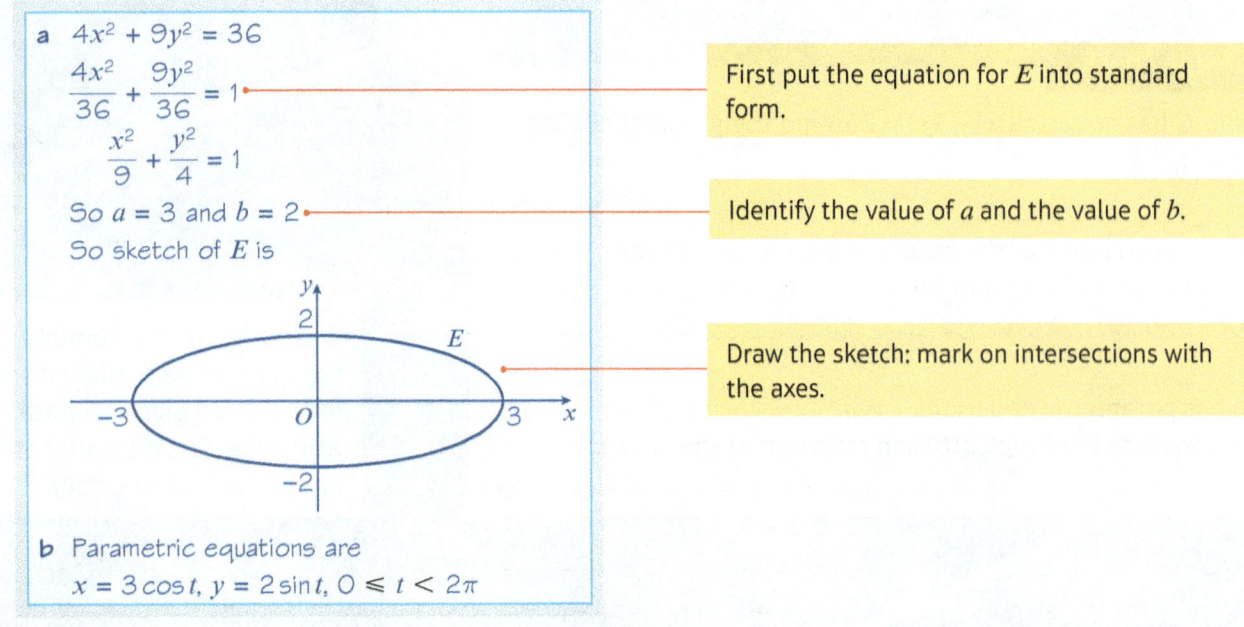

a $4x^2 + 9y^2 = 36$
$\frac{4x^2}{36} + \frac{9y^2}{36} = 1$
$\frac{x^2}{9} + \frac{y^2}{4} = 1$
So $a = 3$ and $b = 2$
So sketch of E is

First put the equation for E into standard form.

Identify the value of a and the value of b.

Draw the sketch: mark on intersections with the axes.

b Parametric equations are
$x = 3\cos t, y = 2\sin t, 0 \leq t < 2\pi$

CHAPTER 2

Example 2

The ellipse E has parametric equations
$$x = 3\cos\theta, \; y = 5\sin\theta, \; 0 \leq \theta < 2\pi$$

a Sketch E. **b** Find a Cartesian equation of E.

a $-5 \leq y \leq 5$
$-3 \leq x \leq 3$ ← Since $\sin\theta$ and $\cos\theta$ are both between -1 and 1.

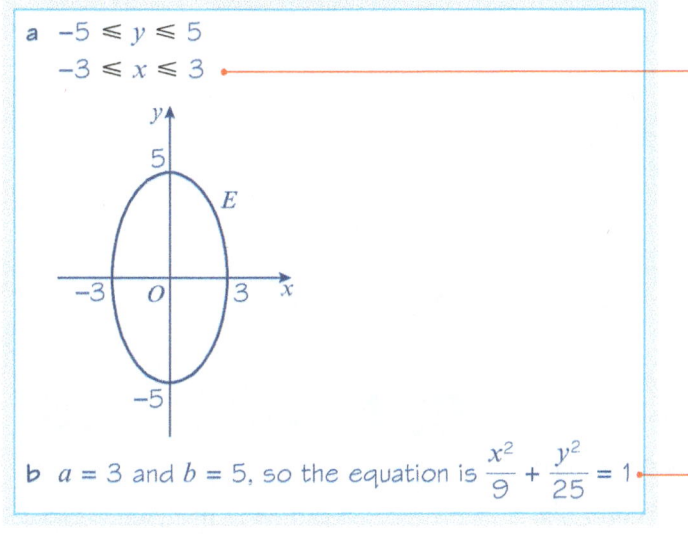

b $a = 3$ and $b = 5$, so the equation is $\dfrac{x^2}{9} + \dfrac{y^2}{25} = 1$ ← Compare with the standard formula.

Exercise 2A — SKILLS — PROBLEM-SOLVING

1 a Sketch the following ellipses showing clearly where the curves cross the coordinate axes.
  i $x^2 + 4y^2 = 16$ **ii** $4x^2 + y^2 = 36$ **iii** $x^2 + 9y^2 = 25$
  b Find parametric equations for these curves.

2 a Sketch ellipses with the following parametric equations.
  i $x = 2\cos\theta, \; y = 3\sin\theta$ **ii** $x = 4\cos\theta, \; y = 5\sin\theta$
  iii $x = \cos\theta, \; y = 5\sin\theta$ **iv** $x = 4\cos\theta, \; y = 3\sin\theta$
  b Find a Cartesian equation for each ellipse.

(P) 3 The diagram shows the circles with equations $x^2 + y^2 = a^2$ and $x^2 + y^2 = b^2$. The line OS makes an angle θ with the positive x-axis and intersects the circles at points P and Q respectively (i.e. in the same order as the things already mentioned). The point R has the same y-coordinate as P and the same x-coordinate as Q, as shown in the diagram.

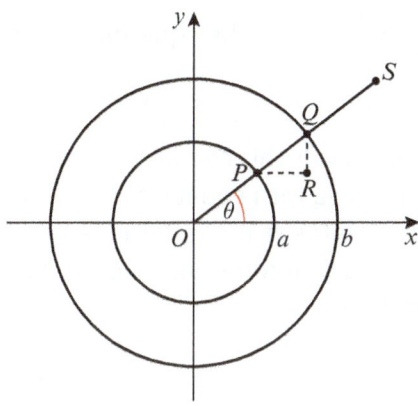

  a Find the coordinates of R in terms of a, b and θ.
  b Hence describe the **locus** of R as θ varies from 0 to 2π, and give its Cartesian equation.
  c Sketch the curve with parametric equations
  $$x = 4\cos t, \; y = \sin t, \; \dfrac{\pi}{2} \leq t \leq \dfrac{3\pi}{2}$$
  showing clearly any points where the curve meets or intersects the coordinate axes.

Challenge

The curve C is formed by rotating the ellipse with equation $\frac{x^2}{a^2} + \frac{y^2}{b^2} = 1$ through 45° anticlockwise (i.e. moving in the opposite direction to the hands of a clock) about the origin.

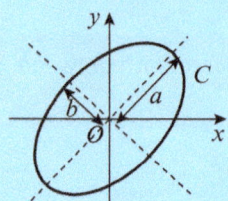

Show that C has equation $\frac{(x+y)^2}{2a^2} + \frac{(x-y)^2}{2b^2} = 1$

Problem-solving

Write the position **vector** of a general point on the original ellipse as $\begin{pmatrix} a\cos t \\ b\sin t \end{pmatrix}$ and then apply a suitable **linear transformation**.

2.2 Hyperbolas

You have previously encountered rectangular hyperbolas with parametric equations $x = ct$, $y = \frac{c}{t}$, $t \in \mathbb{R}$, $t \neq 0$, where c is a positive constant. The Cartesian equation of this rectangular hyperbola is $xy = c^2$. This family of curves have **perpendicular** asymptotes with equations $x = 0$ (the y-axis) and $y = 0$ (the x-axis). A general point P on the curve has coordinates $P\left(ct, \frac{c}{t}\right)$

Links ← Further Pure 1 Section 4.3

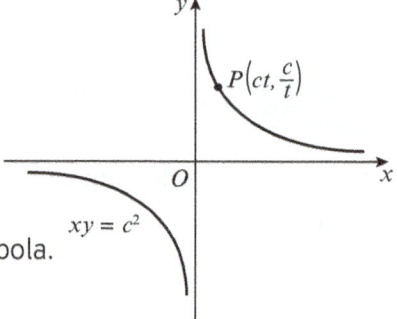

In general, hyperbolas do not need to have perpendicular asymptotes. You can find Cartesian and parametric equations for a standard hyperbola.

- A standard hyperbola has Cartesian equation
$$\frac{x^2}{a^2} - \frac{y^2}{b^2} = 1$$

When $y = 0$, $x^2 = a^2$ and so the curve crosses the x-axis at $(\pm a, 0)$. As x and y tend to infinity, $\frac{x^2}{a^2} \approx \frac{y^2}{b^2}$ and so the equations of the asymptotes are $y = \pm \frac{b}{a} x$

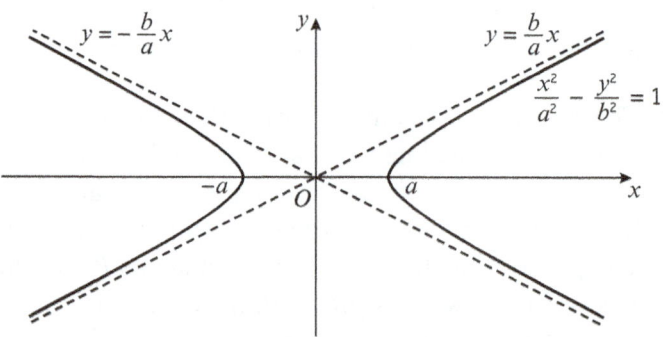

When $a = b$, this creates a rectangular hyperbola with equation $x^2 - y^2 = a^2$ with asymptotes at $y = x$ and $y = -x$
These asymptotes are perpendicular to one another.

Notation The equations of the asymptotes are given in the formula booklet.

Watch out Although $x^2 - y^2 = a^2$ is an example of a rectangular hyperbola because its asymptotes are perpendicular, it is not part of the family of curves of the form $xy = c^2$

In the previous section you saw that the parametric equations of the ellipse were connected to the trigonometric relationship $\cos^2\theta + \sin^2\theta \equiv 1$. You can use the corresponding relationship for the hyperbolic functions to find parametric equations for the hyperbola.

- The standard hyperbola has parametric equations
  $$x = \pm a\cosh t, \; y = b\sinh t, \; t \in \mathbb{R}$$

  Links $\cosh^2 x - \sinh^2 x \equiv 1$
  ← Further Pure 3 Section 1.4

- The standard hyperbola has alternative parametric equations
  $$x = a\sec\theta, \; y = b\tan\theta, \; -\pi \leq \theta < \pi, \; \theta \neq \pm\frac{\pi}{2}$$

- A general point P on a hyperbola has coordinates $(\pm a\cosh t, b\sinh t)$ or $(a\sec\theta, b\tan\theta)$.

Example 3 SKILLS PROBLEM-SOLVING

The hyperbola H has equation $9x^2 - 4y^2 = 36$

a Sketch H.

b Write down the equations of the asymptotes of H.

c Find parametric equations for H.

a Rearrange the equation to get
$$\frac{x^2}{4} - \frac{y^2}{9} = 1$$
So $a = 2$ and $b = 3$

Write the equation in the form $\dfrac{x^2}{a^2} - \dfrac{y^2}{b^2} = 1$ and identify values for a and b.

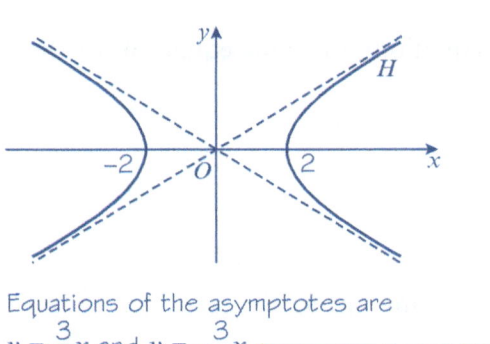

b Equations of the asymptotes are
$y = \frac{3}{2}x$ and $y = -\frac{3}{2}x$

The equations of the asymptotes are $y = \pm\dfrac{b}{a}x$

c Parametric equations are
$x = \pm 2\cosh t, \; y = 3\sinh t, \; t \in \mathbb{R}$

Use $x = a\cosh t$ and $y = b\sinh t$

Example 4

A hyperbola H has parametric equations
$$x = 4\sec t, \; y = \tan t, \; -\pi \leq t < \pi, \; t \neq \pm\frac{\pi}{2}$$

a Find a Cartesian equation for H.

b Sketch H.

c Write down the equations of the asymptotes of H.

a Using $\sec^2 t - \tan^2 t \equiv 1$

$$\left(\frac{x}{4}\right)^2 - y^2 = 1$$

Cartesian equation is

$$\frac{x^2}{16} - y^2 = 1$$

Alternatively, compare with $x = a \sec \theta$ and $y = b \tan \theta$ and use the standard equation.

b $a = 4$ and $b = 1$

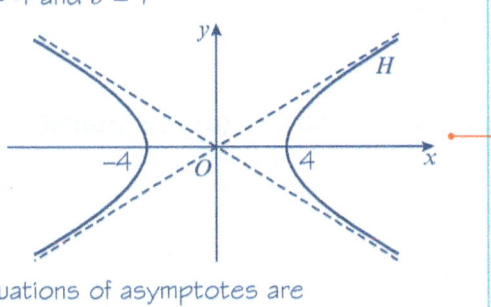

By comparing with $\dfrac{x^2}{a^2} - \dfrac{y^2}{b^2} = 1$ and using $a = 4$ and $b = 1$

c Equations of asymptotes are

$$y = \pm \frac{1}{4}x$$

Use $y = \pm \dfrac{b}{a} x$

Exercise 2B — SKILLS — PROBLEM-SOLVING

1 Sketch the following hyperbolas showing clearly the intersections with the x-axis and the equations of the asymptotes.

a $x^2 - 4y^2 = 16$ **b** $4x^2 - 25y^2 = 100$ **c** $\dfrac{x^2}{8} - \dfrac{y^2}{2} = 1$

2 a Sketch the hyperbolas with the following parametric equations. Give the equations of the asymptotes and show points of intersection with the x-axis.
  i $x = 2\sec\theta$, $y = 3\tan\theta$, $-\pi \leqslant \theta < \pi$, $\theta \neq \pm\dfrac{\pi}{2}$
  ii $x = \pm 4\cosh t$, $y = 3\sinh t$, $t \in \mathbb{R}$
  iii $x = \pm\cosh t$, $y = 2\sinh t$, $t \in \mathbb{R}$
  iv $x = 5\sec\theta$, $y = 7\tan\theta$, $-\pi \leqslant \theta < \pi$, $\theta \neq \pm\dfrac{\pi}{2}$

b Find the Cartesian equation for each of the hyperbolas from part **a**.

Challenge

The rectangular hyperbola with equation $xy = c^2$ is rotated through 45° anticlockwise about the origin. Show that the resulting curve satisfies (i.e. meets the requirements of) the equation $y^2 - x^2 = a^2$ and state the relationship between a and c in this case.

2.3 Eccentricity

You can define the ellipse and hyperbola in terms of their focus–directrix properties. In order to do this, you need to generalise the approach (i.e. apply the approach to a wider situation) used for the parabola in the previous chapter. To do this you need to consider the **eccentricity** of a particular conic section.

Links The parabola with equation $y^2 = 4ax$ is the locus of all the points, P, that are **equidistant** from a fixed point, S, (the **focus**) and a fixed line (the directrix).
← Further Pure 1 Section 4.2

- For all points, P, on a conic section, the ratio of the distance of P from a fixed point (called the focus) and a fixed straight line (called the directrix) is constant. This ratio, e, is known as the eccentricity of the curve.

The diagram shows a fixed point, S, a fixed straight line, the directrix, and a point, P, on a conic section.

For all points, P, on the curve, the ratio $\dfrac{PS}{PM} = e$ is constant.

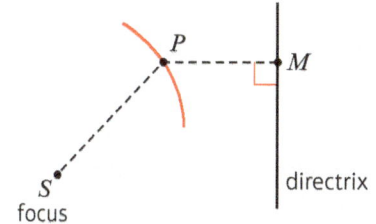

- If $0 < e < 1$, the point P describes an ellipse.
- If $e = 1$, the point P describes a parabola.
- If $e > 1$, the point P describes a hyperbola.

Watch out The special case where $e = 0$ represents a circle, and the special case where e is **infinite** represents a straight line. These are both examples of conic sections, but you will not need to consider them in this chapter.

Example 5 SKILLS REASONING/ARGUMENTATION

Show that, for $0 < e < 1$, the ellipse with focus $(ae, 0)$ and directrix $x = \dfrac{a}{e}$ has equation $\dfrac{x^2}{a^2} + \dfrac{y^2}{b^2} = 1$

Let P be the point with coordinates (x, y).

$$\dfrac{PS}{PM} = e \Rightarrow PS^2 = e^2 PM^2$$

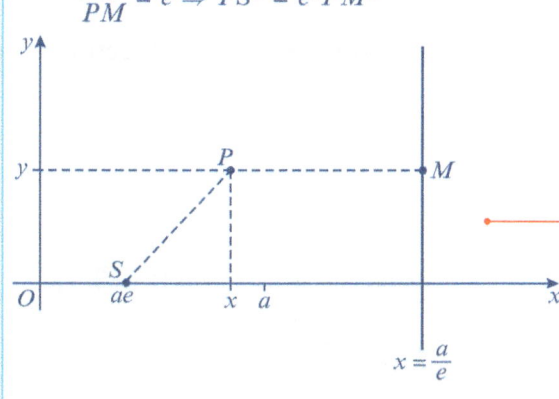

Draw a diagram.

$PS^2 = (x - ae)^2 + y^2$

$PM^2 = \left(\dfrac{a}{e} - x\right)^2 = \dfrac{(a - ex)^2}{e^2}$

Find expressions for PS^2 and PM^2 in terms of a, e and x, y.

So $PS^2 = e^2 PM^2$ gives

$x^2 - 2aex + a^2e^2 + y^2 = a^2 - 2aex + e^2x^2$

$x^2(1 - e^2) + y^2 = a^2(1 - e^2)$

Simplify.

$\dfrac{x^2}{a^2} + \dfrac{y^2}{a^2(1 - e^2)} = 1$

So if $b^2 = a^2(1 - e^2)$ then you have the standard equation of the ellipse.

Problem-solving

This equation only produces an ellipse if $0 < e < 1$. If $e = 0$, then $1 - e^2 = 1$ and the equation reduces to the equation of a circle. If $e > 1$, then $1 - e^2$ is negative and the equation produces a hyperbola.

Because the ellipse is symmetrical about the y-axis, the above derivation will also work for a focus $(-ae, 0)$ with a directrix $x = -\dfrac{a}{e}$.

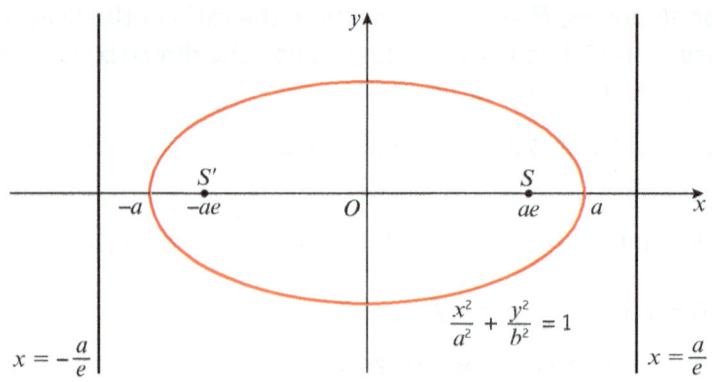

Online Explore the foci and directices of an ellipse using GeoGebra.

For an ellipse with equation $\dfrac{x^2}{a^2} + \dfrac{y^2}{b^2} = 1$, and $a > b$,
- the eccentricity, $0 < e < 1$, is given by $b^2 = a^2(1 - e^2)$
- the foci are at $(\pm ae, 0)$
- the directrices are $x = \pm \dfrac{a}{e}$

Notation

Foci is the plural of focus and **directrices** is the plural of directrix.

Notice that the foci are on the **major axis** which in this case is the x-axis because $a > b$

If the major axis is along the y-axis ($b > a$), then the foci will be on the y-axis at $(0, \pm be)$ and the directrices will have equations $y = \pm \dfrac{b}{e}$. The eccentricity will be given by $a^2 = b^2(1 - e^2)$

Example 6   SKILLS   ANALYSIS

Find the foci of the ellipses with the following equations and give the equations of the directrices.

a $\dfrac{x^2}{9} + \dfrac{y^2}{4} = 1$ 　　　 **b** $\dfrac{x^2}{16} + \dfrac{y^2}{25} = 1$

In each case sketch the ellipse, and show the directrices and foci.

a $\dfrac{x^2}{9} + \dfrac{y^2}{4} = 1$

$b^2 = a^2(1 - e^2)$ gives $4 = 9(1 - e^2)$ so $e^2 = \dfrac{5}{9}$

So $e = \dfrac{\sqrt{5}}{3}$

Foci are at $(\pm\sqrt{5}, 0)$

Directrices are $x = \pm \dfrac{9}{\sqrt{5}}$

Note that $a = 3$ and $b = 2$.
Since $a > b$ use $b^2 = a^2(1 - e^2)$

Use $(\pm ae, 0)$

Use $x = \pm \dfrac{a}{e}$

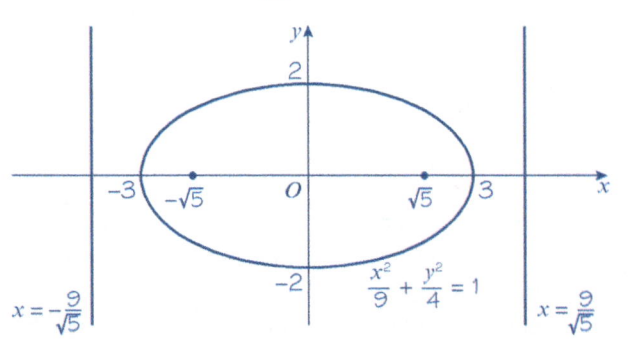

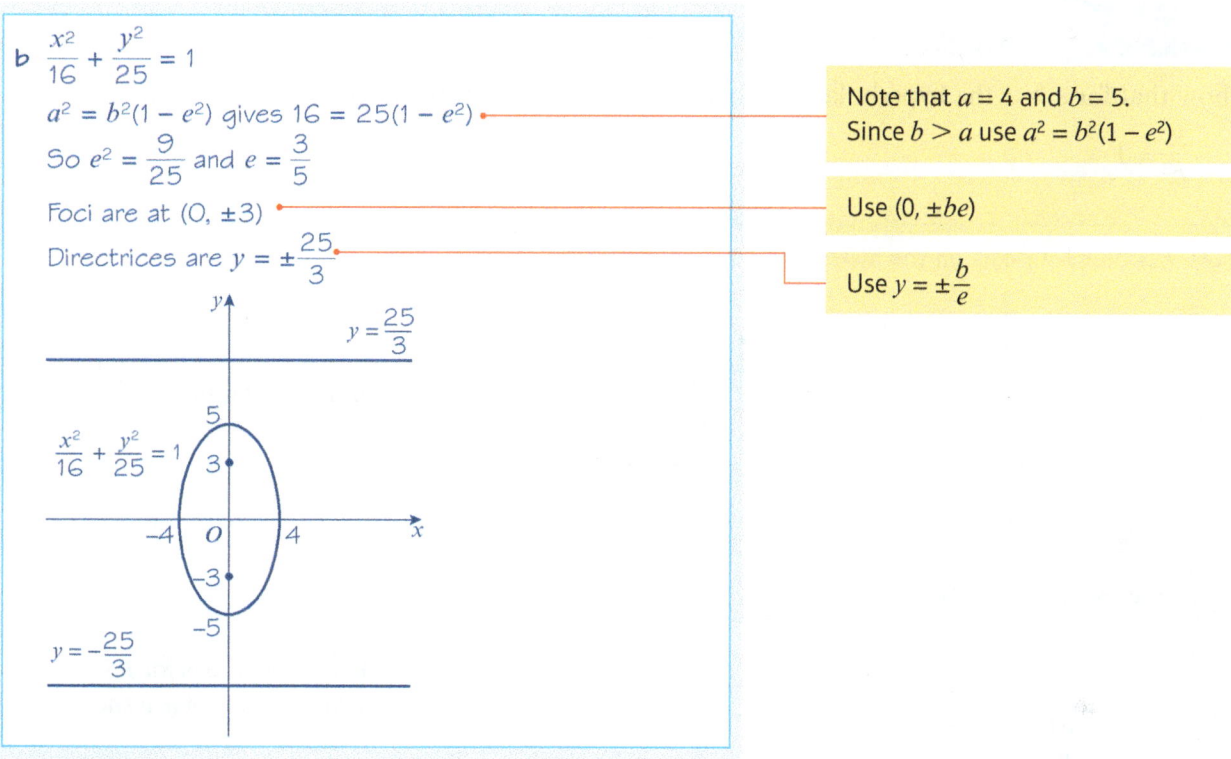

b $\frac{x^2}{16} + \frac{y^2}{25} = 1$

$a^2 = b^2(1 - e^2)$ gives $16 = 25(1 - e^2)$

So $e^2 = \frac{9}{25}$ and $e = \frac{3}{5}$

Foci are at $(0, \pm 3)$

Directrices are $y = \pm \frac{25}{3}$

Note that $a = 4$ and $b = 5$. Since $b > a$ use $a^2 = b^2(1 - e^2)$

Use $(0, \pm be)$

Use $y = \pm \frac{b}{e}$

Example 7

The ellipse with equation $\frac{x^2}{a^2} + \frac{y^2}{b^2} = 1$ has foci at $S(ae, 0)$ and $S'(-ae, 0)$.

Show that if P is any point on the ellipse then $PS + PS' = 2a$

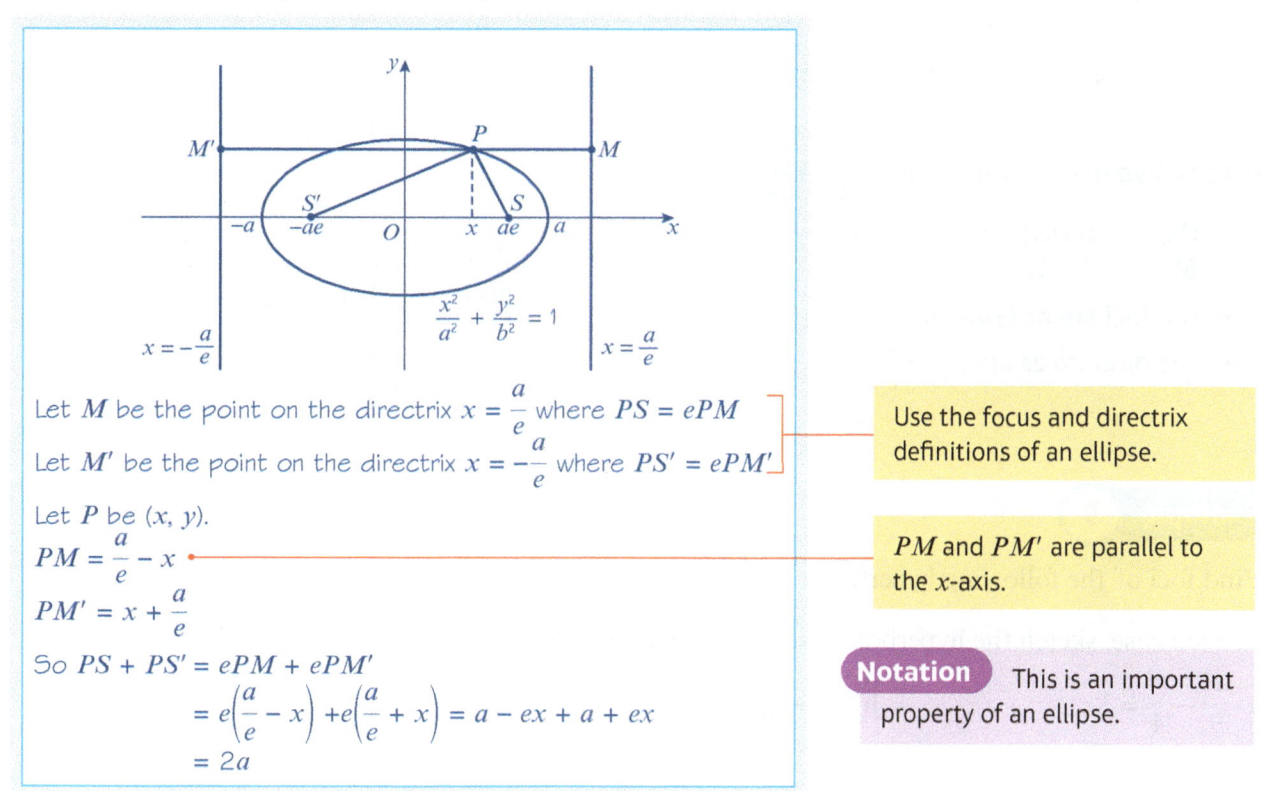

Let M be the point on the directrix $x = \frac{a}{e}$ where $PS = ePM$

Let M' be the point on the directrix $x = -\frac{a}{e}$ where $PS' = ePM'$

Use the focus and directrix definitions of an ellipse.

Let P be (x, y).

$PM = \frac{a}{e} - x$

$PM' = x + \frac{a}{e}$

PM and PM' are parallel to the x-axis.

So $PS + PS' = ePM + ePM'$

$= e\left(\frac{a}{e} - x\right) + e\left(\frac{a}{e} + x\right) = a - ex + a + ex$

$= 2a$

Notation This is an important property of an ellipse.

Example 8 — SKILLS: REASONING/ARGUMENTATION

Show that for $e > 1$ the hyperbola with foci at $(\pm ae, 0)$ and directrices at $x = \pm \dfrac{a}{e}$ has equation $\dfrac{x^2}{a^2} - \dfrac{y^2}{b^2} = 1$

Let $P(x, y)$ be a point on the hyperbola.

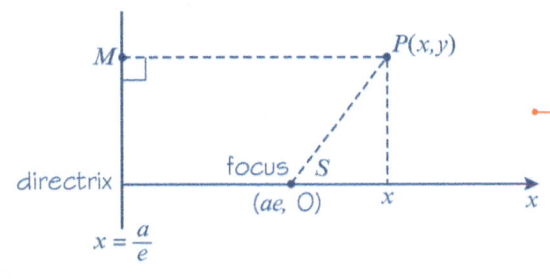

Draw a diagram.

$\dfrac{PS}{PM} = e \Rightarrow PS^2 = e^2 PM^2$

$PS^2 = (x - ae)^2 + y^2$

$PM^2 = \left(x - \dfrac{a}{e}\right)^2 = \dfrac{(ex - a)^2}{e^2}$

Find expressions for PS^2 and PM^2 in terms of a, e and x, y.

So $PS^2 = e^2 PM^2$ gives

$x^2 - 2aex + a^2 e^2 + y^2 = e^2 x^2 - 2aex + a^2$

Simplify.

$a^2(e^2 - 1) = x^2(e^2 - 1) - y^2$

$1 = \dfrac{x^2}{a^2} - \dfrac{y^2}{a^2(e^2 - 1)}$

$e > 1$ so $a^2(e^2 - 1)$ will be positive.

So if $b^2 = a^2(e^2 - 1)$ you have the standard equation of a hyperbola.

■ For a hyperbola with equation $\dfrac{x^2}{a^2} - \dfrac{y^2}{b^2} = 1$,
  - the eccentricity, $e > 1$, is given by $b^2 = a^2(e^2 - 1)$
  - the foci are at $(\pm ae, 0)$
  - the directrices are $x = \pm \dfrac{a}{e}$

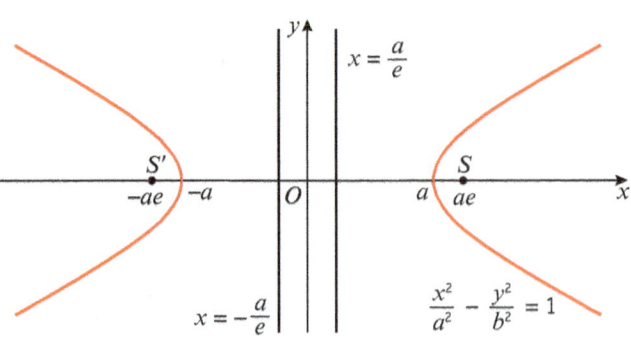

Example 9

Find foci of the following hyperbolas.

In each case, sketch the hyperbola and show the directrices.

  $\dfrac{x^2}{9} - \dfrac{y^2}{4} = 1$ **b** $\dfrac{x^2}{16} - \dfrac{y^2}{25} = 1$

a $\dfrac{x^2}{9} - \dfrac{y^2}{4} = 1$ so $a = 3$ and $b = 2$

Eccentricity is given by $b^2 = a^2(e^2 - 1)$

$4 = 9(e^2 - 1)$

So $\dfrac{4}{9} + 1 = e^2$

$\Rightarrow e = \sqrt{\dfrac{13}{9}} = \dfrac{\sqrt{13}}{3}$

Foci are at $(\pm\sqrt{13}, 0)$

Directrices are $x = \pm\dfrac{9}{\sqrt{13}}$

Asymptotes are $y = \pm\dfrac{b}{a}x \Rightarrow y = \pm\dfrac{2}{3}x$

Compare the equation with $\dfrac{x^2}{a^2} - \dfrac{y^2}{b^2} = 1$ and identify a and b.

Use $b^2 = a^2(e^2 - 1)$

Use $(\pm ae, 0)$

Use $x = \pm\dfrac{a}{e}$

Online Explore the foci and directices of a hyperbola using GeoGebra.

Draw in the asymptotes to make the graph easier to draw.

$y = \dfrac{2}{3}x$

$y = -\dfrac{2}{3}x$

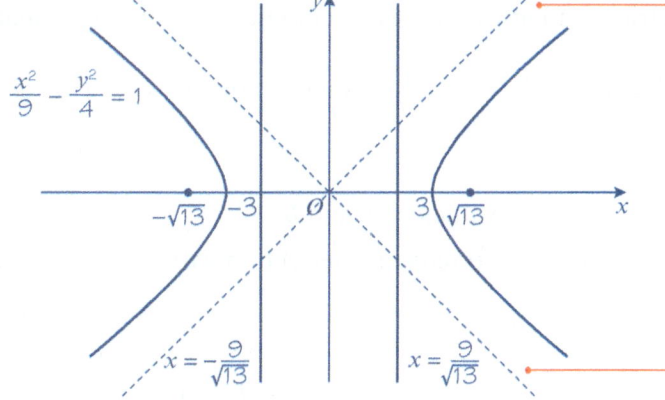

b $\dfrac{x^2}{16} - \dfrac{y^2}{25} = 1$, so $a = 4$ and $b = 5$

Eccentricity is given by $b^2 = a^2(e^2 - 1)$.

$25 = 16(e^2 - 1)$

$\dfrac{25}{16} + 1 = e^2$ so $e = \sqrt{\dfrac{41}{16}} = \dfrac{\sqrt{41}}{4}$

Foci are at $(\pm\sqrt{41}, 0)$

Directrices are $x = \pm\dfrac{16}{\sqrt{41}}$

Asymptotes are $y = \pm\dfrac{b}{a}x \Rightarrow y = \pm\dfrac{5}{4}x$

Compare the equation with $\dfrac{x^2}{a^2} - \dfrac{y^2}{b^2} = 1$ and identify a and b.

Use $b^2 = a^2(e^2 - 1)$

Use $(\pm ae, 0)$

Problem-solving

In this example $b > a$. However, unlike with an ellipse, the foci do not move to the y-axis. Setting $x = 0$ in the general equation of a hyperbola would give $-\dfrac{y^2}{b^2} = 1$ which is never satisfied for real values of y.

Draw in the asymptotes to make the graph easier to draw.

$y = \dfrac{5}{4}x$

$y = -\dfrac{5}{4}x$

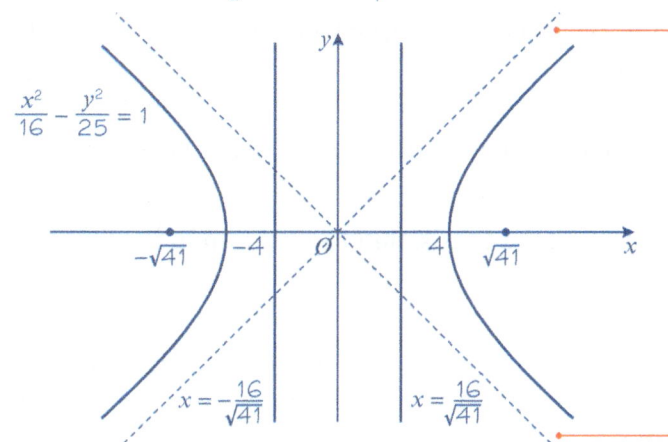

Exercise 2C SKILLS ANALYSIS

1. Find the eccentricity of the following ellipses.
   a $\dfrac{x^2}{9} + \dfrac{y^2}{5} = 1$
   b $\dfrac{x^2}{16} + \dfrac{y^2}{9} = 1$
   c $\dfrac{x^2}{4} + \dfrac{y^2}{8} = 1$

2. Find the foci and directrices of the following ellipses.
   a $\dfrac{x^2}{4} + \dfrac{y^2}{3} = 1$
   b $\dfrac{x^2}{16} + \dfrac{y^2}{7} = 1$
   c $\dfrac{x^2}{5} + \dfrac{y^2}{9} = 1$

3. An ellipse with equation $\dfrac{x^2}{a^2} + \dfrac{y^2}{b^2} = 1$ has focus $(3, 0)$ and the equation of the directrix is $x = 12$
   a Explain why $a > b$
   b Find
      i the eccentricity of the ellipse
      ii the values of a and b.
   c Sketch the ellipse, showing the directrices and any points of intersection with the coordinate axes.

4. An ellipse with equation $\dfrac{x^2}{a^2} + \dfrac{y^2}{b^2} = 1$ has focus $(0, 2)$ and the equation of the directrix is $y = 8$
   a Explain why $b > a$
   b Find
      i the eccentricity of the ellipse
      ii the values of a and b.
   c Sketch the ellipse, showing the directrices and any points of intersection with the coordinate axes.

5. Find the eccentricities of the following hyperbolas.
   a $\dfrac{x^2}{5} - \dfrac{y^2}{3} = 1$
   b $\dfrac{x^2}{9} - \dfrac{y^2}{7} = 1$
   c $\dfrac{x^2}{9} - \dfrac{y^2}{16} = 1$

6. Sketch the following hyperbolas, showing clearly the positions of their foci and directrices.
   a $\dfrac{x^2}{4} - \dfrac{y^2}{8} = 1$
   b $\dfrac{x^2}{16} - \dfrac{y^2}{9} = 1$
   c $\dfrac{x^2}{4} - \dfrac{y^2}{5} = 1$

7. a For each of the following hyperbolas, find the eccentricity and show that the foci are at $(\pm 5, 0)$.
      i $\dfrac{x^2}{24} - y^2 = 1$
      ii $x^2 - \dfrac{y^2}{24} = 1$
      iii $\dfrac{x^2}{16} - \dfrac{y^2}{9} = 1$
      iv $\dfrac{x^2}{9} - \dfrac{y^2}{16} = 1$
   b Hence sketch all four hyperbolas on the same graph, showing the foci and labelling each curve with its eccentricity.

(E/P) 8. The latus rectum of an ellipse is a **chord** perpendicular to the major axis that passes through a focus. Show that the length of the latus rectum of the ellipse with equation $\dfrac{x^2}{a^2} + \dfrac{y^2}{b^2} = 1$, where $a > b$, is $\dfrac{2b^2}{a}$ **(5 marks)**

(E/P) 9. The distance between the foci of an ellipse is 16 and the distance between the directrices is 25.
   a Find the eccentricity of the ellipse. **(3 marks)**
   b Given that both the foci of the ellipse lie on the y-axis, find its equation in the form $\dfrac{x^2}{a^2} + \dfrac{y^2}{b^2} = 1$ **(2 marks)**

(E/P) 10. The point P lies on the ellipse with equation $x^2 + 4y^2 = 36$, and A and B are the points $-3\sqrt{3}, 0$ and $3\sqrt{3}, 0$ respectively. Prove that $PA + PB = 12$ **(4 marks)**

E/P 11 Ellipse E has equation $\frac{x^2}{a^2} + \frac{y^2}{b^2} = 1$, such that $a > b$. The foci of E are at S and S' and the point P is $(0, b)$.

Show that $\cos(PSS') = e$, the eccentricity of E. **(6 marks)**

E/P 12 The ellipse E has foci at S and S'. The point P on E is such that angle PSS' is a right angle and angle $PS'S = 30°$

Show that the eccentricity of the ellipse, e, is $\frac{1}{\sqrt{3}}$ **(6 marks)**

2.4 Tangents and normals to an ellipse

You can use parametric **differentiation** or implicit differentiation to find the equations of the tangent and **normal** to an ellipse at a given point. It is often simpler to derive the equations rather than memorising formulae.

Watch out If you are asked to **prove** a result you will need to show enough working to demonstrate your process for finding the **gradient**.

Example 10 — SKILLS ANALYSIS

Find the equation of the tangent to the ellipse with equation $\frac{x^2}{9} + \frac{y^2}{4} = 1$ at the point $P(3\cos t, 2\sin t)$.

$y = 2\sin t, \; x = 3\cos t$

$\frac{dy}{dx} = \frac{\frac{dy}{dt}}{\frac{dx}{dt}} = \frac{2\cos t}{-3\sin t}$ ← Find the gradient.

Problem-solving

You could also differentiate the equation implicitly: $\frac{2}{9}x + \frac{1}{2}y\frac{dy}{dx} = 0$ and therefore $\frac{dy}{dx} = -\frac{4x}{9y}$

$y - 2\sin t = \frac{2\cos t}{-3\sin t}(x - 3\cos t)$ ← Write down the equation of the tangent using $y - y_1 = m(x - x_1)$

$3y\sin t - 6\sin^2 t = -2x\cos t + 6\cos^2 t$

$3y\sin t + 2x\cos t = 6(\cos^2 t + \sin^2 t)$ ← Simplify.

$3y\sin t + 2x\cos t = 6$ ← Use $\cos^2 t + \sin^2 t \equiv 1$

Example 11 — SKILLS CRITICAL THINKING

Show that the equation of the normal to the ellipse with equation $\frac{x^2}{a^2} + \frac{y^2}{b^2} = 1$ at the point $P(a\cos t, b\sin t)$ is $ax\sin t - by\cos t = (a^2 - b^2)\cos t\sin t$

$\frac{dy}{dx} = \frac{b\cos t}{-a\sin t}$ ← Find the gradient.

Gradient of normal is $\frac{a\sin t}{b\cos t}$ ← Use the perpendicular gradient rule.

Equation is $y - b\sin t = \frac{a\sin t}{b\cos t}(x - a\cos t)$ ← Use $y - y_1 = m(x - x_1)$ and simplify.

$by\cos t - b^2\cos t\sin t = ax\sin t - a^2\cos t\sin t$

$ax\sin t - by\cos t = (a^2 - b^2)\cos t\sin t$

- An equation of the normal to the ellipse with equation $\dfrac{x^2}{a^2} + \dfrac{y^2}{b^2} = 1$ at the point $P(a\cos t, b\sin t)$ is $ax\sin t - by\cos t = (a^2 - b^2)\cos t\sin t$

You can use a similar method to find the general equation of a tangent to an ellipse.

- An equation of the tangent to the ellipse with equation $\dfrac{x^2}{a^2} + \dfrac{y^2}{b^2} = 1$ at the point $P(a\cos t, b\sin t)$ is $bx\cos t + ay\sin t = ab$

Links The derivation of this result is left as an exercise. → Exercise 2D Q3

Example 12

The point $P\left(2, \dfrac{3\sqrt{3}}{2}\right)$ lies on the ellipse E with parametric equations $x = 4\cos\theta$, $y = 3\sin\theta$, $0 \leq \theta < 2\pi$

a Find the value of θ at the point P.

The normal to the ellipse at P cuts the x-axis at the point A.

b Find the coordinates of the point A.

a $4\cos\theta = 2 \Rightarrow \cos\theta = \dfrac{1}{2}$ so $\theta = \dfrac{\pi}{3}, \dfrac{5\pi}{3}$

$3\sin\theta = 3\dfrac{\sqrt{3}}{2} \Rightarrow \sin\theta = \dfrac{\sqrt{3}}{2}$ so $\theta = \dfrac{\pi}{3}, \dfrac{2\pi}{3}$

So $\theta = \dfrac{\pi}{3}$

b $\dfrac{dy}{dx} = \dfrac{3\cos\theta}{-4\sin\theta}$

So gradient of normal is $\dfrac{4\sin\theta}{3\cos\theta}$

At P the gradient of the normal is

$4 \times \dfrac{\frac{\sqrt{3}}{2}}{3 \times \frac{1}{2}} = \dfrac{4\sqrt{3}}{3}$

Equation of normal at P is

$y - 3\dfrac{\sqrt{3}}{2} = \dfrac{4\sqrt{3}}{3}(x - 2)$

Cuts x-axis at $-9\sqrt{3} = 8\sqrt{3}(x - 2)$

So A is $\left(\dfrac{7}{8}, 0\right)$

Set $a\cos\theta$ as the x-coordinate and $b\sin\theta$ as the y-coordinate and solve to find θ. Choose the value of θ in the given range that satisfies both equations.

Use the general point to find the gradient.

Use the perpendicular gradient rule then substitute the value of θ.

This can be found by implicit differentiation on the Cartesian equation $\dfrac{x^2}{16} + \dfrac{y^2}{9} = 1$. Differentiating:

$\dfrac{2}{16}x + \dfrac{2}{9}y\dfrac{dy}{dx} = 0$ so $\dfrac{dy}{dx} = -\dfrac{9x}{16y}$ and

using the coordinates of P, $\dfrac{dy}{dx} = \dfrac{-18}{16 \times 3\frac{\sqrt{3}}{2}} = \dfrac{-3}{4\sqrt{3}}$

so normal gradient is $\dfrac{4\sqrt{3}}{3}$

Let $y = 0$ and solve to find x.

Example 13 SKILLS REASONING/ARGUMENTATION

Show that the condition for $y = mx + c$ to be a tangent to the ellipse $\dfrac{x^2}{a^2} + \dfrac{y^2}{b^2} = 1$ is $b^2 + a^2m^2 = c^2$

The line meets the ellipse when $\dfrac{x^2}{a^2} + \dfrac{(mx+c)^2}{b^2} = 1$

So $b^2x^2 + a^2m^2x^2 + 2a^2mxc + a^2c^2 = a^2b^2$

$x^2(b^2 + a^2m^2) + 2a^2mcx + a^2(c^2 - b^2) = 0$

Substitute $mx + c$ for y.

Multiply out and rearrange as a quadratic equation in x.

To be a tangent there must be only one real root. Therefore the discriminant of this quadratic is 0.

$(2a^2mc)^2 = 4(b^2 + a^2m^2)a^2(c^2 - b^2)$

So $4a^4m^2c^2 = 4a^2(b^2c^2 - b^4 + a^2m^2c^2 - a^2b^2m^2)$

$a^2m^2c^2 = b^2c^2 - b^4 + a^2m^2c^2 - a^2b^2m^2$

$b^4 + a^2b^2m^2 = b^2c^2$

$b^2 + a^2m^2 = c^2$

Use the properties of the discriminant.
← Pure 1 Section 2.5

Multiply out and simplify.

Cancel b^2

Problem-solving

This is a general result about tangents to ellipses. Unless you are asked to prove it, you could quote it in your exam.

Example 14

The ellipse C has equation $\dfrac{x^2}{5^2} + \dfrac{y^2}{3^2} = 1$. The line l is normal to the ellipse at P and passes through the point Q, where C cuts the y-axis, as shown in the diagram.

Find the exact coordinates of the point R where l cuts the positive x-axis.

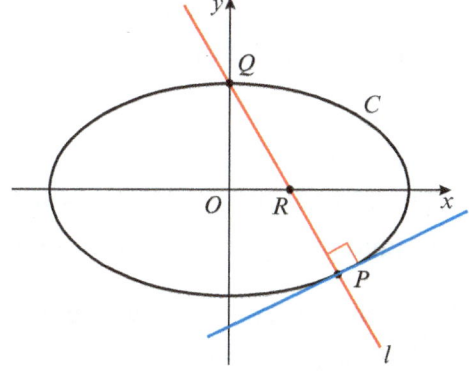

$\dfrac{x^2}{5^2} + \dfrac{y^2}{3^2} = 1$ so $a = 5$ and $b = 3$

$ax\sin\theta - by\cos\theta = (a^2 - b^2)\cos\theta\sin\theta$

$5x\sin\theta - 3y\cos\theta = 16\cos\theta\sin\theta$

Q cuts the y-axis at $(0, 3)$

$-9\cos\theta = 16\cos\theta\sin\theta$

$-9 = 16\sin\theta$

$\sin\theta = -\dfrac{9}{16}$

$\cos\theta = \sqrt{1 - \sin^2\theta}$

$= \dfrac{5\sqrt{7}}{16}$

So the equation of l is:

$5\left(-\dfrac{9}{16}\right)x - 3\left(\dfrac{5\sqrt{7}}{16}\right)y = 16\left(\dfrac{5\sqrt{7}}{16}\right)\left(-\dfrac{9}{16}\right)$

$-3x - \sqrt{7}y = -3\sqrt{7}$

When $y = 0$

$-3x = -3\sqrt{7}$

$x = \sqrt{7}$

So l cuts the x-axis at $(\sqrt{7}, 0)$.

Deduce the values for a and b from the general equation of an ellipse, $\dfrac{x^2}{a^2} + \dfrac{y^2}{b^2} = 1$

State the general equation for the normal of an ellipse and substitute $a = 5$ and $b = 3$

The ellipse cuts the y-axis at $(0, \pm b)$ and $b = 3$

Substitute $x = 0$, $y = 3$ into the general equation for the normal to an ellipse.

Use your value of $\sin\theta$ to find the value of $\cos\theta$.

Problem-solving

The identity $\cos^2\theta + \sin^2\theta \equiv 1$ gives $\cos\theta = \pm\dfrac{5\sqrt{7}}{16}$ However, from the diagram you can see that P is in the fourth **quadrant**, so $\cos\theta$ must be positive.

Substitute your exact values for $\sin\theta$ and $\cos\theta$ to find the equation of l.

Substitute $y = 0$ to find the points where l cuts the x-axis.

Exercise 2D

1 Find the equations of tangents and normals to the following ellipses at the points given.

   a $\dfrac{x^2}{4} + y^2 = 1$ at $(2\cos\theta, \sin\theta)$
   b $\dfrac{x^2}{25} + \dfrac{y^2}{9} = 1$ at $(5\cos\theta, 3\sin\theta)$

2 Find equations of tangents and normals to the following ellipses at the points given.

   a $\dfrac{x^2}{9} + \dfrac{y^2}{1} = 1$ at $\left(\sqrt{5}, \dfrac{2}{3}\right)$
   b $\dfrac{x^2}{16} + \dfrac{y^2}{4} = 1$ at $(-2, \sqrt{3})$

(P) 3 Show that the equation of the tangent to the ellipse $\dfrac{x^2}{a^2} + \dfrac{y^2}{b^2} = 1$ at the point $(a\cos t, b\sin t)$ is $bx\cos t + ay\sin t = ab$

4 a Show that the line $y = x + \sqrt{5}$ is a tangent to the ellipse with equation $\dfrac{x^2}{4} + \dfrac{y^2}{1} = 1$
   b Find the point of contact of this tangent.

5 a Find an equation of the normal to the ellipse with equation $\dfrac{x^2}{9} + \dfrac{y^2}{4} = 1$ at the point $P(3\cos\theta, 2\sin\theta)$.

This normal crosses the x-axis at the point $\left(-\dfrac{5}{6}, 0\right)$.

   b Find the value of θ and the exact coordinates of the possible positions of P.

6 The line $y = 2x + c$ is a tangent to $x^2 + \dfrac{y^2}{4} = 1$

Find the possible values of c.

7 The line with equation $y = mx + 3$ is a tangent to $x^2 + \dfrac{y^2}{5} = 1$

Find the possible values of m.

(E) 8 The line $y = mx + 4$ ($m > 0$) is a tangent to the ellipse E with equation $\dfrac{x^2}{3} + \dfrac{y^2}{4} = 1$ at the point P.

   a Find the value of m.   **(4 marks)**
   b Find the coordinates of the point P.   **(2 marks)**

The normal to E at P crosses the y-axis at the point A.

   c Find the coordinates of A.   **(5 marks)**

The tangent to E at P crosses the y-axis at the point B.

   d Find the area of triangle APB.   **(5 marks)**

(E) 9 The ellipse E has equation $\dfrac{x^2}{9} + \dfrac{y^2}{4} = 1$

   a Show that the gradient of the tangent to E at the point $P(3\cos\theta, 2\sin\theta)$ is $-\dfrac{2}{3}\cot\theta$ **(4 marks)**
   b Show that the point $Q\left(\dfrac{9}{5}, -\dfrac{8}{5}\right)$ lies on E.   **(2 marks)**
   c Find the gradient of the tangent to E at Q.   **(1 mark)**

The tangents to E at the points P and Q are perpendicular.

   d Find the value of $\tan\theta$ and hence the exact coordinates of the two possible positions of P.   **(4 marks)**

(P) 10 The line $y = mx + c$ is a tangent to both of the ellipses $\dfrac{x^2}{9} + \dfrac{y^2}{46} = 1$ and $\dfrac{x^2}{25} + \dfrac{y^2}{14} = 1$

Find the possible values of m and c.

E/P 11 The ellipse E has equation $\dfrac{x^2}{8^2} + \dfrac{y^2}{4^2} = 1$. The line l_1 is tangent to E at the point $P(8\cos\theta, 4\sin\theta)$ and the line l_2 is normal to E at the point $P(8\cos\theta, 4\sin\theta)$. Line l_1 cuts the x-axis at A and line l_2 cuts the y-axis at B. Find the equation of the line AB. **(6 marks)**

E/P 12 The ellipse E has equation $\dfrac{x^2}{5^2} + \dfrac{y^2}{3^2} = 1$. The line l_1 is tangent to E at the point $P(5\cos\theta, 3\sin\theta)$.
  a Use calculus to show that an equation for l_1 is $3x\cos\theta + 5y\sin\theta = 15$ **(5 marks)**
  The line l_1 cuts the y-axis at Q. The line l_2 passes through the point Q, perpendicular to l_1.
  b Find the equation of the line l_2. **(3 marks)**
  c Given that l_2 cuts the x-axis at $(-4, 0)$, show that $\cos\theta = \dfrac{4}{5}$ **(3 marks)**

E/P 13 The ellipse E has equation $\dfrac{x^2}{4} + \dfrac{y^2}{16} = 1$. The line l_1 is tangent to E at the point $P(2\cos t, 4\sin t)$.
  a Use calculus to show that an equation for l_1 is $2x\cos t + y\sin t = 4$ **(5 marks)**
  The line l_2 passes through the origin and is perpendicular to l_1. The lines l_1 and l_2 intersect at the point Q.
  b Show that the coordinates of Q are $\left(\dfrac{8\cos t}{4\cos^2 t + \sin^2 t}, \dfrac{4\sin t}{4\cos^2 t + \sin^2 t}\right)$ **(4 marks)**

2.5 Tangents and normals to a hyperbola

You can find the equations of the tangent and normal to a hyperbola at a given point.

Example 15 SKILLS ANALYSIS

Find the equation of the tangent to the hyperbola with equation $\dfrac{x^2}{9} - \dfrac{y^2}{4} = 1$ at the point $(6, 2\sqrt{3})$.

Differentiating, $\dfrac{2}{9}x - \dfrac{2}{4}y\dfrac{dy}{dx} = 0$ — Use implicit differentiation.

At $(6, 2\sqrt{3})$,

$\dfrac{12}{9} - \dfrac{4\sqrt{3}}{4}\dfrac{dy}{dx} = 0 \Rightarrow \dfrac{dy}{dx} = \dfrac{4\sqrt{3}}{9}$

Equation of tangent is

$y - 2\sqrt{3} = \dfrac{4\sqrt{3}}{9}(x - 6)$ — Use $y - y_1 = m(x - x_1)$

or $y = \dfrac{4\sqrt{3}}{9}x - \dfrac{2\sqrt{3}}{3}$

Example 16

Show that the equation of the tangent to the hyperbola with equation $\dfrac{x^2}{a^2} - \dfrac{y^2}{b^2} = 1$ at the point $(a\cosh t, b\sinh t)$ can be written as $bx\cosh t - ay\sinh t = ab$

$x = a\cosh t, y = b\sinh t$

$\dfrac{dy}{dx} = \dfrac{\frac{dy}{dt}}{\frac{dx}{dt}} = \dfrac{b\cosh t}{a\sinh t}$ ⟵ Use the chain rule to find $\dfrac{dy}{dx}$

See Chapter 3 for differentiation of $\sinh t$ and $\cosh y$

Equation of tangent is

$y - b\sinh t = \dfrac{b\cosh t}{a\sinh t}(x - a\cosh t)$ ⟵ Use $y - y_1 = m(x - x_1)$

$ay\sinh t - ab\sinh^2 t = bx\cosh t - ab\cosh^2 t$

$ay\sinh t + ab(\cosh^2 t - \sinh^2 t) = bx\cosh t$

$ay\sinh t + ab = bx\cosh t$ ⟵ Use $\cosh^2 t - \sinh^2 t \equiv 1$

$bx\cosh t - ay\sinh t = ab$

- An equation of the tangent to the hyperbola with equation $\dfrac{x^2}{a^2} - \dfrac{y^2}{b^2} = 1$ at the point $P(a\cosh t, b\sinh t)$ is $ay\sinh t + ab = bx\cosh t$

You can use the alternative form of a general point on a hyperbola to find a different general equation of a tangent to a hyperbola.

- An equation of the tangent to the hyperbola with equation $\dfrac{x^2}{a^2} - \dfrac{y^2}{b^2} = 1$ at the point $P(a\sec\theta, b\tan\theta)$ is $bx\sec\theta - ay\tan\theta = ab$

Links The derivation of this result is left as an exercise. → **Exercise 2E Q3**

Example 17   SKILLS   REASONING/ARGUMENTATION

Show that an equation of the normal to the hyperbola with equation $\dfrac{x^2}{a^2} - \dfrac{y^2}{b^2} = 1$ at $(a\sec\theta, b\tan\theta)$ is $by + ax\sin\theta = (a^2 + b^2)\tan\theta$

$y = b\tan\theta, x = a\sec\theta$

$\dfrac{dy}{dx} = \dfrac{\frac{dy}{d\theta}}{\frac{dx}{d\theta}} = \dfrac{b\sec^2\theta}{a\sec\theta\tan\theta} = \dfrac{b}{a\sin\theta}$ ⟵ Use the chain rule to find $\dfrac{dy}{dx}$

So gradient of normal is $-\dfrac{a\sin\theta}{b}$ ⟵ Use the perpendicular gradient rule.

Equation of the normal is

$y - b\tan\theta = -\dfrac{a\sin\theta}{b}(x - a\sec\theta)$ ⟵ Use $y - y_1 = m(x - x_1)$

$by - b^2\tan\theta = -ax\sin\theta + a^2\tan\theta$

So $by + ax\sin\theta = (a^2 + b^2)\tan\theta$

- An equation of the normal to the hyperbola with equation $\dfrac{x^2}{a^2} - \dfrac{y^2}{b^2} = 1$ at the point $P(a\sec\theta, b\tan\theta)$ is $by + ax\sin\theta = (a^2 + b^2)\tan\theta$

You can use the other form of a general point on a hyperbola to find a different general equation of a normal to a hyperbola.

FURTHER COORDINATE SYSTEMS — CHAPTER 2

- An equation of the normal to the hyperbola with equation $\dfrac{x^2}{a^2} - \dfrac{y^2}{b^2} = 1$ at the point $P(a\cosh t, b\sinh t)$ is $ax\sinh t + by\cosh t = (a^2 + b^2)\sinh t \cosh t$

Links The derivation of this result is left as an exercise → Exercise 2E Q4

Example 18

Show that the condition for the line $y = mx + c$ to be a tangent to the hyperbola $\dfrac{x^2}{a^2} - \dfrac{y^2}{b^2} = 1$ is that m and c satisfy $b^2 + c^2 = a^2 m^2$

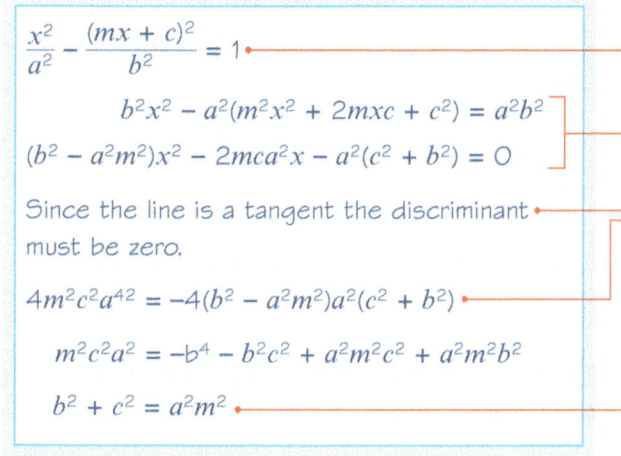

Substitute $mx + c$ for y in the equation of the hyperbola.

Multiply out and collect terms as a quadratic in x.

Use discriminant properties.

Cancel $4a^2$

Cancel b^2

Problem-solving

This is a general result about tangents to hyperbolas. Unless you are asked to prove it, you could quote it in your exam.

Example 19

The tangent to the hyperbola with equation $\dfrac{x^2}{9} - \dfrac{y^2}{4} = 1$ at the point $(3\cosh t, 2\sinh t)$ crosses the y-axis at the point $(0, -1)$. Find the value of t.

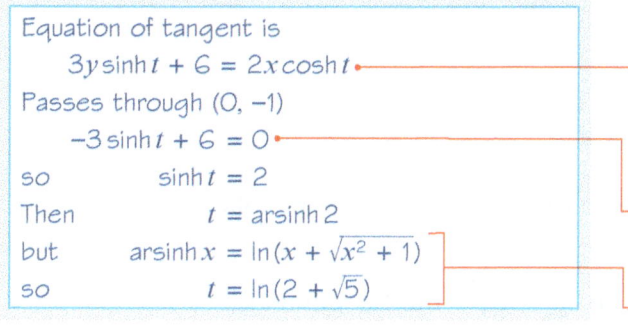

Remember that for a hyperbola with equation $\dfrac{x^2}{a^2} - \dfrac{y^2}{b^2} = 1$, the equation of the tangent at point $(a\cosh t, b\sinh t)$ is $ay\sinh t + ab = bx\cosh t$
Here $a = 3$ and $b = 2$

Substitute $x = 0$ and $y = -1$

Use the formula for $\operatorname{arsinh}(x)$ from the formula booklet.

Example 20

The hyperbola H has equation $\dfrac{x^2}{36} - \dfrac{y^2}{9} = 1$

The line l_1 is the tangent to H at the point $P(6\cosh t, 3\sinh t)$. The line l_2 passes through the origin and is perpendicular to l_1. The lines l_1 and l_2 intersect at the point Q.

Show that the coordinates of the point Q are $\left(\dfrac{6\cosh t}{4\sinh^2 t + \cosh^2 t}, -\dfrac{12\sinh t}{4\sinh^2 t + \cosh^2 t}\right)$

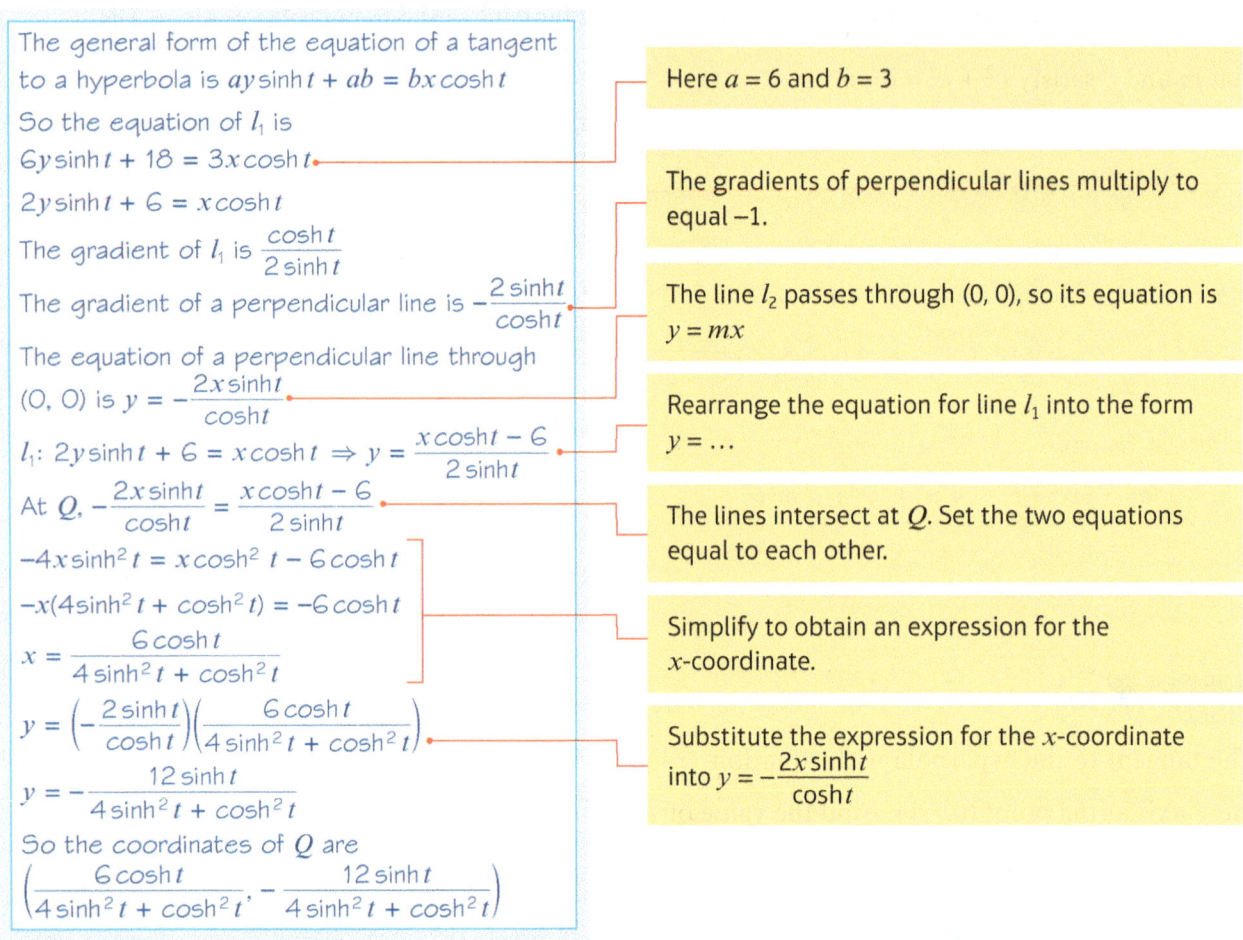

Exercise 2E   SKILLS   ANALYSIS, REASONING/ARGUMENTATION

1 Find the equations of the tangents and normals to the hyperbolas with the following equations at the points indicated.

  a $\dfrac{x^2}{16} - \dfrac{y^2}{2} = 1$ at the point $(12, 4)$

  b $\dfrac{x^2}{36} - \dfrac{y^2}{12} = 1$ at the point $(12, 6)$

  c $\dfrac{x^2}{25} - \dfrac{y^2}{3} = 1$ at the point $(10, 3)$

2 Find the equations of the tangents and normals to the hyperbolas with the following equations at the points indicated.

  a $\dfrac{x^2}{25} - \dfrac{y^2}{4} = 1$ at the point $(5\cosh t, 2\sinh t)$

  b $\dfrac{x^2}{1} - \dfrac{y^2}{9} = 1$ at the point $(\sec t, 3\tan t)$

3 Show that the equation of the tangent to the hyperbola $\frac{x^2}{a^2} - \frac{y^2}{b^2} = 1$ at the point $(a\sec t, b\tan t)$ is $bx\sec t - ay\tan t = ab$

4 Show that the equation of the normal to the hyperbola $\frac{x^2}{a^2} - \frac{y^2}{b^2} = 1$ at the point $(a\cosh t, b\sinh t)$ is $ax\sinh t + by\cosh t = (a^2 + b^2)\sinh t\cosh t$

5 The point $P(4\cosh t, 3\sinh t)$, $t \neq 0$, lies on the hyperbola with equation $\frac{x^2}{16} - \frac{y^2}{9} = 1$
The tangent at P crosses the y-axis at the point A.
  a Find, in terms of t, the coordinates of A.
The normal to the hyperbola at P crosses the y-axis at B.
  b Find, in terms of t, the coordinates of B.
  c Find, in terms of t, the area of triangle APB.

6 The tangents from the points P and Q on the hyperbola with equation $\frac{x^2}{4} - \frac{y^2}{9} = 1$ meet at the point $(1, 0)$. Find the exact coordinates of P and Q.

7 The line $y = 2x + c$ is a tangent to the hyperbola $\frac{x^2}{10} - \frac{y^2}{4} = 1$. Find the possible values of c.

8 The line $y = mx + 12$ is a tangent to the hyperbola $\frac{x^2}{49} - \frac{y^2}{25} = 1$ at the point P.
Find the possible values of m.

9 The line with equation $y = mx + c$ is a tangent to both of the hyperbolas $\frac{x^2}{4} - \frac{y^2}{15} = 1$ and $\frac{x^2}{9} - \frac{y^2}{95} = 1$
Find the possible values of m and c.

10 The line $y = -x + c$, $c > 0$, touches the hyperbola $\frac{x^2}{25} - \frac{y^2}{16} = 1$ at the point P.
  a Find the value of c.   **b** Find the exact coordinates of P.

11 The hyperbola H has equation $\frac{x^2}{a^2} - \frac{y^2}{b^2} = 1$
  a Use calculus to show that the equation of the normal to H at the point $(a\cosh t, b\sinh t)$, $t \neq 0$, may be written in the form $ax\sinh t + by\cosh t = (a^2 + b^2)\sinh t\cosh t$. **(4 marks)**
The line l_1 is the normal to H at the point $(a\cosh t, b\sinh t)$. Given that l_1 meets the x-axis at the point P.
  b find, in terms of a, b and t, the coordinates of P. **(2 marks)**
The line l_2 is the tangent to H at the point $(a, 0)$. Given that l_1 and l_2 meet at the point Q,
  c find, in terms of a, b and t, the coordinates of Q. **(2 marks)**

12 The hyperbola H has equation $\frac{x^2}{49} - \frac{y^2}{25} = 1$
The line l_1 is the tangent to H at the point $(7\sec\theta, 5\tan\theta)$.
  a Use calculus to show that an equation of l_1 is $7y\sin\theta = 5x - 35\cos\theta$ **(5 marks)**
The line l_2 passes through the origin and is perpendicular to l_1. The lines l_1 and l_2 intersect at the point Q.
  b Show that the coordinates of the point Q are $\left(\frac{175\cos\theta}{25 + 49\sin^2\theta}, \frac{-245\sin\theta\cos\theta}{25 + 49\sin^2\theta}\right)$ **(5 marks)**

E/P 13 P and Q are two distinct points on the hyperbola described by the equation $x^2 - 4y^2 = 16$
The line l passes through the point P and the point Q. The tangent to the hyperbola at P and the tangent to the hyperbola at Q intersect at the point (m, n). Show that an equation of the line l is $mx - 4ny = 16$ **(9 marks)**

P 14 Show that there are exactly two tangents to the hyperbola $\dfrac{x^2}{4^2} - \dfrac{y^2}{2^2} = 1$ passing through the point $(6, 4)$ and find each of their equations.

E/P 15 The point P lies on the hyperbola H with equation $x^2 - y^2 = 1$. The tangent to H at P cuts the asymptotes of P at the points A and B.
  a Prove that P is the **midpoint** of the line segment AB. **(6 marks)**
  b Prove that $OA \times OB$ remains constant as the position of P varies on H. **(3 marks)**

2.6 Loci

Each of the conic sections can be defined as a locus of points. For example, the parabola is the locus of points equidistant from a fixed point and a fixed straight line. You can use the properties of the conic sections, and the general points on each curve, to find other loci associated with these curves.

Example 21   SKILLS   CRITICAL THINKING

The tangent to the ellipse with equation $\dfrac{x^2}{a^2} + \dfrac{y^2}{b^2} = 1$ at the point $P(a\cos t, b\sin t)$ crosses the x-axis at A and the y-axis at B.

Find an equation for the locus of the midpoint of AB as P moves round the ellipse.

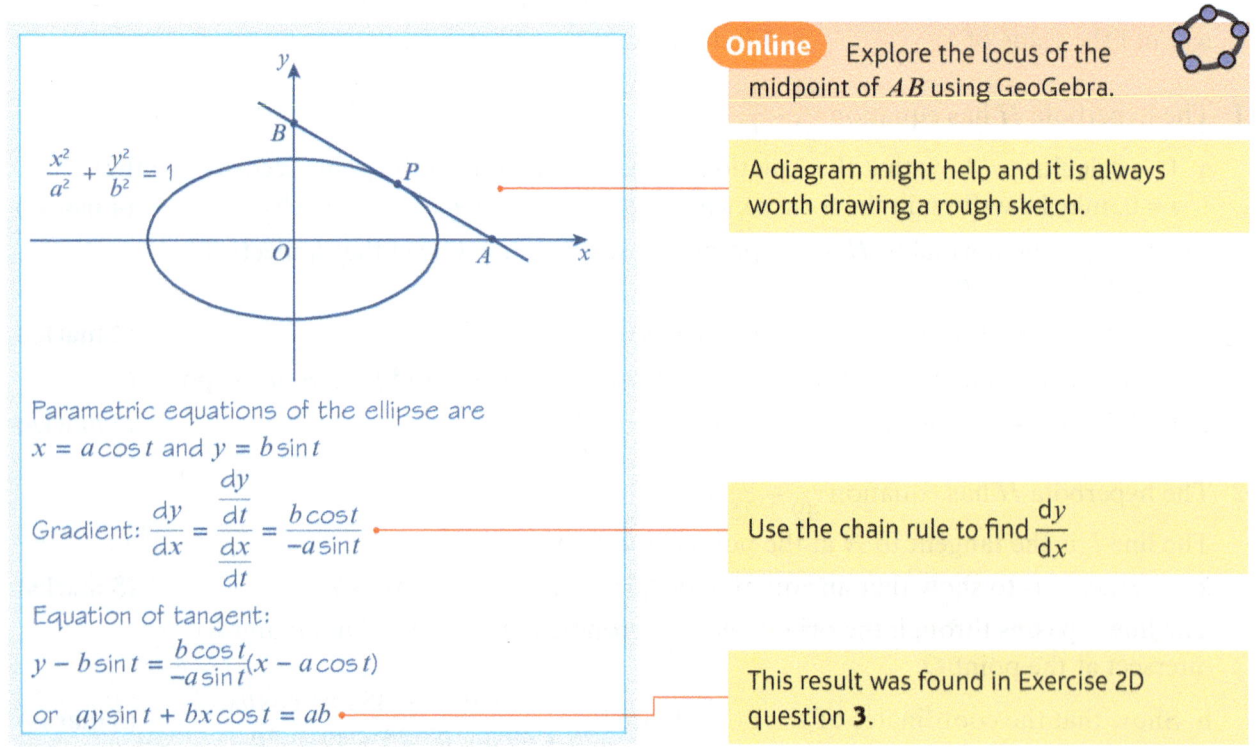

Online Explore the locus of the midpoint of AB using GeoGebra.

A diagram might help and it is always worth drawing a rough sketch.

Parametric equations of the ellipse are
$x = a\cos t$ and $y = b\sin t$

Gradient: $\dfrac{dy}{dx} = \dfrac{\frac{dy}{dt}}{\frac{dx}{dt}} = \dfrac{b\cos t}{-a\sin t}$

Use the chain rule to find $\dfrac{dy}{dx}$

Equation of tangent:
$y - b\sin t = \dfrac{b\cos t}{-a\sin t}(x - a\cos t)$
or $ay\sin t + bx\cos t = ab$

This result was found in Exercise 2D question **3**.

A is $(a\sec t, 0)$.
B is $(0, b\csc t)$.

The midpoint of AB has coordinates (X, Y) where
$X = \dfrac{a\sec t}{2}$
$Y = \dfrac{b\csc t}{2}$

Rearranging:
$\cos t = \dfrac{a}{2X}$ and $\sin t = \dfrac{b}{2Y}$

Using $\cos^2 t + \sin^2 t \equiv 1$ gives the locus
$\left(\dfrac{a}{2X}\right)^2 + \left(\dfrac{b}{2Y}\right)^2 = 1$

— Use $x = 0$ to find B and $y = 0$ to find A.

— To find the locus of the midpoint, let the coordinates of the midpoint be (X, Y) and then form parametric equations for X and Y.

— Eliminate (i.e. get rid of) the parameter (t in this case) to find an equation in X and Y.

Problem-solving

In some questions, you may be asked to show that the locus has a particular shape, so you may need to rearrange the final equation into an appropriate form.

You might also need to use properties of the parabola and rectangular hyperbola when solving loci questions. This table summarises the results.

	Parabola	Rectangular hyperbola
Standard Cartesian equation	$y^2 = 4ax$	$xy = c^2$
Parametric equations	$x = at^2, y = 2at$	$x = ct, y = \dfrac{c}{t}$
General point, P	$(at^2, 2at)$	$\left(ct, \dfrac{c}{t}\right)$
Equation of tangent at P	$ty = x + at^2$	$x + t^2 y = 2ct$
Equation of normal at P	$y + tx = 2at + at^3$	$t^3 x - ty = c(t^4 - 1)$

Example 22 · SKILLS · REASONING/ARGUMENTATION

The normal at $P(ap^2, 2ap)$ and the normal at $Q(aq^2, 2aq)$ to the parabola with equation $y^2 = 4ax$ meet at R.

a Find the coordinates of R.

The chord PQ passes through the focus $(a, 0)$ of the parabola.

b Show that $pq = -1$.

c Show that the locus of R is a parabola with equation $y^2 = a(x - 3a)$

a To find R, find the intersections of the normals.
Normal at P is $y + px = 2ap + ap^3$
Normal at Q is $y + qx = 2aq + aq^3$

— Use the standard result for the equation of a normal to a parabola at $(at^2, 2at)$: $y + tx = 2at + at^3$

Subtracting,
$(p - q)x = 2a(p - q) + a(p^3 - q^3)$
$(p - q)x = 2a(p - q) + a(p - q)(p^2 + pq + q^2)$
$x = 2a + a(p^2 + pq + q^2)$
$y = 2ap + ap^3 - 2ap - ap^3 - ap^2q - apq^2$
$= -apq(p + q)$
So R is $(2a + a(p^2 + pq + q^2), -apq(p + q))$

Problem-solving

The factorisations of $(p^3 \pm q^3) = (p \pm q)(p^2 \mp pq + q^2)$ are particularly useful in this type of problem and should be learned.

Substitute for x to find y.

b Chord PQ has gradient
$$\frac{2a(p - q)}{a(p^2 - q^2)} = \frac{2(p - q)}{(p - q)(p + q)} = \frac{2}{p + q}$$

Use $\frac{y_1 - y_2}{x_1 - x_2}$

Equation of chord is
$$y - 2ap = \frac{2}{p + q}(x - ap^2)$$
$$\Rightarrow y(p + q) = 2x + 2apq$$

Since the chord passes through $(a, 0)$,
$0 = 2a + 2apq$
$\Rightarrow pq = -1$

Problem-solving

Notice that if you let $p = q$ in the equation of the chord you get the equation of the tangent at Q. This is sometimes a useful technique to use.

c Using $pq = -1$ the coordinates of R become
$(a + a(p^2 + q^2), a(p + q))$
Let R be (X, Y), then
$X = a + a(p^2 + q^2)$
$Y = a(p + q)$
So $X = a + a((p + q)^2 - 2pq)$
and using $pq = -1$
$X = 3a + a(p + q)^2$
But $p + q = \frac{Y}{a}$
So $X = 3a + a\left(\frac{Y}{a}\right)^2$
$\Rightarrow Y^2 = a(X - 3a)$

The following technique is particularly useful when tackling questions of this sort.

Since $(p + q)^2 = p^2 + q^2 + 2pq$
then $p^2 + q^2 = (p + q)^2 - 2pq$
Using $pq = -1$ gives $p^2 + q^2 = (p + q)^2 + 2$

Now use Y to eliminate p and q.

Rearrange to the specified form.

Exercise 2F SKILLS REASONING/ARGUMENTATION

 1 The tangent at $P(ap^2, 2ap)$ and the tangent at $Q(aq^2, 2aq)$ to the parabola with equation $y^2 = 4ax$ meet at R.
  a Find the coordinates of R.
  The chord PQ passes through the focus $(a, 0)$ of the parabola.
  b Show that the locus of R lies on the line $x = -a$
  Given instead that the chord PQ has gradient 2,
  c find the locus of R.

E/P 2 The hyperbola H has equation $\dfrac{x^2}{a^2} - \dfrac{y^2}{b^2} = 1$

The line l_1 is tangent to H at the point $P(a \sec t, b \tan t)$.

  a Use calculus to show that an equation for l_1 is $bx \sec t - ay \tan t = ab$ **(4 marks)**

The line l_1 cuts the x-axis at A and the y-axis at B.

  b Show that the locus of the midpoint of AB is $\dfrac{a^2}{4x^2} - \dfrac{b^2}{4y^2} = 1$ **(5 marks)**

E/P 3 The hyperbola H has equation $\dfrac{x^2}{a^2} - \dfrac{y^2}{b^2} = 1$. The line l_1 is normal to H at the point $P(a \sec t, b \tan t)$.

  a Use calculus to show that an equation for l_1 is $ax \sin t + by = (a^2 + b^2) \tan t$ **(4 marks)**

The line l_1 cuts the x-axis at A and the y-axis at B.

  b Show that the locus of the midpoint of AB is $4a^2 x^2 = (a^2 + b^2)^2 + 4b^2 y^2$ **(5 marks)**

E/P 4 The ellipse E has equation $\dfrac{x^2}{25} + \dfrac{y^2}{9} = 1$. The line l_1 is normal to E at the point $P(5 \cos \theta, 3 \sin \theta)$.

  a Use calculus to show that an equation for l_1 is $3y \cos \theta = 5x \sin \theta - 16 \sin \theta \cos \theta$ **(4 marks)**

The line l_1 cuts the x-axis at M and the y-axis at N.

  b Show that the locus of the midpoint of MN is $\dfrac{25x^2}{64} + \dfrac{9y^2}{64} = 1$ **(5 marks)**

E/P 5 The tangent at the point $P\left(cp, \dfrac{c}{p}\right)$ and the tangent at the point $Q\left(cq, \dfrac{c}{q}\right)$ to the rectangular hyperbola $xy = c^2$ intersect at the point R.

  a Show that R is $\left(\dfrac{2cpq}{p+q}, \dfrac{2c}{p+q}\right)$ **(4 marks)**

  b Show that the chord PQ has equation $ypq + x = c(p + q)$ **(3 marks)**

  c Find the locus of R, given that:

    i the chord PQ has gradient 2   **(2 marks)**

    ii the chord PQ passes through the point $(1, 0)$ **(2 marks)**

    iii the chord PQ passes through the point $(0, 1)$.   **(2 marks)**

P 6 **a** Find the gradient of the parabola with equation $y^2 = 4ax$ at the point $P(at^2, 2at)$.

  b Hence show that the equation of the tangent at this point is $x - ty + at^2 = 0$

The tangent meets the y-axis at T, and O is the origin.

  c Show that the coordinates of the centre of the circle through O, P and T are $\left(\dfrac{at^2}{2} + a, \dfrac{at}{2}\right)$

E/P 7 The chord PQ to the rectangular hyperbola $xy = c^2$ passes through the point $(0, 1)$.

Find the equation of the locus of the midpoint of PQ as P and Q vary.   **(7 marks)**

E/P 8 The point P lies on the ellipse with equation $\dfrac{x^2}{4} + \dfrac{y^2}{16} = 1$. The point N is the foot of the perpendicular from point P to the line $y = 6$. M is the midpoint of PN.

  a Find an equation for the locus of M as P moves around the ellipse.   **(4 marks)**

  b Show that this locus is a circle and state its centre and radius.   **(3 marks)**

> **Challenge**
>
> The points A and B lie on an ellipse with equation $\dfrac{x^2}{a^2} + \dfrac{y^2}{b^2} = 1$ such that the chord AB has gradient k. Show that the locus of the midpoints of all possible such chords AB has equation $ka^2y + b^2x = 0$, and describe this locus.

Chapter review 2

SKILLS PROBLEM-SOLVING

1 The ellipse E has parametric equations $x = 4\cos\theta$, $y = 9\sin\theta$
  a Find a Cartesian equation of the ellipse.
  b Sketch the ellipse, labelling any points of intersection with the coordinate axes.
  c Find the equation of the normal to the ellipse at $P(4\cos\theta, 9\sin\theta)$.

2 The hyperbola H has parametric equations $x = \pm 2\cosh t$, $y = 5\sinh t$
  a Find a Cartesian equation of the hyperbola.
  b Sketch the hyperbola, giving the equations of the asymptotes and show points of intersection of the hyperbola with the x-axis.
  c Find the equation of the tangent to the hyperbola at $Q(2\cosh t, 5\sinh t)$.

(E/P) 3 A hyperbola of the form $\dfrac{x^2}{a^2} - \dfrac{y^2}{b^2} = 1$ has asymptotes with equations $y = \pm mx$ and passes through the point $(a, 0)$.
  a Find an equation of the hyperbola in terms of x, y, a and m. **(4 marks)**
  A point P on this hyperbola is equidistant from one of the hyperbola's asymptotes and the x-axis.
  b Prove that, for all values of m, P lies on the curve with equation
    $(x^2 - y^2)^2 = 4x^2(x^2 - a^2)$ **(3 marks)**

(E/P) 4 a Prove that the gradient of the chord joining the point $P\left(cp, \dfrac{c}{p}\right)$ and the point $Q\left(cq, \dfrac{c}{q}\right)$ on the rectangular hyperbola with equation $xy = c^2$ is $-\dfrac{1}{pq}$ **(5 marks)**
  The points P, Q and R lie on a rectangular hyperbola, such that the angle QPR is a right angle.
  b Prove that the angle between QR and the tangent at P is also a right angle. **(5 marks)**

(E/P) 5 a Show that an equation of the tangent to the rectangular hyperbola with equation $xy = c^2$ (with $c > 0$) at the point $\left(ct, \dfrac{c}{t}\right)$ is $t^2y + x = 2ct$ **(4 marks)**
  Tangents are drawn from the point $(-3, 3)$ to the rectangular hyperbola with equation $xy = 16$.
  b Find the coordinates of the points of contact of these tangents with the hyperbola. **(4 marks)**

(E/P) 6 The point P lies on the ellipse with equation $9x^2 + 25y^2 = 225$, and A and B are the points $(-4, 0)$ and $(4, 0)$ respectively.
  a Prove that $PA + PB = 10$ **(4 marks)**
  b Prove also that the normal at P bisects the angle APB. **(6 marks)**

7 A curve is given parametrically by $x = ct$, $y = \frac{c}{t}$

    a Show that an equation of the tangent to the curve at the point $\left(ct, \frac{c}{t}\right)$ is $t^2y + x = 2ct$

    (4 marks)

    The point P is the foot of the perpendicular from the origin to this tangent.

    b Show that the locus of P is the curve with equation $(x^2 + y^2)^2 = 4c^2xy$ **(6 marks)**

8 The points $P(ap^2, 2ap)$ and $Q(aq^2, 2aq)$ lie on the parabola with equation $y^2 = 4ax$
The angle $POQ = 90°$, where O is the origin.

    a Prove that $pq = -4$ **(4 marks)**

    Given that the normal at P to the parabola has equation

        $y + xp = ap^3 + 2ap$

    b write down an equation of the normal to the parabola at Q. **(1 mark)**

    c Show that these two normals meet at the point R, with coordinates

        $(ap^2 + aq^2 - 2a, 4a(p + q))$ **(3 marks)**

    d Show that, as p and q vary, the locus of R has equation $y^2 = 16ax - 96a^2$ **(4 marks)**

9 Show that, for all values of m, the straight lines with equations $y = mx \pm \sqrt{b^2 + a^2m^2}$ are tangents to the ellipse with equation $\frac{x^2}{a^2} + \frac{y^2}{b^2} = 1$ **(6 marks)**

10 The chord PQ, where P and Q are points on $xy = c^2$, has gradient 1.
Show that the locus of the point of intersection of the tangents from P and Q is the line $y = -x$ **(6 marks)**

11 The ellipse E has equation $\frac{x^2}{36} + \frac{y^2}{16} = 1$. The line l_1 is tangent to E at the point $P(6\cos\theta, 4\sin\theta)$.

    a Use calculus to show that an equation for l_1 is $2x\cos\theta + 3y\sin\theta = 12$ **(4 marks)**

    The line l_1 cuts the x-axis at A and the y-axis at B.

    b Show that the locus of the midpoint of AB is $\frac{9}{x^2} + \frac{4}{y^2} = 1$ **(5 marks)**

12 The ellipse E has equation $\frac{x^2}{169} + \frac{y^2}{25} = 1$. The line l_1 is tangent to E at the point $P(13\cos\theta, 5\sin\theta)$.

    a Use calculus to show that an equation for l_1 is $5x\cos\theta + 13y\sin\theta = 65$ **(5 marks)**

    The line l_1 cuts the y-axis at A. The line l_2 passes through the point A, perpendicular to l_1.

    b Find the equation of the line l_2. **(3 marks)**

    c Given that l_2 cuts the x-axis at the focus of the ellipse $(-ae, 0)$, show that $\cos\theta = e$ **(3 marks)**

13 The hyperbola H has equation $\frac{x^2}{16} - \frac{y^2}{64} = 1$

    The line l_1 is normal to H at the point $P(4\sec\theta, 8\tan\theta)$.

    a Use calculus to show that an equation for l_1 is $x\sin\theta + 2y = 20\tan\theta$. **(4 marks)**

    The line l_1 cuts the x-axis at A and the y-axis at B.

    b Show that the locus of the midpoint of AB is also a hyperbola and find the equation of this hyperbola. **(6 marks)**

14 The ellipse E has equation $\frac{x^2}{a^2} + \frac{y^2}{b^2} = 1$. The line l_1 is normal to E at the point $P(a\cos t, b\sin t)$.

  a Use calculus to show that an equation for l_1 is $ax\sin t - by\cos t = (a^2 - b^2)\cos t \sin t$. (4 marks)

  The line l_1 cuts the x-axis at M and the y-axis at N.

  b Show that the locus of the midpoint of MN is $4b^2 y^2 + 4a^2 x^2 = (a^2 - b^2)^2$ (5 marks)

 15 The ellipse E with equation $\frac{x^2}{5^2} + \frac{y^2}{3^2}$ has foci at S and S'. Prove that for any point P on the ellipse, $PS + PS' = 10$ (5 marks)

(P) 16 The hyperbola H has equation $\frac{x^2}{9} - \frac{y^2}{16} = 1$. The tangents to the hyperbola at points P and Q both meet one directrix of H at a single point A with y-coordinate 0, and the other directrix of H at points B and C. Find the area of triangle ABC.

> **Challenge**
>
> Let P be a point on an ellipse with eccentricity e. The normal to the ellipse at P meets the major axis at Q. Prove that $QS = ePS$, where S is a focus.
>
>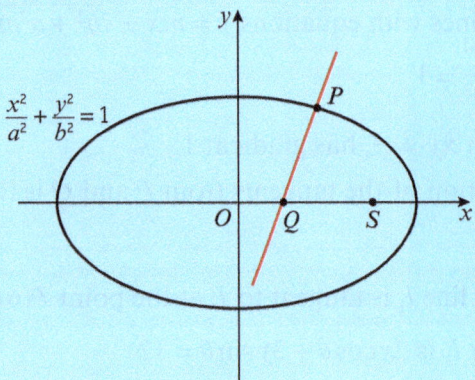

Summary of key points

1. A standard **ellipse** has Cartesian equation $\frac{x^2}{a^2} + \frac{y^2}{b^2} = 1$
   - The standard ellipse has parametric equations $x = a\cos t$, $y = b\sin t$, $0 \le t < 2\pi$
   - A general point P on an ellipse has coordinates $(a\cos t, b\sin t)$.

2. A standard **hyperbola** has Cartesian equation $\frac{x^2}{a^2} - \frac{y^2}{b^2} = 1$
   - The standard hyperbola has parametric equations $x = \pm a\cosh t$, $y = b\sinh t$, $t \in \mathbb{R}$
   - The standard hyperbola has alternative parametric equations $x = a\sec\theta$, $y = b\tan\theta$, $-\pi \le \theta < \pi$, $\theta \ne \pm\frac{\pi}{2}$
   - A general point P on a hyperbola has coordinates $(\pm a\cosh t, b\sinh t)$ or $(a\sec\theta, b\tan\theta)$.

3. For all points, P, on a conic section, the ratio of the distance of P from a fixed point (called the **focus**) and a fixed straight line (called the **directrix**) is constant. This ratio, e, is known as the **eccentricity** of the curve.
   - If $0 < e < 1$, the point P describes an ellipse.
   - If $e = 1$, the point P describes a parabola.
   - If $e > 1$, the point P describes a hyperbola.

4. For an ellipse with equation $\frac{x^2}{a^2} + \frac{y^2}{b^2} = 1$, and $a > b$
   - the eccentricity, $0 < e < 1$, is given by $b^2 = a^2(1 - e^2)$
   - the foci are at $(\pm ae, 0)$
   - the directrices are $x = \pm \frac{a}{e}$

5. For a hyperbola with equation $\frac{x^2}{a^2} - \frac{y^2}{b^2} = 1$
   - the eccentricity, $e > 1$, is given by $b^2 = a^2(e^2 - 1)$
   - the foci are at $(\pm ae, 0)$
   - the directrices are $x = \pm \frac{a}{e}$

6. An equation of the tangent to the ellipse with equation $\frac{x^2}{a^2} + \frac{y^2}{b^2} = 1$ at the point $P(a\cos t, b\sin t)$ is $bx\cos t + ay\sin t = ab$

7. An equation of the normal to the ellipse with equation $\frac{x^2}{a^2} + \frac{y^2}{b^2} = 1$ at the point $P(a\cos t, b\sin t)$ is $ax\sin t - by\cos t = (a^2 - b^2)\cos t \sin t$

8. - An equation of the tangent to the hyperbola with equation $\frac{x^2}{a^2} - \frac{y^2}{b^2} = 1$ at the point $P(a\cosh t, b\sinh t)$ is $ay\sinh t + ab = bx\cosh t$
   - An equation of the tangent to the hyperbola with equation $\frac{x^2}{a^2} - \frac{y^2}{b^2} = 1$ at the point $P(a\sec\theta, b\tan\theta)$ is $bx\sec\theta - ay\tan\theta = ab$

9. - An equation of the normal to the hyperbola with equation $\frac{x^2}{a^2} - \frac{y^2}{b^2} = 1$ at the point $P(a\cosh t, b\sinh t)$ is $ax\sinh t + by\cosh t = (a^2 + b^2)\sinh t\cosh t$
   - An equation of the normal to the hyperbola with equation $\frac{x^2}{a^2} - \frac{y^2}{b^2} = 1$ at the point $P(a\sec\theta, b\tan\theta)$ is $by + ax\sin\theta = (a^2 + b^2)\tan\theta$

3 DIFFERENTIATION

Learning objectives

After completing this chapter you should be able to:
- Find the derivatives of the hyperbolic functions and expressions involving them → pages 47–48
- Find the derivatives of inverse hyperbolic functions → pages 49–50
- Find the derivatives of inverse trigonometric functions → pages 50–52

3.1
3.2

Prior knowledge check

1. Given that $y = \sin^2 x$, show that $\dfrac{dy}{dx} = \sin 2x$
 ← Pure 3 Section 6.4

2. Show that $\dfrac{d(\tan x)}{dx} = \sec^2 x$ ← Pure 3 Section 6.4

3. Given that $y = x^2 \cos 3x$, find $\dfrac{dy}{dx}$
 ← Pure 3 Sections 6.3, 6.4

Hyperbolic functions arise in many problems in mathematics and physics which involve circular functions. For example, the hyperbolic sine function arises in the gravitational potential of a cylinder and the hyperbolic cosine function arises in the shape of a hanging cable.

DIFFERENTIATION   CHAPTER 3

3.1 Differentiating hyperbolic functions

You can differentiate the hyperbolic functions.

The formulae marked (*) are given to you in the formula booklet.

- $\dfrac{d}{dx}(\sinh x) = \cosh x$ (*)

- $\dfrac{d}{dx}(\cosh x) = \sinh x$ (*)

- $\dfrac{d}{dx}(\tanh x) = \text{sech}^2 x$ (*)

- $\dfrac{d}{dx}(\coth x) = -\text{cosech}^2 x$

- $\dfrac{d}{dx}(\text{cosech}\, x) = -\coth x\, \text{cosech}\, x$

- $\dfrac{d}{dx}(\text{sech}\, x) = -\tanh x\, \text{sech}\, x$

Watch out The rules for $\sinh x$ and $\tanh x$ are the same as the corresponding rules for $\sin x$ and $\tan x$. However, the **derivative** of $\cosh x$ is **positive** $\sinh x$.

Example 1

Show that $\dfrac{d}{dx}(\cosh x) = \sinh x$

$\cosh x = \dfrac{e^x + e^{-x}}{2}$ — Use the definition of $\cosh x$.

So $\dfrac{d}{dx}(\cosh x) = \dfrac{d}{dx}\left(\dfrac{e^x + e^{-x}}{2}\right)$

$= \dfrac{e^x - e^{-x}}{2}$ — Differentiate with respect to x: $\dfrac{d}{dx}(e^{-x}) = -e^{-x}$

$\dfrac{e^x - e^{-x}}{2} = \sinh x$ — By definition.

So $\dfrac{d}{dx}(\cosh x) = \sinh x$

Example 2

Differentiate $\cosh 3x$ with respect to x.

$\dfrac{d}{dx}(\cosh 3x) = 3\sinh 3x$ — Use the chain rule.

Example 3 — SKILLS: ANALYSIS

Differentiate $x^2 \cosh 4x$ with respect to x.

$$\frac{d}{dx}(x^2 \cosh 4x) = \frac{d}{dx}(x^2) \cosh 4x + x^2 \frac{d}{dx}(\cosh 4x)$$
$$= 2x \cosh 4x + x^2 \times 4 \sinh 4x$$
$$= 2x \cosh 4x + 4x^2 \sinh 4x$$

Use the product rule.

Example 4 — SKILLS: REASONING; ARGUMENTATION

Given that $y = A \cosh 3x + B \sinh 3x$, where A and B are constants, prove that $\frac{d^2y}{dx^2} = 9y$

$$\frac{dy}{dx} = 3A \sinh 3x + 3B \cosh 3x$$ — Differentiate y.

$$\frac{d^2y}{dx^2} = 9A \cosh 3x + 9B \sinh 3x$$ — Differentiate again.

$$= 9(A \cosh 3x + B \sinh 3x)$$ — Factorise.

$$= 9y$$ — $y = A \cosh 3x + B \sinh 3x$

Exercise 3A — SKILLS: ANALYSIS

In questions 1–16, differentiate with respect to x.

1. $\sinh 2x$
2. $\cosh 5x$
3. $\tanh 2x$
4. $\sinh 3x$
5. $\coth 4x$
6. $\operatorname{sech} 2x$ **Hint** $\operatorname{sech} 2x = \dfrac{1}{\cosh 2x}$
7. $e^{-x} \sinh x$
8. $x \cosh 3x$
9. $\dfrac{\sinh x}{3x}$
10. $x^2 \cosh 3x$
11. $\sinh 2x \cosh 3x$
12. $\ln(\cosh x)$
13. $\sinh x^3$
14. $\cosh^2 2x$
15. $e^{\cosh x}$
16. $\operatorname{cosech} x$ **Hint** $\operatorname{cosech} x = \dfrac{1}{\sinh x}$

17. If $y = a \cosh nx + b \sinh nx$, where a and b are constants, prove that $\dfrac{d^2y}{dx^2} = n^2 y$

18. Find the stationary values of the curve with equation $y = 12 \cosh x - \sinh x$

19. Given that $y = \cosh 3x \sinh x$, find $\dfrac{d^2y}{dx^2}$

20. Find the equation of the tangent and normal to the hyperbola $\dfrac{x^2}{256} - \dfrac{y^2}{16} = 1$ at the point $(16 \cosh q, 4 \sinh q)$.

3.2 Differentiating inverse hyperbolic functions

You can also differentiate the inverse hyperbolic functions.

- $\dfrac{d}{dx}(\operatorname{arsinh} x) = \dfrac{1}{\sqrt{x^2 + 1}}$

- $\dfrac{d}{dx}(\operatorname{arcosh} x) = \dfrac{1}{\sqrt{x^2 - 1}}, \; x > 1$

- $\dfrac{d}{dx}(\operatorname{artanh} x) = \dfrac{1}{1 - x^2}, \; |x| < 1$

Example 5

Given $y = x \operatorname{arcosh} x$, find $\dfrac{dy}{dx}$

$\dfrac{dy}{dx} = \operatorname{arcosh} x + x \times \dfrac{1}{\sqrt{x^2 - 1}}$ ← Use the product rule. ← Pure 3 Section 6.4

$= \operatorname{arcosh} x + \dfrac{x}{\sqrt{x^2 - 1}}$

Example 6 SKILLS REASONING; ARGUMENTATION

Given $y = (\operatorname{arcosh} x)^2$, prove that $(x^2 - 1)\left(\dfrac{dy}{dx}\right)^2 = 4y$

$\dfrac{dy}{dx} = 2\operatorname{arcosh} x \times \dfrac{1}{\sqrt{x^2 - 1}}$ ← Use the chain rule.

$\Rightarrow \sqrt{x^2 - 1}\,\dfrac{dy}{dx} = 2\operatorname{arcosh} x$ ← Multiply by $\sqrt{x^2 - 1}$

$\Rightarrow (x^2 - 1)\left(\dfrac{dy}{dx}\right)^2 = 4(\operatorname{arcosh} x)^2$ ← Square both sides.

But $y = (\operatorname{arcosh} x)^2$

so $(x^2 - 1)\left(\dfrac{dy}{dx}\right)^2 = 4y$

Exercise 3B SKILLS ANALYSIS

1 Differentiate:
   a $\operatorname{arcosh} 2x$
   b $\operatorname{arsinh}(x + 1)$
   c $\operatorname{artanh} 3x$
   d $\operatorname{arsech} x$
   e $\operatorname{arcosh} x^2$
   f $\operatorname{arcosh} 3x$
   g $x^2 \operatorname{arcosh} x$
   h $\operatorname{arsinh} \dfrac{x}{2}$
   i $e^{x^3} \operatorname{arsinh} x$
   j $\operatorname{arsinh} x \operatorname{arcosh} x$
   k $\operatorname{arcosh} x \operatorname{sech} x$
   l $x \operatorname{arcosh} 3x$

2 Prove that:
   a $\dfrac{d}{dx}(\operatorname{arcosh} x) = \dfrac{1}{\sqrt{x^2 - 1}}$
   b $\dfrac{d}{dx}(\operatorname{artanh} x) = \dfrac{1}{1 - x^2}$

E/P 3 Given that $y = \text{artanh}\left(\dfrac{e^x}{2}\right)$, prove that

$$(4 - e^{2x})\dfrac{dy}{dx} = 2e^x \qquad \text{(5 marks)}$$

E/P 4 Given that $y = \text{arsinh}\, x$, show that

$$(1 + x^2)\dfrac{d^3y}{dx^3} + 3x\dfrac{d^2y}{dx^2} + \dfrac{dy}{dx} = 0 \qquad \text{(7 marks)}$$

P 5 If $y = (\text{arcosh}\, x)^2$, find $\dfrac{d^2y}{dx^2}$

E/P 6 Find the equation of the tangent at the point where $x = \dfrac{12}{13}$ on the curve with equation $y = \text{artanh}\, x$. (7 marks)

3.3 Differentiating inverse trigonometric functions

You can also differentiate the inverse trigonometric functions.

- $\dfrac{d}{dx}(\arcsin x) = \dfrac{1}{\sqrt{1 - x^2}}$

- $\dfrac{d}{dx}(\arccos x) = \dfrac{1}{\sqrt{1 - x^2}}$

- $\dfrac{d}{dx}(\arctan x) = \dfrac{1}{1 + x^2}$

Example 7

Show that $\dfrac{d}{dx}(\arcsin x) = \dfrac{1}{\sqrt{1 - x^2}}$

Let $y = \arcsin x$

then $\sin y = x$

$\cos y \dfrac{dy}{dx} = 1$ — Differentiate.

$\dfrac{dy}{dx} = \dfrac{1}{\cos y}$ — Divide by $\cos y$.

$= \dfrac{1}{\sqrt{1 - \sin^2 y}}$ — Use $\cos^2 y = 1 - \sin^2 y$

but $\sin y = x$

so $\dfrac{dy}{dx} = \dfrac{1}{\sqrt{1 - x^2}}$

Example 8

Given $y = \arcsin x^2$ find $\dfrac{dy}{dx}$

$\sin y = x^2$

$\cos y \dfrac{dy}{dx} = 2x$ — Differentiate.

$\dfrac{dy}{dx} = \dfrac{2x}{\cos y}$ — Divide by $\cos y$.

$\dfrac{dy}{dx} = \dfrac{2x}{\sqrt{1 - \sin^2 y}}$ — Use $\cos^2 y = 1 - \sin^2 y$

but $\sin y = x^2$

so $\dfrac{dy}{dx} = \dfrac{2x}{\sqrt{1 - x^4}}$

Example 9 — SKILLS ANALYSIS

Given $y = \tan^{-1}\left(\dfrac{1-x}{1+x}\right)$, find $\dfrac{dy}{dx}$

$\tan y = \left(\dfrac{1-x}{1+x}\right)$

$\sec^2 y \dfrac{dy}{dx} = \dfrac{-(1+x) - 1(1-x)}{(1+x)^2}$ — Differentiate using quotient rule on $\left(\dfrac{1-x}{1+x}\right)$

← Pure 3 Section 6.5

$\dfrac{dy}{dx} = \dfrac{1}{\sec^2 y} \times -\left[\dfrac{1 + x + 1 - x}{(1+x)^2}\right]$ — Divide by $\sec^2 y$.

$= \dfrac{1}{1 + \tan^2 y} \times -\left[\dfrac{2}{(1+x)^2}\right]$ — Use $1 + \tan^2 y = \sec^2 y$ and simplify brackets.

but $\tan y = \left(\dfrac{1-x}{1+x}\right)$

$\dfrac{dy}{dx} = \dfrac{1}{1 + \left(\dfrac{1-x}{1+x}\right)^2} \times -\left[\dfrac{2}{(1+x)^2}\right]$

$= \dfrac{(1+x)^2}{(1+x)^2 + (1-x)^2} \times -\dfrac{2}{(1+x)^2}$ — Cancel $(1+x)^2$

$= \dfrac{-2}{2 + 2x^2}$

$= -\dfrac{1}{1 + x^2}$

Exercise 3C

1. Given that $y = \arccos x$ prove that
$$\frac{dy}{dx} = -\frac{1}{\sqrt{1-x^2}}$$

2. Differentiate with respect to x
   - **a** $\arccos 2x$
   - **b** $\arctan \frac{x}{2}$
   - **c** $\arcsin 3x$
   - **d** $\text{arccot}\, x$
   - **e** $\text{arcsec}\, x$
   - **f** $\text{arccosec}\, x$
   - **g** $\arcsin\left(\frac{x}{x-1}\right)$
   - **h** $\arccos x^2$
   - **i** $e^x \arccos x$
   - **j** $\arcsin x \cos x$
   - **k** $x^2 \arccos x$
   - **l** $e^{\arctan x}$

3. If $\tan y = x \arctan x$, find $\frac{dy}{dx}$

4. Given that $y = \arcsin x$ prove that
$$(1+x^2)\frac{d^2y}{dx^2} - x\frac{dy}{dx} = 0$$
(8 marks)

5. Find an equation of the tangent to the curve with equation $y = \arcsin 2x$ at the point where $x = \frac{1}{4}$
(7 marks)

Challenge

$y = \cos x \cosh x$ then show that $\frac{d^4y}{dx^4} = -4y$

Chapter review 3

SKILLS REASONING; ARGUMENTATION

1. Given $y = \cosh 2x$, find $\frac{dy}{dx}$

2. Differentiate with respect to x.
   - **a** $\text{arsinh}\, 3x$
   - **b** $\text{arsinh}\, x^2$
   - **c** $\text{arcosh}\, \frac{x}{2}$
   - **d** $x^2 \text{arcosh}\, 2x$

3. Given that $y = \arctan x$ prove that $\frac{dy}{dx} = \frac{1}{1+x^2}$

4. Given that $y = (\arctan x)^2$ prove that $(1+x^2)\frac{d^2y}{dx^2} + x\frac{dy}{dx} - 2 = 0$

5. Given $y = 5\cosh x - 3\sinh x$
   - **a** find $\frac{dy}{dx}$
   - **b** find the minimum turning points.

6. Given that $y = (\arcsin x)^2$ show that $(1-x^2)\frac{d^2y}{dx^2} - x\frac{dy}{dx} - 2 = 0$
(7 marks)

7 Differentiate arcosh (sinh 2x).

(P) 8 Given that $y = x - \arctan x$ prove that $\dfrac{d^2y}{dx^2} = 2x\left(1 - \dfrac{dy}{dx}\right)^2$

9 Differentiate $\arcsin \dfrac{x}{\sqrt{1 + x^2}}$

(E) 10 Show that the curve with equation $y = \text{sech } x$ has $\dfrac{d^2y}{dx^2} = 0$ at the point where $x = \pm \ln p$ and state a value of p. **(9 marks)**

(P) 11 Find the equation of the tangent and normal to the hyperbola $\dfrac{x^2}{a^2} - \dfrac{y^2}{b^2} = 1$ at the point $(a \cosh q, b \sinh q)$.

Summary of key points

1. $\dfrac{d}{dx}(\sinh x) = \cosh x$

   $\dfrac{d}{dx}(\cosh x) = \sinh x$

   $\dfrac{d}{dx}(\tanh x) = \text{sech}^2 x$

   $\dfrac{d}{dx}(\coth x) = -\text{cosech}^2 x$

   $\dfrac{d}{dx}(\text{cosech } x) = -\coth x \, \text{cosech } x$

   $\dfrac{d}{dx}(\text{sech } x) = -\tanh x \, \text{sech } x$

   $\dfrac{d}{dx}(\text{arsinh } x) = \dfrac{1}{\sqrt{x^2 + 1}}$

   $\dfrac{d}{dx}(\text{arcosh } x) = \dfrac{1}{\sqrt{x^2 - 1}}$

   $\dfrac{d}{dx}(\text{artanh } x) = \dfrac{1}{\sqrt{1 - x^2}}$

   $\dfrac{d}{dx}(\arcsin x) = \dfrac{1}{\sqrt{1 - x^2}}$

   $\dfrac{d}{dx}(\arccos x) = \dfrac{1}{\sqrt{1 - x^2}}$

   $\dfrac{d}{dx}(\arctan x) = \dfrac{1}{1 + x^2}$

4 INTEGRATION

4.1
4.2
4.3
4.4
4.5
4.6

Learning objectives

After completing this chapter you should be able to:
- Use standard integrals → pages 55–57
- Integrate hyperbolic functions → pages 58–61
- Use trigonometric and hyperbolic substitutions → pages 61–67
- Integrate functions involving quadratic surds → pages 67–71
- Integrate inverse trigonometric and inverse hyperbolic functions → pages 71–73
- Derive and use reduction formulae → pages 73–78
- Use integration to calculate:
  the length of an arc of a curve → pages 79–82
  the area of a surface of revolution → pages 82–87

Prior knowledge check

1 Show that $\int \cos^2 x \, dx = \frac{\sin 2x}{4} + \frac{x}{2} + c$ ← Pure 4 Section 6.4

2 Given $f(x) = e^{\frac{1}{2}} + \frac{1}{x}, x > 0$, find, giving your answer to 2 decimal places, the area of the region bounded by the curve $f(x)$, the x-axis, and the lines $x = 2$ and $x = 4$ ← Pure 2 Section 8.2, Pure 3 Section 7.1

3 Find $\int e^x \sin x \, dx$ ← Pure 4 Section 6.4

Simple 3D shapes have simple formulae for volumes and areas. Integration is used in a wide variety of practical applications, examples being finding the curved surface area of a car windscreen or of a parabolic mirror for a telescope.

INTEGRATION — CHAPTER 4

4.1 Standard integrals

As integration is the reverse process to differentiation, the results found in Chapter 3 mean that you can add the following to your list of standard integrals.

You should know which results are given in your formula booklet; those included in the Edexcel booklet are denoted (i.e. shown) by (*) throughout this chapter.

- **You should be familiar with the following results:**

1. $\int \sinh x \, dx = \cosh x + C$ (*)   since $\frac{d}{dx}(\cosh x) = \sinh x$

2. $\int \cosh x \, dx = \sinh x + C$ (*)   since $\frac{d}{dx}(\sinh x) = \cosh x$

3. $\int \tanh x = \ln \cosh x$ (*)

4. $\int \operatorname{sech}^2 x \, dx = \tanh x + C$ since $\frac{d}{dx}(\tanh x) = \operatorname{sech}^2 x$

5. $\int \operatorname{cosech}^2 x \, dx = -\coth x + C$ since $\frac{d}{dx}(\coth x) = -\operatorname{cosech}^2 x$

6. $\int \operatorname{sech} x \tanh x \, dx = -\operatorname{sech} x + C$ since $\frac{d}{dx}(\operatorname{sech} x) = -\operatorname{sech} x \tanh x$

7. $\int \operatorname{cosech} x \coth x \, dx = -\operatorname{cosech} x + C$ since $\frac{d}{dx}(\operatorname{cosech} x) = -\operatorname{cosech} x \coth x$

8. $\int \frac{1}{\sqrt{(1-x^2)}} \, dx = \arcsin x + C, \; |x| < 1$ since $\frac{d}{dx}(\arcsin x) = \frac{1}{\sqrt{(1-x^2)}}$

9. $\int \frac{1}{1+x^2} \, dx = \arctan x + C$ since $\frac{d}{dx}(\arctan x) = \frac{1}{1+x^2}$

10. $\int \frac{1}{\sqrt{(1+x^2)}} \, dx = \operatorname{arsinh} x + C$ since $\frac{d}{dx}(\operatorname{arsinh} x) = \frac{1}{\sqrt{(1+x^2)}}$

11. $\int \frac{1}{\sqrt{(x^2-1)}} \, dx = \operatorname{arcosh} x + C, \; x > 1$ since $\frac{d}{dx}(\operatorname{arcosh} x) = \frac{1}{\sqrt{(x^2-1)}}$

Notation   If x is replaced by a **linear** function of x in results **1, 2, 4–7**, they can be generalised using
$\int f'(ax+b) \, dx = \frac{1}{a} f(ax+b) + C$

Example 1   SKILLS ANALYSIS

Find   **a** $\int \cosh(4x-1) \, dx$ **b** $\int \operatorname{cosech} 3x \coth 3x \, dx$

a $\int \cosh(4x-1) \, dx = \frac{1}{4} \sinh(4x-1) + c$

b $\int \operatorname{cosech} 3x \coth 3x \, dx = -\frac{1}{3} \operatorname{cosech} 3x + c$

Results **8** to **11** will be generalised in Sections 4.3 and 4.4, but at this stage it is important to recognise their structure. There are many integrals that have the same denominators as those in results **8** to **11**, that you found in Pure 4, for example $\int \dfrac{x}{1+x^2}\,dx$ and $\int \dfrac{x}{\sqrt{1+x^2}}\,dx$

The results of these integrals were found using one of these two general results:

$$\int f'(x)\,[f(x)]^n\,dx = \dfrac{[f(x)]^{n+1}}{n+1} + C, \ n \neq -1 \quad (1)$$

$$\int \dfrac{f'(x)}{f(x)}\,dx = \ln|f(x)| + C \quad (2)$$

You should be confident in recognising these forms (they will occur frequently in this chapter); the next example is included as revision.

Example 2

Integrate with respect to x

a $\dfrac{4x}{1+x^2}$ **b** $\dfrac{5x}{\sqrt{1+x^2}}$

a $\int \dfrac{4x}{1+x^2}\,dx = 2\int \dfrac{2x}{1+x^2}\,dx$

$= 2\ln|(1+x^2)| + c$

$= 2\ln(1+x^2) + c$

This of the form $k\int \dfrac{f'(x)}{f(x)}\,dx$ so use **(2)**.

Watch out An integral with a denominator of $1+x^2$ does not automatically imply arctan x.

b $\int \dfrac{5x}{\sqrt{1+x^2}}\,dx = 5\int x(1+x^2)^{-\frac{1}{2}}\,dx$

$= \dfrac{5}{2}\int 2x(1+x^2)^{-\frac{1}{2}}\,dx$

$= \dfrac{5}{2}\left\{\dfrac{(1+x^2)^{\frac{1}{2}}}{\left(\frac{1}{2}\right)}\right\} + c$

$= 5\sqrt{1+x^2} + c$

This is of the form $k\int f'(x)[f(x)]^n\,dx$ so use **(1)**.

Result is $k\dfrac{[f(x)]^{n+1}}{n+1} + c$ with $f(x) = 1+x^2,\ n = -\dfrac{1}{2}$

Watch out An integral with a denominator of $\sqrt{1+x^2}$ does not automatically imply arsinh x.

It may be possible to reduce more complicated looking integrals into two parts, one of which is one of those listed in **8** to **11** and one of which you already know how to integrate.

Example 3

Find $\int \dfrac{2+5x}{\sqrt{x^2+1}}\,dx$

$\int \dfrac{2+5x}{\sqrt{x^2+1}}\,dx = \int \dfrac{5x}{\sqrt{x^2+1}}\,dx + \int \dfrac{2}{\sqrt{x^2+1}}\,dx$

$= 5\int x(x^2+1)^{-\frac{1}{2}}\,dx + 2\int \dfrac{1}{\sqrt{x^2+1}}\,dx$

$= 5\sqrt{x^2+1} + 2\,\text{arsinh}\,x + C$

Splitting the numerator gives two recognisable integrals.

See Example 2**b**.

Use result **9**.

Exercise 4A SKILLS ANALYSIS

1 Integrate with respect to x

    a $\sinh x + 3\cosh x$ **b** $5\operatorname{sech}^2 x$

    c $\dfrac{1}{\sinh^2 x}$ **d** $\cosh x - \dfrac{1}{\cosh^2 x}$

    e $\dfrac{\sinh x}{\cosh^2 x}$ **f** $\dfrac{3}{\sinh x \tanh x}$

    g $\operatorname{sech} x(\operatorname{sech} x + \tanh x)$ **h** $(\operatorname{sech} x + \operatorname{cosech} x)(\operatorname{sech} x + \operatorname{cosech} x)$

2 Find

    a $\displaystyle\int \sinh 2x\,dx$ **b** $\displaystyle\int \cosh\left(\dfrac{x}{3}\right)dx$

    c $\displaystyle\int \operatorname{sech}^2(2x-1)\,dx$ **d** $\displaystyle\int \operatorname{cosech}^2 5x\,dx$

    e $\displaystyle\int \operatorname{cosech} 2x \coth 2x\,dx$ **f** $\displaystyle\int \operatorname{sech}\left(\dfrac{x}{\sqrt{2}}\right)\tanh\left(\dfrac{x}{\sqrt{2}}\right)dx$

    g $\displaystyle\int \left\{5\sinh 5x - 4\cosh 4x + 3\operatorname{sech}^2\left(\dfrac{x}{2}\right)\right\}dx$

3 Write down the results of the following.

    a $\displaystyle\int \dfrac{1}{1+x^2}\,dx$ **b** $\displaystyle\int \dfrac{1}{\sqrt{1+x^2}}\,dx$

    c $\displaystyle\int \dfrac{1}{1+x}\,dx$ **d** $\displaystyle\int \dfrac{2x}{1+x^2}\,dx$

    e $\displaystyle\int \dfrac{1}{\sqrt{1-x^2}}\,dx,\ |x|<1$ **f** $\displaystyle\int \dfrac{1}{\sqrt{x^2-1}}\,dx$

    g $\displaystyle\int \dfrac{3x}{\sqrt{x^2-1}}\,dx$ **h** $\displaystyle\int \dfrac{3}{(1+x)^2}\,dx$

4 Find

    a $\displaystyle\int \dfrac{2x+1}{\sqrt{1-x^2}}\,dx$ **b** $\displaystyle\int \dfrac{1+x}{\sqrt{x^2-1}}\,dx$ **c** $\displaystyle\int \dfrac{x-3}{\sqrt{1+x^2}}\,dx$

5 **a** Show that $\dfrac{x^2}{1+x^2} = 1 - \dfrac{1}{1+x^2}$

    b Hence find $\displaystyle\int \dfrac{x^2}{1+x^2}\,dx$

4.2 Integration

You need to be able to integrate expressions involving hyperbolic functions.

- **The method for integrating hyperbolic expressions is usually the same as that applied to the corresponding trigonometric expressions.**

Many hyperbolic functions can be integrated by recognising that they are of the form

$$\int f'(x)[f(x)]^n \, dx \quad \text{or} \quad \int \frac{f'(x)}{f(x)} \, dx$$

Example 4

Find

a $\int \text{sech}^6 x \tanh x \, dx$

b $\int \cosh^5 2x \sinh 2x \, dx$

c $\int \tanh x \, dx$

d $\int \frac{\text{sech}^2 x}{2 + 5 \tanh x} \, dx$

a $\int \text{sech}^6 x \tanh x \, dx = \int \text{sech}^5 x (-\text{sech} x \tanh x) \, dx$

$= -\frac{1}{6} \text{sech}^6 x + C$

Use $\int [f(x)]^n f'(x) \, dx = \frac{[f(x)]^{n+1}}{n+1} + C$ with $f(x) = \text{sech } x$ and $n = 5$

b $\int \cosh^5 2x \sinh 2x \, dx = \frac{1}{2} \int (\cosh 2x)^5 (2 \sinh 2x) \, dx$

$= \frac{1}{12} \cosh^6 2x + C$

Use $\int [f(x)]^n f'(x) \, dx = \frac{[f(x)]^{n+1}}{n+1} + C$ with $f(x) = \cosh 2x$ and $n = 5$

c $\int \tanh x \, dx = \int \frac{\sinh x}{\cosh x} \, dx = \ln \cosh x + C \, (*)$

Use $\int \frac{f'(x)}{f(x)} \, dx = \ln|f(x)| + C$ with $f(x) = \cosh x$ (modulus signs are not required because $\cosh x > 0$ for all x)

d $\int \frac{\text{sech}^2 x}{2 + 5 \tanh x} \, dx = \frac{1}{5} \int \frac{5 \text{sech}^2 x}{2 + 5 \tanh x} \, dx$

$= \frac{1}{5} \ln|2 + 5 \tanh x| + C$

You can arrange so it is of the form

$$k \int \frac{f'(x)}{f(x)} \, dx$$

Hint Remember that you can always check your result; differentiating should give the integrand (i.e. the expression to be integrated).

In Pure 4, you saw how using trigonometric identities often transformed a trigonometric expression that you could not integrate directly into one that you could. The same technique can be used with hyperbolic functions.

Example SKILLS ANALYSIS

Find

a $\int \tanh^2 x \, dx$ **b** $\int \cosh^2 3x \, dx$ **c** $\int \sinh^3 x \, dx$

a $\int \tanh^2 x \, dx = \int (1 - \text{sech}^2 x) \, dx$
$= x - \tanh x + C$

Using $1 - \tanh^2 x = \text{sech}^2 x$ gives two standard integrals.

b $\int \cosh^2 3x \, dx = \int \left(\frac{1 + \cosh 6x}{2}\right) dx$
$= \frac{1}{2}\left(x + \frac{\sinh 6x}{6}\right) + C$
$= \frac{1}{2}x + \frac{1}{12}\sinh 6x + C$

Using $\cosh 2A = 2\cosh^2 A - 1$ with $A = 3x$

c $\int \sinh^3 x \, dx = \int \sinh^2 x \sinh x \, dx$
$= \int (\cosh^2 x - 1) \sinh x \, dx$
$= \int \cosh^2 x \sinh x \, dx - \int \sinh x \, dx$
$= \frac{1}{3}\cosh^3 x - \cosh x + C$

For small odd values of n, you can use
$\int \sinh^n x \, dx = \int \sinh^{n-1} x \sinh x \, dx$
$\left(\int \cosh^n x \, dx, n \text{ odd, can be found similarly}\right)$

Sometimes however, the method used for trigonometric functions may break down, or may not be the simplest.

In such cases you can use the exponential definition of the hyperbolic functions.

Example SKILLS PROBLEM-SOLVING

Find $\int e^{2x} \sinh x \, dx$

Watch out $\int e^{2x} \sinh x \, dx$ can be found by using integration by parts twice. However, using integration by parts on $\int e^{ax} \sinh ax \, dx$ breaks down.

$\int e^{2x} \sinh x \, dx = \int e^{2x} \left(\frac{e^x - e^{-x}}{2}\right) dx$
$= \frac{1}{2} \int (e^{3x} - e^x) \, dx$
$= \frac{1}{2}\left(\frac{e^{3x}}{3} - e^x\right) + C$
$= \frac{1}{6}(e^{3x} - 3e^x) + C$

Using the definition of $\sinh x$

Example 7

Find $\int \operatorname{sech} x \, dx$

Writing $\operatorname{sech} x$ as $\dfrac{1}{\cosh x}$ and using the exponential form of $\cosh x$ gives

$\int \operatorname{sech} x \, dx = \int \dfrac{2}{e^x + e^{-x}} \, dx$

$= \int \dfrac{2e^x}{e^{2x} + 1} \, dx$ ← Multiply numerator and denominator by e^x

Use the substitution $u = e^x$, then $\dfrac{du}{dx} = e^x$ so $e^x \, dx$ can be replaced by du.

$\int \operatorname{sech} x \, dx = \int \dfrac{2}{u^2 + 1} \, du = 2 \int \dfrac{1}{u^2 + 1} \, du$ ← This is now standard form 8 with **variable** u.

$= 2 \arctan u + C$

$= 2 \arctan e^x + C$

Watch out $\int \sec x$ can be found by noting that $\dfrac{d}{dx}(\ln[\sec x + \tan x])$

$= \dfrac{1}{\sec x + \tan x} \times (\sec x \tan x + \sec^2 x)$

$= \sec x$

So that $\sec x \, dx = \ln(\sec x + \tan x) + C$ but $\int \operatorname{sech} x \, dx$ needs a different approach.

Exercise 4B SKILLS PROBLEM-SOLVING

1 Find

  a $\int \sinh^3 x \cosh x \, dx$ **b** $\int \tanh 4x \, dx$ **c** $\int \tanh^5 x \operatorname{sech}^2 x \, dx$

  d $\int \operatorname{cosech}^7 x \coth x \, dx$ **e** $\int \sqrt{\cosh 2x} \sinh 2x \, dx$ **f** $\int \operatorname{sech}^{10} 3x \tanh 3x \, dx$

2 Find

  a $\int \dfrac{\sinh x}{2 + 3 \cosh x} \, dx$ **b** $\int \dfrac{1 + \tanh x}{\cosh^2 x} \, dx$ **c** $\int \dfrac{5 \cosh x + 2 \sinh x}{\cosh x} \, dx$

3 a Show that $\int \coth x \, dx = \ln \sinh x + C$.

  b Show that $\int_1^2 \coth 2x \, dx = \ln \sqrt{\left(e^2 + \dfrac{1}{e^2}\right)}$

4 Use integration by parts to find:

  a $\int x \sinh 3x \, dx$ **b** $\int x \operatorname{sech}^2 x \, dx$

5 Find

  a $\int e^x \cosh x \, dx$ **b** $\int e^{-2x} \sinh 3x \, dx$ **c** $\int \cosh x \cosh 3x \, dx$

(P) 6 By writing $\cosh 3x$ in exponential form, find $\int \cosh^2 3x \, dx$ and show that it is equivalent to the result found in Example 5b.

7 Evaluate $\int_0^1 \dfrac{1}{\sinh x + \cosh x} \, dx$, giving your answer in terms of e.

8 Use appropriate identities to find

    a $\int \sinh^2 x \, dx$ **b** $\int (\text{sech}\, x - \tanh x)^2 \, dx$ **c** $\int \dfrac{\cosh^2 3x}{\sinh^2 3x} \, dx$

    d $\int \sinh^2 x \cosh^2 x \, dx$ **e** $\int \cosh^5 x \, dx$ **f** $\int \tanh^3 2x \, dx$

9 Show that $\int_0^{\ln 2} \cosh^2 \left(\dfrac{x}{2} \right) dx = \dfrac{1}{8}(3 + \ln 16)$

10 The region bounded by the curve $y = \sinh x$, the line $x = 1$ and the positive x-axis is rotated through 360° about the x-axis. Show that the volume of the solid of revolution formed is $\dfrac{\pi}{8e^2}(e^4 - 4e^2 - 1)$. **(9 marks)**

11 Using the result for $\int \text{sech}\, x \, dx$ given in Example 7, find

    a $\int \dfrac{2}{\cosh x} \, dx$ **b** $\int \text{sech}\, 2x \, dx$ **c** $\int \sqrt{1 - \tanh^2 \left(\dfrac{x}{2} \right)} \, dx$

12 Using the substitution $u = x^2$, or otherwise, find:

    a $\int x \cosh^2(x^2) \, dx$ **b** $\int \dfrac{x}{\cosh^2(x^2)} \, dx$

4.3 Trigonometric and hyperbolic substitutions

■ **You need to be able to use trigonometric and hyperbolic substitutions in integration.**

The standard results below, which you met in Section 4.1, can be derived directly by using a substitution.

8 $\int \dfrac{1}{\sqrt{(1 - x^2)}} \, dx = \arcsin x + C, \, |x| < 1$

9 $\int \dfrac{1}{1 + x^2} \, dx = \arctan x + C$

10 $\int \dfrac{1}{\sqrt{(1 + x^2)}} \, dx = \text{arsinh}\, x + C$

11 $\int \dfrac{1}{\sqrt{(x^2 - 1)}} \, dx = \text{arcosh}\, x + C, \, x > 1$

In these cases, **algebraic** substitutions such as $u = 1 - x^2$ for result **8**, do not help, but an appropriate trigonometric or hyperbolic substitution can be used. The suggested substitutions below are made so that the two termed expressions in the denominator are transformed into one, by use of a relevant identity.

■ For an integral involving $\sqrt{(1 - x^2)}$ try $x = \sin \theta$ or $x = \tanh u$ *As $1 - \sin^2 A = \cos^2 A$*
                                                                                                                  $1 - \tanh^2 u = \text{sech}^2 u$

■ For an integral involving $1 + x^2$ try $x = \tan \theta$ or $x = \sinh u$ *As $1 + \tan^2 A = \sec^2 A$*
                                                                                                                   $1 + \sinh^2 u = \cosh^2 u$

■ For an integral involving $\sqrt{(1 + x^2)}$ try $x = \sinh u$ or $x = \tan \theta$

■ For an integral involving $\sqrt{(x^2 - 1)}$ try $x = \cosh u$ or $x = \sec \theta$ *As $\cosh^2 u - 1 = \sinh^2 A$*
                                                                                                                   $\sec^2 A - 1 = \tan^2 A$

Example 8 — SKILLS: CRITICAL THINKING

By using an appropriate substitution show that $\int \dfrac{1}{1+x^2}\,dx = \arctan x + C$

Use the substitution $x = \tan A$,
then $\dfrac{dx}{dA} = \sec^2 A$
so 'dx' can be replaced by '$\sec^2 A\, dA$' and
then $\int \dfrac{1}{1+x^2}\,dx = \int \dfrac{1}{1+\tan^2 A}\sec^2 A\, dA$
$\phantom{\int \dfrac{1}{1+x^2}\,dx} = \int 1\, dA$
$\phantom{\int \dfrac{1}{1+x^2}\,dx} = A + C$
$\phantom{\int \dfrac{1}{1+x^2}\,dx} = \arctan x + C$

Hint The first substitution in each suggested pair is the one more likely to prove the better choice.

Use $1 + \tan^2 A = \sec^2 A$

If the substitution $x = \sinh u$ is used, the resulting integral reduces to $\int \operatorname{sech} u\, du$.

- For integrals of the form $\int \dfrac{1}{\sqrt{(a^2 - x^2)}}\,dx$, $|x| < a$, $\int \dfrac{1}{a^2 + x^2}\,dx$, $\int \dfrac{1}{\sqrt{(a^2 + x^2)}}\,dx$ and $\int \dfrac{1}{\sqrt{(x^2 - a^2)}}\,dx$, the substitutions $x = a \sin\theta$, $x = a \tan\theta$, $x = a \sinh u$ and $x = a \cosh u$ respectively are suggested.

Example 9

By using an appropriate substitution in each case,

a find $\int \dfrac{1}{\sqrt{x^2 - a^2}}\,dx$, $x > a$

b show that $\int \dfrac{1}{4 + x^2}\,dx = \dfrac{1}{2}\arctan\left(\dfrac{x}{2}\right) + C$

a As $\cosh^2 u - 1 = \sinh^2 u$, it follows that
$a^2 \cosh^2 u - a^2 = a^2 \sinh^2 u$
So, using the substitution $x = a \cosh u$,
$\dfrac{dx}{du} = a \sinh u$ so 'dx' can be replaced by '$a \sinh u\, du$', and

$\int \dfrac{1}{\sqrt{x^2 - a^2}}\,dx = \int \dfrac{1}{\sqrt{a^2 \cosh^2 u - a^2}}\, a \sinh u\, du$
$\phantom{\int \dfrac{1}{\sqrt{x^2 - a^2}}\,dx} = \int \dfrac{1}{a \sinh} a \sinh u\, du$
$\phantom{\int \dfrac{1}{\sqrt{x^2 - a^2}}\,dx} = u + C$
$\phantom{\int \dfrac{1}{\sqrt{x^2 - a^2}}\,dx} = \operatorname{arcosh}\left(\dfrac{x}{a}\right) + C$

As $x = a \cosh u$, $\cosh u = \dfrac{x}{a}$
so $u = \operatorname{arcosh}\left(\dfrac{x}{a}\right)$

b Let $x = 2\tan\theta$,
then $4 + x^2 = 4 + 4\tan^2\theta = 4(1 + \tan^2\theta) = 4\sec^2\theta$
and $\dfrac{dx}{d\theta} = 2\sec^2\theta$ so 'dx' can be replaced by '$2\sec^2\theta\, d\theta$'

so $\displaystyle\int \dfrac{1}{4 + x^2}\,dx = \int \dfrac{1}{4\sec^2\theta}\,2\sec^2\theta\, d\theta$

$= \dfrac{1}{2}\displaystyle\int 1\, d\theta$

$= \dfrac{1}{2}\theta + C$

$= \dfrac{1}{2}\arctan\left(\dfrac{x}{2}\right) + C$

As $x = 2\tan\theta$, $\tan\theta = \dfrac{x}{2}$
so $\theta = \arctan\left(\dfrac{x}{2}\right)$

The following results are given in the Edexcel formula booklet.

12 $\displaystyle\int \dfrac{1}{\sqrt{(a^2 - x^2)}}\,dx = \arcsin\left(\dfrac{x}{a}\right), |x| < a$ (*)

13 $\displaystyle\int \dfrac{1}{a^2 + x^2}\,dx = \dfrac{1}{a}\arctan\left(\dfrac{x}{a}\right)$ (*)

14 $\displaystyle\int \dfrac{1}{\sqrt{(a^2 + x^2)}}\,dx = \operatorname{arsinh}\left(\dfrac{x}{a}\right)$ (*)

15 $\displaystyle\int \dfrac{1}{\sqrt{(x^2 - a^2)}}\,dx = \operatorname{arcosh}\left(\dfrac{x}{a}\right), x > a$ (*)

Hint When $a = 1$, these become the results **8** to **11**.

Only **13** has the factor $\dfrac{1}{a}$ in the result.

You may still be asked to find a result by using a suitable substitution, but usually you will be able to use the results **12** to **15** for integrals of this type.

Example 10   SKILLS   CRITICAL THINKING

a Find $\displaystyle\int \dfrac{4}{5 + x^2}\,dx$ **b** Show that $\displaystyle\int_5^8 \dfrac{1}{\sqrt{x^2 - 16}}\,dx = \ln\left(\dfrac{2 + \sqrt{3}}{2}\right)$

a $\displaystyle\int \dfrac{4}{5 + x^2}\,dx = 4\int \dfrac{1}{5 + x^2}\,dx$

$= 4\left\{\dfrac{1}{\sqrt{5}}\arctan\left(\dfrac{x}{\sqrt{5}}\right)\right\} + C$

$= \dfrac{4}{\sqrt{5}}\arctan\left(\dfrac{x}{\sqrt{5}}\right) + C$

Using result **13** with $a = \sqrt{5}$

b $\displaystyle\int_5^8 \dfrac{1}{\sqrt{x^2 - 16}}\,dx = \left[\operatorname{arcosh}\left(\dfrac{x}{4}\right)\right]_5^8$

Using **13** with $a = 4$

$= \operatorname{arcosh}\left(\dfrac{8}{4}\right) - \operatorname{arcosh}\left(\dfrac{5}{4}\right)$

$= \ln(2 + \sqrt{3}) - \ln\left(\dfrac{5}{4} + \sqrt{\dfrac{9}{16}}\right)$

Using $\operatorname{arcosh} x = \ln(x + \sqrt{x^2 - 1})$ (*)

$= \ln(2 + \sqrt{3}) - \ln 2$

Using $\operatorname{arcosh} x = \ln(x + \sqrt{x^2 - 1})$ (*)

$= \ln\left(\dfrac{2 + \sqrt{3}}{2}\right)$

Using $\ln a - \ln b = \ln\left(\dfrac{a}{b}\right)$

CHAPTER 4 INTEGRATION

- Integrals of the form

$$\int \frac{1}{\sqrt{(a^2 - b^2x^2)}} dx, \int \frac{1}{(a^2 + b^2x^2)} dx, \int \frac{1}{\sqrt{(a^2 + b^2x^2)}} dx \text{ and } \int \frac{1}{\sqrt{(b^2x^2 - a^2)}} dx$$

can be easily manipulated to use the results **12** to **15**

Example 11

a $\int \frac{1}{25 + 9x^2} dx$

b Evaluate $\int_{-\frac{\sqrt{3}}{4}}^{\frac{\sqrt{3}}{4}} \frac{1}{\sqrt{3 - 4x^2}} dx$, leaving your answer in terms of π.

a $\int \frac{1}{25 + 9x^2} dx = \int \frac{1}{9\left(\frac{25}{9} + x^2\right)} dx$ — You need to write $25 + 9x^2$ in the form $k(a^2 + x^2)$

$= \frac{1}{9}\left\{\left(\frac{1}{\left(\frac{5}{3}\right)}\right)\arctan\left(\frac{x}{\left(\frac{5}{3}\right)}\right)\right\} + C$ — Using result **13** with $a = \frac{5}{3}$

$= \frac{1}{15}\left\{\arctan\left(\frac{3x}{5}\right)\right\} + C$

b $\int_{-\frac{\sqrt{3}}{4}}^{\frac{\sqrt{3}}{4}} \frac{1}{\sqrt{3 - 4x^2}} dx = \int_{-\frac{\sqrt{3}}{4}}^{\frac{\sqrt{3}}{4}} \frac{1}{\sqrt{4\left(\frac{3}{4} - x^2\right)}} dx$ — You need to write $3 - 4x^2$ in the form $k(a^2 + x^2)$

$= \frac{1}{2} \int_{-\frac{\sqrt{3}}{4}}^{\frac{\sqrt{3}}{4}} \frac{1}{\sqrt{\left(\frac{3}{4} - x^2\right)}} dx$ — Using result **12** with $a = \frac{\sqrt{3}}{2}$

$= \frac{1}{2}\left[\arcsin\frac{2x}{\sqrt{3}}\right]_{-\frac{\sqrt{3}}{4}}^{\frac{\sqrt{3}}{4}}$ — $-\frac{\pi}{2} < \arcsin x < \frac{\pi}{2}$

$= \frac{1}{2}\left[\arcsin\left(\frac{1}{2}\right)\right] - \frac{1}{2}\left[\arcsin\left(-\frac{1}{2}\right)\right]$

$= \frac{\pi}{12} - \left(-\frac{\pi}{12}\right)$ — So $\arcsin\left(\frac{1}{2}\right) = \frac{\pi}{6}$
  $\arcsin\left(-\frac{1}{2}\right) = -\frac{\pi}{6}$

$= \frac{\pi}{6}$

INTEGRATION CHAPTER 4

The substitutions suggested in the previous pages can be used in a wide range of integrals.

Example 12 — SKILLS: REASONING/ARGUMENTATION

Show that $\int \sqrt{1 + x^2}\, dx = \frac{1}{2}\operatorname{arsinh} x + \frac{1}{2}x\sqrt{1 + x^2} + C$

Using $x = \sinh u$, then $\frac{dx}{du} = \cosh u$ so 'dx' can be replaced by '$\cosh u\, du$'

So
$$\int \sqrt{1 + x^2}\, dx = \int \sqrt{1 + \sinh^2 u}\, \cosh u\, du$$
$$= \int \cosh^2 u\, du$$ — Using $\cosh u = 2\cosh^2 u - 1$
$$= \frac{1}{2}\int (1 + \cosh 2u)\, du$$
$$= \frac{1}{2}\left(u + \frac{\sinh 2u}{2}\right) + C$$
$$= \frac{1}{2}(u + \sinh u \cosh u) + C$$ — You need to be able to use $x = \sinh u$, so use $\sinh 2u = 2\sinh u \cosh u$
$$= \frac{1}{2}\operatorname{arsinh} x + \frac{1}{2}x\sqrt{1 + x^2} + C$$ — As $u = \operatorname{arsinh} x$ and $\cosh u = \sqrt{1 + \sinh^2 u}$

Example 13

By using a hyperbolic substitution, evaluate $\int_0^6 \frac{x^3}{\sqrt{x^2 + 9}}\, dx$ **Hint** You could use integration by parts.

Use the substitution $x = 3\sinh u$ then $\frac{dx}{du} = 3\cosh u$, and 'dx' can be replaced by '$3\cosh u\, du$',

so
$$\int_0^6 \frac{x^3}{\sqrt{x^2 + 9}}\, dx = \int_0^{\operatorname{arsinh}2} \frac{27\sinh^3 u}{3\cosh u}\, 3\cosh u\, du$$
$$= 27\int_0^{\operatorname{arsinh}2} \sinh^3 u\, du$$
$$= 27\left[\frac{1}{3}\cosh^3 u - \cosh u\right]_0^{\operatorname{arsinh}2}$$
$$= 27\left[\frac{5\sqrt{5}}{3} - \sqrt{5}\right] - 27\left[\frac{1}{3} - 1\right]$$
$$= 18\sqrt{5} + 18$$
$$= 18(\sqrt{5} + 1) \text{ or } 58.2 \text{ (3 s.f.)}$$

You need to reduce $x^2 + 9$ to a single term; using $x = 3\sinh u$ gives
$9\sinh^2 u + 9 = 9(\sinh^2 u + 1) = 9\cosh^2 u$

Limits: When $x = 6$, $\sinh u = 2 \Rightarrow u = \operatorname{arsinh}2$
When $x = 0$, $\sinh u = 0 \Rightarrow u = 0$

See Example 5c.

As $\sinh u = 2$
$\cosh u = \sqrt{\{1 + (2)^2\}} = \sqrt{5}$

Exercise 4C SKILLS CRITICAL THINKING

Unless a substitution is given or asked for, use the standard results **8** to **15**. Give numerical answers to 3 significant figures, unless otherwise stated.

1. Use the substitution $x = a \tan \theta$ to show that $\int \dfrac{1}{a^2 + x^2} dx = \dfrac{1}{a} \arctan\left(\dfrac{x}{a}\right) + C$

2. Use the substitution $x = \cos \theta$ to show that $\int \dfrac{1}{\sqrt{1 - x^2}} dx = -\arccos x + C$

3. Use suitable substitutions to find

   a $\int \dfrac{3}{\sqrt{4 - x^2}} dx$ b $\int \dfrac{1}{\sqrt{x^2 - 9}} dx$ c $\int \dfrac{4}{5 + x^2} dx$ d $\int \dfrac{1}{\sqrt{4x^2 + 25}} dx$

4. Write down the results for the following:

   a $\int \dfrac{1}{\sqrt{25 - x^2}} dx$ b $\int \dfrac{3}{\sqrt{x^2 + 9}} dx$ c $\int \dfrac{1}{\sqrt{x^2 - 2}} dx$ d $\int \dfrac{1}{16 + x^2} dx$

5. Find

   a $\int \dfrac{1}{\sqrt{4x^2 - 12}} dx$ b $\int \dfrac{1}{4 + 3x^2} dx$ c $\int \dfrac{1}{\sqrt{9x^2 + 16}} dx$ d $\int \dfrac{1}{\sqrt{3 - 4x^2}} dx$

6. Evaluate

   a $\int_1^3 \dfrac{2}{1 + x^2} dx$ b $\int_1^2 \dfrac{3}{\sqrt{1 + 4x^2}} dx$ c $\int_{-1}^2 \dfrac{1}{\sqrt{21 - 3x^2}} dx$

7. Evaluate the following, giving your answers in terms of π or as a single natural logarithm, whichever is appropriate:

   a $\int_0^4 \dfrac{1}{\sqrt{x^2 + 16}} dx$ b $\int_{13}^{15} \dfrac{1}{\sqrt{x^2 - 144}} dx$ c $\int_{\sqrt{2}}^{\sqrt{3}} \dfrac{1}{\sqrt{4 - x^2}} dx$

E/P 8. The curve C has equation $y = \dfrac{2}{\sqrt{2x^2 + 9}}$. The region R is bounded by C, the coordinate axes and the lines $x = -1$ and $x = 3$.

   a Find the area of R. (4 marks)

   The region R is rotated through 360° about the x-axis.

   b Find the volume of the solid generated. (5 marks)

E 9. A circle C has centre the origin and radius r.

   a Show that the area of C can be written as $4 \int_0^r \sqrt{r^2 - x^2} \, dx$ (3 marks)

   b Hence show that the area of C is πr^2. (5 marks)

10. a Use the substitution $x = \dfrac{2}{3} \tan \theta$ to find $\int \dfrac{x^2}{9x^2 + 4} dx$

    b Use the substitution $x = \sinh^2 u$ to find $\int \sqrt{\dfrac{x}{x + 1}} dx$, $x > 0$

11 By splitting up each integral into two separate integrals, or otherwise, find

  a $\int \dfrac{x-2}{\sqrt{x^2-4}}\,dx$ **b** $\int \dfrac{2x-1}{\sqrt{2-x^2}}\,dx$ **c** $\int \dfrac{2+3x}{1+3x^2}\,dx$

12 Use the method of partial fractions to find $\int \dfrac{x^2+4x+10}{x^3+5x}\,dx$, $x>0$

13 Show that $\int_0^1 \dfrac{2}{(x+1)(x^2+1)}\,dx = \dfrac{1}{4}(\pi + 2\ln 2)$

14 By using the substitution $u = x^2$, evaluate $\int_2^3 \dfrac{2x}{\sqrt{x^4-1}}\,dx$

(P) 15 By using the substitution $x = \dfrac{1}{2}\sin\theta$, show that $\int_0^{\frac{1}{4}} \dfrac{x^2}{\sqrt{1-4x^2}}\,dx = \dfrac{1}{192}(2\pi - 3\sqrt{3})$

(E/P) 16 a Use the substitution $x = 2\cosh u$ to show that
$$\int \sqrt{x^2-4}\,dx = \dfrac{1}{2}x\sqrt{x^2-4} - 2\,\text{arcosh}\left(\dfrac{x}{2}\right) + C$$ **(5 marks)**

  b Find the area enclosed between the hyperbola with equation $\dfrac{x^2}{4} - \dfrac{y^2}{9} = 1$ and the line $x = 4$ **(6 marks)**

(P) 17 a Show that $\int \dfrac{1}{2\cosh x - \sinh x}\,dx$ can be written as $\int \dfrac{2e^x}{e^{2x}+3}\,dx$

  b Hence, by using the substitution $u = e^x$, find $\int \dfrac{1}{2\cosh x - \sinh x}\,dx$

18 Using the substitution $u = \dfrac{2}{3}\sinh x$, evaluate $\int_0^1 \dfrac{\cosh x}{\sqrt{4\sinh^2 x + 9}}\,dx$

(P) 19 a Find $\int \dfrac{dx}{a^2-x^2}$, $|x| < a$, by using

  i partial fractions

  ii the substitution $x = a\tanh\theta$

  b Deduce the logarithmic form of $\text{artanh}\left(\dfrac{x}{a}\right)$

20 Using the substitution $x = \sec\theta$, find

  a $\int \dfrac{1}{x\sqrt{x^2-1}}\,dx$ **b** $\int \dfrac{\sqrt{x^2-1}}{x}\,dx$

4.4 Integrating expressions

■ You can use the method of **completing the square** to reduce integrals of the type $\int \dfrac{1}{px^2+qx+r}\,dx$ and $\int \dfrac{1}{\sqrt{px^2+qx+r}}\,dx$ to one of the forms in the results below, results **12** to **15** of which you have met previously in Section 4.3 on page 65.

12 $\int \dfrac{1}{\sqrt{(a^2-x^2)}}\,dx = \arcsin\left(\dfrac{x}{a}\right)$, $|x| < a$ (*)

13 $\int \dfrac{1}{a^2+x^2}\,dx = \dfrac{1}{a}\arctan\left(\dfrac{x}{a}\right)$ (*)

14 $\int \dfrac{1}{\sqrt{(a^2 + x^2)}} \, dx = \text{arsinh}\left(\dfrac{x}{a}\right)$ (*)

15 $\int \dfrac{1}{\sqrt{(x^2 - a^2)}} \, dx = \text{arcosh}\left(\dfrac{x}{a}\right), \; x > a$ (*)

16 $\int \dfrac{1}{a^2 - x^2} \, dx = \dfrac{1}{2a} \ln\left|\dfrac{a + x}{a - x}\right|, \; |x| < a$ (*)

Notation Results **16** and **17** are found using partial fractions.

17 $\int \dfrac{1}{x^2 - a^2} \, dx = \dfrac{1}{2a} \ln\left|\dfrac{x - a}{x + a}\right|$ (*)

Hint If $px^2 + qx + r$ factorises, then integrals of the form $\int \dfrac{1}{px^2 + qx + r} \, dx$ are best obtained by using the method of partial fractions which is covered in Pure 4 Section 2.1.

Example 14

Find **a** $\int \dfrac{1}{x^2 - 8x + 8} \, dx$ **b** $\int \dfrac{1}{2x^2 + 4x + 11} \, dx$

a $x^2 - 8x + 8 = (x - 4)^2 - 8$

So $\int \dfrac{1}{x^2 - 8x + 8} \, dx = \int \dfrac{1}{(x - 4)^2 - 8} \, dx$

Put $u = x - 4$, then 'du' = 'dx'

So $\int \dfrac{1}{x^2 - 8x + 8} \, dx = \int \dfrac{1}{u^2 - 8} \, du$

$= \dfrac{1}{4\sqrt{2}} \ln\left|\dfrac{u - 2\sqrt{2}}{u + 2\sqrt{2}}\right| + C$

$= \dfrac{\sqrt{2}}{8} \ln\left|\dfrac{x - 4 - 2\sqrt{2}}{x - 4 + 2\sqrt{2}}\right| + C$

b $2x^2 + 4x + 11 = 2\left(x^2 + 2x + \dfrac{11}{2}\right)$

$= 2\left\{(x + 1)^2 + \dfrac{9}{2}\right\}$

So $\int \dfrac{1}{2x^2 + 4x + 11} \, dx = \int \dfrac{1}{2\left\{(x + 1)^2 + \dfrac{9}{2}\right\}} \, dx$

Put $u = x + 1$, then 'du' = 'dx' so that

$\int \dfrac{1}{2x^2 + 4x + 11} \, dx = \dfrac{1}{2} \int \dfrac{1}{u^2 + \dfrac{9}{2}} \, du$

$= \dfrac{1}{2} \dfrac{1}{\left(\dfrac{3}{\sqrt{2}}\right)} \arctan \dfrac{u}{\left(\dfrac{3}{\sqrt{2}}\right)} + C$

$= \dfrac{1}{3\sqrt{2}} \arctan\left(\dfrac{\sqrt{2}(x + 1)}{3}\right) + C$

— First complete the square to express in the form $(x + b)^2 + c$

— Reduce to one of the forms in results **12** to **17**. Select the appropriate standard result.

— Using result **17** with $a = \sqrt{8} = 2\sqrt{2}$

— First complete the square to express in the form $(x + b)^2 + c$

— Use result **13** with $a = \sqrt{\dfrac{9}{2}} = \dfrac{3}{\sqrt{2}}$

— As $u = x + 1$

Example 15 — SKILLS: ANALYSIS

Find $\int \dfrac{1}{\sqrt{12x + 2x^2}}\,dx$

$12x + 2x^2 = 2(x^2 - 6x)$
$ = 2\{(x+3)^2 - 9\}$ — First complete the square.

So $\int \dfrac{1}{\sqrt{12x + 2x^2}}\,dx = \int \dfrac{1}{\sqrt{2\{(x+3)^2 - 9\}}}\,dx$

Put $u = x + 3$, then $du = dx$ and — Choose the substitution.

$\int \dfrac{1}{\sqrt{12x + 2x^2}}\,dx = \dfrac{1}{\sqrt{2}} \int \dfrac{1}{\sqrt{u^2 - 9}}\,du$ — Select the standard form. Use result **15** with $a = 3$

$= \dfrac{1}{\sqrt{2}} \operatorname{arcosh}\left(\dfrac{u}{3}\right) + C$

$= \dfrac{1}{\sqrt{2}} \operatorname{arcosh}\left(\dfrac{x+3}{3}\right) + C$ — As $u = x + 3$

You may be asked to find any of these integrals by using a trigonometric or hyperbolic substitution, so you should be prepared for that situation.

Example 16

Use a suitable trigonometric substitution to find $\int \dfrac{1}{\sqrt{4 - 2x - x^2}}\,dx$

$4 - 2x - x^2 = -(x^2 + 2x - 4)$
$ = -\{(x+1)^2 - 5\}$ — You still need to complete the square.
$ = 5 - (x+1)^2$

So $\int \dfrac{1}{\sqrt{4 - 2x - x^2}}\,dx = \int \dfrac{1}{\sqrt{5 - (x+1)^2}}\,dx$

Use the substitution $x + 1 = \sqrt{5} \sin\theta$ — For $\sqrt{a^2 - x^2}$ you use $X = a \sin\theta$

Then as $\dfrac{dx}{d\theta} = \sqrt{5} \cos\theta$, dx can be replaced by $\sqrt{5} \cos\theta\,d\theta$

So $\int \dfrac{1}{\sqrt{4 - 2x - x^2}}\,dx = \int \dfrac{1}{\sqrt{5 - 5\sin^2\theta}} \sqrt{5} \cos\theta\,d\theta$

$= \int \dfrac{1}{\sqrt{5}\cos\theta} \sqrt{5}\cos\theta\,d\theta$

$= \int 1\,d\theta$

$= \theta + C$

$= \arcsin\left(\dfrac{x+1}{\sqrt{5}}\right) + C$ — As $x + 1 = \sqrt{5}\sin\theta$

You may also be directed to use a particular substitution.

Example 17

Use the substitution $x = \frac{1}{2}(3 + 4\cosh u)$ to find $\displaystyle\int \frac{1}{\sqrt{4x^2 - 12x - 7}}\,dx$

For $x = \frac{1}{2}(3 + 4\cosh u)$, $\frac{dx}{du} = \frac{1}{2}(4\sinh u) = 2\sinh u$ so dx can be replaced by $2\sinh u\,du$, and

$$4x^2 - 12x - 7 = 4\left\{\frac{1}{4}(9 + 24\cosh u + 16\cosh^2 u)\right\} - 6(3 + 4\cosh u) - 7$$

$$= 9 + 24\cosh u + 16\cosh^2 u - 18 - 24\cosh u - 7$$

$$= 16\cosh^2 u - 16$$

$$= 16\sinh^2 u \quad \text{Using } \cosh^2 u - \sinh^2 u = 1$$

So $\displaystyle\int \frac{1}{\sqrt{4x^2 - 12x - 7}}\,dx = \int \frac{1}{4\sinh u}\,2\sinh u\,du$

$$= \frac{1}{2}\int 1\,du$$

$$= \frac{1}{2}u + C$$

$$= \frac{1}{2}\operatorname{arcosh}\left(\frac{2x-3}{4}\right) + C \quad \text{As } x = \frac{1}{2}(3 + 4\cosh u),\ \cosh u = \left(\frac{2x-3}{4}\right)$$

Exercise 4D SKILLS ANALYSIS

1 Find

a $\displaystyle\int \frac{1}{\sqrt{5 - 4x - x^2}}\,dx$

b $\displaystyle\int \frac{1}{\sqrt{x^2 - 4x - 12}}\,dx$

c $\displaystyle\int \frac{1}{\sqrt{x^2 + 6x + 10}}\,dx$

d $\displaystyle\int \frac{1}{\sqrt{x(x-2)}}\,dx$

e $\displaystyle\int \frac{1}{\sqrt{2x^2 + 4x + 7}}\,dx$

f $\displaystyle\int \frac{1}{\sqrt{-4x^2 - 12x}}\,dx$

g $\displaystyle\int \frac{1}{\sqrt{14 - 12x - 2x^2}}\,dx$

h $\displaystyle\int \frac{1}{\sqrt{9x^2 - 8x + 1}}\,dx$

2 Find

a $\displaystyle\int \frac{1}{\sqrt{4x^2 - 12x + 10}}\,dx$

b $\displaystyle\int \frac{1}{\sqrt{4x^2 - 12x + 4}}\,dx$

3 Evaluate the following, giving answers to 3 significant figures.

a $\displaystyle\int_1^2 \frac{1}{\sqrt{x^2 + 2x + 5}}\,dx$

b $\displaystyle\int_1^3 \frac{1}{x^2 + x + 1}\,dx$

c $\displaystyle\int_0^1 \frac{1}{\sqrt{2 + 3x - 2x^2}}\,dx$

4 a Evaluate $\displaystyle\int_1^3 \frac{1}{\sqrt{x^2 - 2x + 2}}\,dx$, giving your answer as a single natural logarithm.

b Evaluate $\displaystyle\int_1^2 \frac{1}{\sqrt{1 + 6x - 3x^2}}\,dx$, giving your answer in the form $k\pi$.

5 Show that $\int_{1}^{3} \frac{1}{\sqrt{3x^2 - 6x + 7}} dx = \frac{1}{\sqrt{3}} \ln(2+\sqrt{3})$

6 Using a suitable hyperbolic or trigonometric substitution find

   a $\int \frac{1}{\sqrt{x^2 + 4x + 5}} dx$ b $\int \frac{1}{\sqrt{-x^2 + 4x + 5}} dx$

7 Using the substitution $x = \frac{1}{5}(\sqrt{3} \tan\theta - 1)$, obtain $\int_{-0.2}^{0} \frac{1}{25x^2 + 10x + 4} dx$, giving your answer in terms of π.

8 Evaluate $\int_{3}^{4} \frac{1}{\sqrt{(x-2)(x+4)}} dx$, giving your answer in the form $\ln(a + b\sqrt{c})$, where a, b and c are **integers** to be found.

9 Using the substitution $x = 1 + \sinh\theta$, show that $\int \frac{1}{(x^2 - 2x + 2)^{\frac{3}{2}}} dx = \frac{x-1}{\sqrt{x^2 - 2x + 2}} + C$

10 Use the substitution $x = 2\sin\theta - 1$ to find $\int \frac{x}{\sqrt{3 - 2x - x^2}} dx$

Challenge

SKILLS
CREATIVITY

1 By means of a suitable substitution, or otherwise, find:

   a $\int x \cosh^2(x^2) \, dx$

   b $\int \frac{x}{\cosh^2(x^2)} dx$

4.5 Integrating inverse trigonometric and hyperbolic functions

You previously integrated the single function $\ln x$ by writing

$\int \ln x \, dx = \int (\ln x \times 1) \, dx$ and then using integration by parts

with $u = \ln x$ and $\frac{dv}{dx} = 1$

Links ← Pure 4 Section 6.4

You can use the same technique to integrate inverse trigonometric and hyperbolic functions.

Example 18 SKILLS ANALYSIS

Find $\int \operatorname{artanh} x \, dx$

Let $I = \int \operatorname{artanh} x \, dx = \int \operatorname{artanh} x \times 1 \, dx$

Using $u = \operatorname{artanh} x$ and $\frac{dv}{dx} = 1$

$\frac{du}{dx} = \frac{1}{1 - x^2}$ and $v = x$

So $I = x\operatorname{artanh} x - \int \dfrac{x}{1-x^2}\,dx$ — Using $\int u\dfrac{dv}{dx}\,dx = uv - \int v\dfrac{du}{dx}\,dx$

$= x\operatorname{artanh} x - \left(-\dfrac{1}{2}\right)\int \dfrac{(-2x)}{1-x^2}\,dx$ — This is of the form $\int \dfrac{f'(x)}{f(x)}\,dx$, where $f(x) = 1 - x^2$

$= x\operatorname{artanh} x + \dfrac{1}{2}\ln|1-x^2| + C$

This could be written in terms of ln by using $\operatorname{arctan} x = \dfrac{1}{2}\ln\left(\dfrac{1+x}{1-x}\right)$ (*).

Example 19

Evaluate $\displaystyle\int_{-\frac{1}{2}}^{\frac{\sqrt{3}}{2}} \arcsin x\,dx$

Let $I = \int \arcsin x\,dx = \int \arcsin x \times 1\,dx$

Using $u = \arcsin x$ and $\dfrac{dv}{dx} = 1$

$\dfrac{du}{dx} = \dfrac{1}{\sqrt{1-x^2}}$ and $v = x$

So $I = x\arcsin x - \int \dfrac{x}{\sqrt{1-x^2}}\,dx$ — Using $\int u\dfrac{dv}{dx}\,dx = uv - \int v\dfrac{du}{dx}\,dx$

$= x\arcsin x - \left(-\dfrac{1}{2}\right)\int (-2x)(1-x^2)^{-\frac{1}{2}}\,dx$ — This is of the form $\int f'(x)[f(x)]^n\,dx$

$= x\arcsin x + \dfrac{1}{2}\dfrac{(1-x^2)^{\frac{1}{2}}}{\left(\frac{1}{2}\right)} + C$ — This is $\dfrac{[f(x)]^{n+1}}{n+1}$, with $n = -\dfrac{1}{2}$

$= x\arcsin x + \sqrt{1-x^2} + C$

So $\displaystyle\int_{-\frac{1}{2}}^{\frac{\sqrt{3}}{2}} \arcsin x\,dx = \left[x\arcsin x + \sqrt{1-x^2}\right]_{-\frac{1}{2}}^{\frac{\sqrt{3}}{2}}$

$= \left[\dfrac{\sqrt{3}}{2}\arcsin\left(\dfrac{\sqrt{3}}{2}\right) + \dfrac{1}{2}\right] - \left[-\dfrac{1}{2}\arcsin\left(-\dfrac{1}{2}\right) + \sqrt{\dfrac{3}{4}}\right]$

$= \dfrac{\sqrt{3}\,\pi}{6} + \dfrac{1}{2} - \dfrac{\pi}{12} - \dfrac{\sqrt{3}}{2}$

$= 0.279$ (3 s.f.) — Remember that $-\dfrac{\pi}{2} \leqslant \arcsin x \leqslant \dfrac{\pi}{2}$

Exercise 4E   SKILLS   REASONING; ARGUMENTATION

1 a Show that $\int \operatorname{arsinh} x\,dx = x\operatorname{arsinh} x - \sqrt{1+x^2} + C$

  b Evaluate $\displaystyle\int_0^1 \operatorname{arsinh} x\,dx$, giving your answer to 3 significant figures.

  c Using the substitution $u = 2x + 1$ and the result in **a**, or otherwise, find $\int \operatorname{arsinh}(2x+1)\,dx$

2 Show that $\int \arctan 3x\,dx = x\arctan 3x - \dfrac{1}{6}\ln(1+9x^2) + C$

INTEGRATION
CHAPTER 4

 3 **a** Show that $\int \operatorname{arcosh} x \, dx = x \operatorname{arcosh} x - \sqrt{x^2 - 1} + C$

b Hence show that $\int_1^2 \operatorname{arcosh} x = \ln(7 + 4\sqrt{3}) - \sqrt{3}$

4 **a** Show that $\int \arctan x \, dx = x \arctan x - \frac{1}{2}\ln(1 + x^2) + C$ **(5 marks)**

b Hence show that $\int_{-1}^{\sqrt{3}} \arctan x \, dx = \frac{(4\sqrt{3} - 3)\pi}{12} - \frac{1}{2}\ln 2$ **(4 marks)**

The curve C has equation $y = 2 \arctan x$. The region R is enclosed by C, the y-axis, the line $y = \pi$ and the line $x = 3$

c Find the area of R, giving your answer to 3 significant figures. **(5 marks)**

5 Evaluate

a $\int_0^{\frac{\sqrt{2}}{2}} \arcsin x \, dx$

b $\int_0^1 x \arctan x \, dx$

giving your answers in terms of π.

6 Using the result that if $y = \operatorname{arcsec} x$, then $\dfrac{dy}{dx} = \dfrac{1}{x\sqrt{x^2 - 1}}$, show that

$\int \operatorname{arcsec} x \, dx = x \operatorname{arcsec} x - \ln\{x + \sqrt{x^2 - 1}\} + C$

4.6 Deriving and using reduction formulae

Often a method used to integrate a function involving n, (usually a power, where n is small) is not viable as n becomes large. For example, the methods used to find $\int \sin^2 x \, dx$ using the double angle formula for $\cos 2x$ to give $\int \sin^2 x \, dx = \int (1 - \cos 2x) \, dx$, to find $\int \sin^3 x \, dx$ by writing as $\int \sin x(1 - \cos^2 x) \, dx$, and to find $\int x^2 e^x \, dx$, by using integration by parts, become increasingly difficult to manage when applied to $\int \sin^n x \, dx$ and $\int x^n e^x \, dx$ as n increases. In such cases, it may be possible (usually by using integration by parts) to relate the given integral in n to a similar integral in $n - 1$ (or $n - 2$, or lower); this relation is called a **reduction formula**. By repeated use of the reduction formula, the given integral may be reduced to a form where only simple integration is required.

Example 20 — SKILLS CRITICAL THINKING

Given that $I_n = \int x^n e^x \, dx$, where n is a positive integer,

a show that $I_n = x^n e^x - n I_{n-1}$, $n \geq 1$

b find $\int x^4 e^x \, dx$.

a Let $u = x^n$ and $\dfrac{dv}{dx} = e^x$

so that $\dfrac{du}{dx} = nx^{n-1}$ and $v = e^x$

Then $I_n = \int x e^x \, dx = x^n e^x - \int n x^{n-1} e^x \, dx$

$\quad = x^n e^x - n \int x^{n-1} e^x \, dx$

so $I_n = x^n e^x - n I_{n-1}$

b $\int x^4 e^x \, dx = I_4$

Using the reduction formula

$I_4 = x^4 e^x - 4 I_3$

$\quad = x^4 e^x - 4(x^3 e^x - 3 I_2)$

$\quad = x^4 e^x - 4 x^3 e^x + 12(x^2 e^x - 2 I_1)$

$\quad = x^4 e^x - 4 x^3 e^x + 12 x^2 e^x - 24(x e^x - I_0)$

$\quad = x^4 e^x - 4 x^3 e^x + 12 x^2 e^x - 24 x e^x + 24 \int e^x \, dx$

$\quad = x^4 e^x - 4 x^3 e^x + 12 x^2 e^x - 24 x e^x + 24 e^x + C$

- The symbol I_n is introduced to avoid the repeated use of integral signs.
- Use $\int u \dfrac{dv}{dx} dx = uv - \int v \dfrac{du}{dx} dx$
- Replacing n by $n-1$ in $\int I_n = \int x^n e^x \, dx$, gives $\int I_{n-1} = \int x^{n-1} e^x \, dx$ This is the reduction formula for I_n
- As the formula reduces n by 1 each time it is applied, I_n can be written, after sufficient steps, in terms of I_0, i.e. $\int e^x \, dx$, a standard integral.
- Using $I_n = x^n e^x - n I_{n-1}$ with $n = 3$
- Using $I_n = x^n e^x - n I_{n-1}$ with $n = 2$
- Using $I_n = x^n e^x - n I_{n-1}$ with $n = 1$
- As $I_0 = \int e^x \, dx$

Sometimes, after using integration by parts, you may need to use an algebraic or trigonometric identity to produce the reduction formula.

Example 21 — SKILLS — CRITICAL THINKING

Show that, if $I_n = \int_0^1 x^n \sqrt{1 - x} \, dx$, then $I_n = \dfrac{2n}{2n + 3} I_{n-1}$, $n \geq 1$

Let $u = x^n$ and $\dfrac{dv}{dx} = \sqrt{1 - x}$

So $\dfrac{du}{dx} = n x^{n-1}$ and $v = -\dfrac{2}{3}(1 - x)^{\frac{3}{2}}$

Then, integrating by parts,

$I_n = \left[-\dfrac{2}{3} x^n (1 - x)^{\frac{3}{2}} \right]_0^1 + \int_0^1 \dfrac{2}{3} n x^{n-1} (1 - x)^{\frac{3}{2}} \, dx$

$\quad = [0 - 0] + \int_0^1 \dfrac{2}{3} n x^{n-1} (1 - x)^{\frac{3}{2}} \, dx$

Using the identity $(1 - x)^{\frac{3}{2}} \equiv (1 - x)\sqrt{1 - x}$

$I_n = \dfrac{2n}{3} \int_0^1 x^{n-1}(1 - x)\sqrt{1 - x} \, dx$

$\quad = \dfrac{2n}{3} \int_0^1 x^{n-1} \sqrt{1 - x} \, dx - \dfrac{2n}{3} \int_0^1 x^n \sqrt{1 - x} \, dx$

$\quad = \dfrac{2n}{3} I_{n-1} - \dfrac{2n}{3} I_n$

So $(3 + 2n) I_n = 2n I_{n-1}$

$I_n = \dfrac{2n}{2n + 3} I_{n-1}$

- If $n \geq 1$, $[\cdots\cdots]_0^1 = 0$
- You need to write this so that the term $\sqrt{1-x}$ appears.
- Collecting up terms in I_n

INTEGRATION   CHAPTER 4

Example 22

Given that $I_n = \int_0^{\frac{\pi}{2}} \sin^n x \, dx$, $n \geq 0$,

a derive the reduction formula $nI_n = (n-1)I_{n-2}$, $n \geq 2$

b deduce the values of

   i $\int_0^{\frac{\pi}{2}} \sin^5 dx$ **ii** $\int_0^{\frac{\pi}{2}} \sin^6 dx$

a First write $\sin^n x$ as $\sin^{n-1} x \sin x$

Using integration by parts on $\int \sin^{n-1} x \sin x \, dx$

Let $u = \sin^{n-1} x$ and $\dfrac{dv}{dx} = \sin x$

$\Rightarrow \dfrac{du}{dx} = (n-1)\sin^{n-2} x \cos x$ and $v = -\cos x$

So $I_n = \int_0^{\frac{\pi}{2}} \sin^n x \, dx$

$= \left[-\sin^{n-1} x \cos x\right]_0^{\frac{\pi}{2}} + \int_0^{\frac{\pi}{2}} (n-1)\sin^{n-2} x \cos^2 x \, dx$

$= [0 - 0] + (n-1)\int_0^{\frac{\pi}{2}} \sin^{n-2} x (1 - \sin^2 x) \, dx$

$= (n-1)\int_0^{\frac{\pi}{2}} \sin^{n-2} x \, dx - (n-1)\int_0^{\frac{\pi}{2}} \sin^n x \, dx$

$I_n = (n-1)I_{n-2} - (n-1)I_n$

$\Rightarrow I_n + (n-1)I_n = (n-1)I_{n-2}$

So $nI_n = (n-1)I_{n-2}$, $n \geq 2$ is the reduction formula

b If n is odd

$I_n = \left(\dfrac{n-1}{n}\right)\left(\dfrac{n-3}{n-2}\right)\cdots\left(\dfrac{2}{3}\right)I_1$

$= \left(\dfrac{n-1}{n}\right)\left(\dfrac{n-3}{n-2}\right)\cdots\left(\dfrac{2}{3}\right)\int_0^{\frac{\pi}{2}} \sin x \, dx$

$= \left(\dfrac{n-1}{n}\right)\left(\dfrac{n-3}{n-2}\right)\cdots\left(\dfrac{2}{3}\right)$ **(1)**

If n is even

$I_n = \left(\dfrac{n-1}{n}\right)\left(\dfrac{n-3}{n-2}\right)\cdots\left(\dfrac{3}{4}\right)\left(\dfrac{1}{2}\right)I_0$

$= \left(\dfrac{n-1}{n}\right)\left(\dfrac{n-3}{n-2}\right)\cdots\left(\dfrac{3}{4}\right)\left(\dfrac{1}{2}\right)\int_0^{\frac{\pi}{2}} 1 \, dx$

$= \left(\dfrac{n-1}{n}\right)\left(\dfrac{n-3}{n-2}\right)\cdots\left(\dfrac{3}{4}\right)\left(\dfrac{1}{2}\right)\left(\dfrac{\pi}{2}\right)$ **(2)**

  i $\int_0^{\frac{\pi}{2}} \sin^5 x \, dx = I_5$

$= \left(\dfrac{4}{5}\right)\left(\dfrac{2}{3}\right)(1) = \dfrac{8}{15}$

  ii $\int_0^{\frac{\pi}{2}} \sin^6 x \, dx = I_6$

$= \left(\dfrac{5}{6}\right)\left(\dfrac{3}{4}\right)\left(\dfrac{1}{2}\right)\left(\dfrac{\pi}{2}\right) = \left(\dfrac{5\pi}{32}\right)$

Hint To find $\int \cos^n x \, dx$ you would write $\cos^n x$ as $\cos^{n-1} x \cos x$

Take care with signs.

This is not in a convenient form but you can use $\sin^2 x + \cos^2 x = 1$ to express as powers of $\sin x$ only.

Collect up terms in I_n and I_{n-2}

Using **(1)** with $n = 5$

Using **(2)** with $n = 6$

It is not always necessary to use integration by parts to produce a reduction formula.

Example 23 — SKILLS: CRITICAL THINKING

$I_n = \int \tan^n x \, dx$ where n is a positive integer.

By writing $\tan^n x$ as $\tan^{n-2} x \tan^2 x$, and using $1 + \tan^2 x = \sec^2 x$, establish the reduction formula

$$I_n = \frac{1}{n-1} \tan^{n-1} x - I_{n-2}, \; n \geq 2.$$

$$I_n = \int \tan^{n-2} x \tan^2 x \, dx = \int \tan^{n-2} x (\sec^2 x - 1) \, dx$$
$$= \int \tan^{n-2} x \sec^2 x \, dx - \int \tan^{n-2} x \, dx$$
$$\text{So } I_n = \frac{1}{n-1} \tan^{n-1} x - I_{n-2}$$

Use $\int [f(x)]^n f'(x) \, dx = \dfrac{[f(x)]^{n+1}}{n+1} + C$

Exercise 4F — SKILLS: CRITICAL THINKING

1 Given that $I_n = \int x^n e^{\frac{x}{2}} \, dx$,

   a show that $I_n = 2x^n e^{\frac{x}{2}} - 2n I_{n-1}, \; n \geq 1$;

   b hence find $\int x^3 e^{\frac{x}{2}} \, dx$

2 Given that $I_n = \int_1^e x(\ln x)^n \, dx, \; n \in \mathbb{N}$

   a show that $I_n = \dfrac{e^2}{2} - \dfrac{n}{2} I_{n-1}, \; n \in \mathbb{N}$;

   b hence show that $\int_1^e x(\ln x)^4 \, dx = \dfrac{e^2 - 3}{4}$.

3 In Example 21 you saw that, if $I_n = \int_0^1 x^n \sqrt{1-x} \, dx$, then $I_n = \dfrac{2n}{2n+3} I_{n-1}, \; n \geq 1$.

   Use this reduction formula to evaluate $\int_0^1 (x+1)(x+2)\sqrt{1-x} \, dx$.

E/P 4 Given that $I_n = \int x^n e^{-x} \, dx$, where n is a positive integer,

   a show that $I_n = -x^n e^{-x} + n I_{n-1}, \; n \geq 1$ **(7 marks)**

   b find $\int x^3 e^{-x} \, dx$ **(4 marks)**

   c evaluate $\int_0^1 x^4 e^{-x} \, dx$, giving your answer in terms of e.   **(4 marks)**

E/P 5 $I_n = \int \tanh^n x \, dx$

   a By writing $\tanh^n x = \tanh^{n-2} x \tanh^2 x$ show that for $n \geq 2$,

   $I_n = I_{n-2} - \dfrac{1}{n-1} \tanh^{n-1} x$ **(6 marks)**

   b Find $\int \tanh^5 x \, dx$ **(3 marks)**

   c Show that $\int_0^{\ln 2} \tanh^4 x \, dx = \ln 2 - \dfrac{84}{125}$ **(4 marks)**

INTEGRATION CHAPTER 4

E/P 6 Given that $\int \tan^n x \, dx = \dfrac{1}{n-1} \tan^{n-1} x - \int \tan^{n-2} x \, dx$

    a find $\int \tan^4 x \, dx$ **(4 marks)**

    b evaluate $\int_0^{\frac{\pi}{4}} \tan^5 x \, dx$ **(3 marks)**

    c show that $\int_0^{\frac{\pi}{3}} \tan^6 x \, dx = \dfrac{9\sqrt{3}}{5} - \dfrac{\pi}{3}$ **(4 marks)**

Hint This result was derived in Example 23.

E/P 7 Given that $I_n = \int_1^a (\ln x)^n \, dx$, where $a > 1$ is a constant,

    a show that, for $n \geq 1$, $I_n = a(\ln a)^n - nI_{n-1}$ **(8 marks)**

    b find the exact value of $\int_1^2 (\ln x)^3 \, dx$ **(4 marks)**

    c show that $\int_1^e (\ln x)^6 \, dx = 5(53e - 144)$ **(4 marks)**

8 Using the results given in Example 22, evaluate

    a $\int_0^{\frac{\pi}{2}} \sin^7 x \, dx$

    b $\int_0^{\frac{\pi}{2}} \sin^2 x \cos^4 x \, dx$

    c $\int_0^1 x^5 \sqrt{1 - x^2} \, dx$, using the substitution $x = \sin \theta$

    d $\int_0^{\frac{\pi}{6}} \sin^8 3t \, dt$, using a suitable substitution.

E 9 Given that $I_n = \int \dfrac{\sin^{2n} x}{\cos x} \, dx$

    a write down a similar expression for I_{n+1} and hence show that $I_n - I_{n+1} = \dfrac{\sin^{2n+1} x}{2n + 1}$ **(6 marks)**

    b find $\int \dfrac{\sin^4 x}{\cos x} \, dx$ and hence show that $\int_0^{\frac{\pi}{4}} \dfrac{\sin^4 x}{\cos x} \, dx = \ln(1 + \sqrt{2}) - \dfrac{7\sqrt{2}}{12}$ **(6 marks)**

E/P 10 **a** Given that $I_n = \int_0^1 x(1 - x^3)^n \, dx$, show that $I_n = \dfrac{3n}{3n + 2} I_{n-1}$ **(6 marks)**

    b Use your reduction formula to evaluate I_4 **(4 marks)**

Hint After integrating by parts, write x^4 as $x\{1 - (1 - x^3)\}$.

P 11 Given that $I_n = \int_0^a (a^2 - x^2)^n \, dx$, where a is a positive constant,

    a show that, for $n > 0$, $I_n = \dfrac{2na^2}{2n + 1} I_{n-1}$

    b use the reduction formula to evaluate

        i $\int_0^1 (1 - x^2)^4 \, dx$ **ii** $\int_0^3 (9 - x^2)^3 \, dx$ **iii** $\int_0^2 \sqrt{4 - x^2} \, dx$

    c Check your answer to part **b iii** by using another method.

E/P 12 Given that $I_n = \int_0^4 x^n \sqrt{4 - x} \, dx$,

    a establish the reduction formula $I_n = \dfrac{8n}{2n + 3} I_{n-1}$, $n \geq 1$ **(8 marks)**

    b evaluate $\int_0^4 x^3 \sqrt{4 - x}$, giving your answer correct to 3 significant figures. **(3 marks)**

13 Given that $I_n = \int \cos^n x \, dx$,

   a establish, for $n \geq 2$, the reduction formula $nI_n = \cos^{n-1} x \sin x + (n-1)I_{n-2}$ **(8 marks)**

   Defining $J_n = \int_0^{2\pi} \cos^n x \, dx$,

   b write down a reduction formula relating J_n and J_{n-2}, for $n \geq 2$ **(4 marks)**

   c hence evaluate   **i** J_4 **ii** J_8 **(4 marks)**

   d Show that if n is odd, J_n is always equal to zero.   **(4 marks)**

14 Given $I_n = \int_0^1 x^n \sqrt{(1-x^2)} \, dx, n \geq 0$,

   a show that $(n+2)I_n = (n-1)I_{n-2}, n \geq 2$ **(6 marks)**

   b hence evaluate $\int_0^1 x^7 \sqrt{(1-x^2)} \, dx$ **(4 marks)**

> **Hint** Write $x^n \sqrt{1-x^2}$ as $x^{n-1}\{x\sqrt{1-x^2}\}$ before integrating by parts.

15 Given $I_n = \int x^n \cosh x \, dx$,

   a show that for $n \geq 2$, $I_n = x^n \sinh x - nx^{n-1} \cosh x + n(n-1)I_{n-2}$ **(8 marks)**

   b find $\int x^4 \cosh x \, dx$ **(4 marks)**

   c evaluate $\int_0^1 x^3 \cosh x \, dx$, giving your answer in terms of e.   **(4 marks)**

16 Given that $I_n = \int \frac{\sin nx}{\sin x} \, dx, n > 0$,

   a write down a similar expression for I_{n-2}, and hence show that
$$I_n - I_{n-2} = \frac{2 \sin(n-1)x}{n-1}$$ **(7 marks)**

   b find

   i $\int \frac{\sin 4x}{\sin x} \, dx$ **ii** the exact value of $\int_{\frac{\pi}{6}}^{\frac{\pi}{3}} \frac{\sin 5x}{\sin x} \, dx$ **(6 marks)**

17 Given that $I_n = \int \sinh^n x \, dx, n \in \mathbb{N}$,

   a derive the reduction formula $nI_n = \sinh^{n-1} x \cosh x - (n-1)I_{n-2}, n \geq 2$ **(9 marks)**

   b hence

   i evaluate $\int_0^{\ln 3} \sinh^5 x \, dx$ **ii** show that $\int_0^{\text{arsinh} 1} \sinh^4 x \, dx = \frac{1}{8}(3\ln(1+\sqrt{2}) - \sqrt{2})$

   (7 marks)

> **Challenge**
>
> **a** Derive a reduction formula for $I_n = \int x^a (\ln x)^n \, dx$, where a is a rational number such that $a \neq -1$ and n is an integer such that $n \geq 1$.
>
> **b** Hence find $\int 1\sqrt{x}(\ln x)^3 \, dx$

4.7 Finding the length of an arc of a curve

Suppose that $P(x, y)$ is any point on the curve C, whose equation is $y = f(x)$, and that the length of the arc from a fixed point on C to P is denoted by s. Let $Q(x + \delta x, y + \delta y)$ be a neighbouring point on C, and the length of the arc PQ be δs.

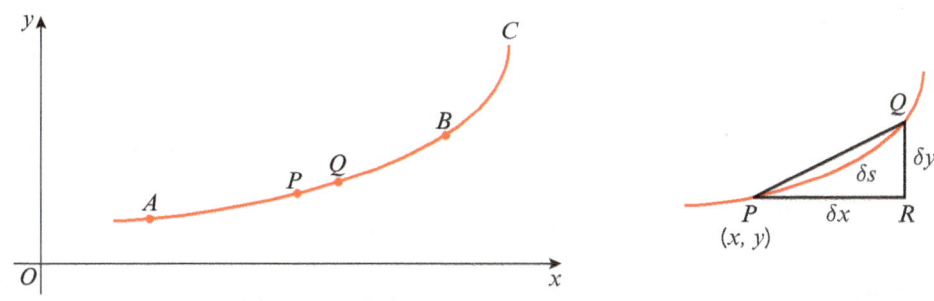

As P and Q are close together, $\delta s \approx$ the length of the chord PQ — See triangle PQR

so $(\delta s)^2 \approx (\delta x)^2 + (\delta y)^2$ **(1)**

$\Rightarrow \left(\dfrac{\delta s}{\delta x}\right)^2 \approx 1 + \left(\dfrac{\delta y}{\delta x}\right)^2$

Remember this: it is the key to deriving all the formulae.

As $\delta x \to 0$ (Q approaches P), $\dfrac{\delta s}{\delta x} \to \dfrac{ds}{dx}$ and $\dfrac{\delta y}{\delta x} \to \dfrac{dy}{dx}$

so, in the limit $\left(\dfrac{ds}{dx}\right)^2 = 1 + \left(\dfrac{dy}{dx}\right)^2$

and $\dfrac{ds}{dx} = \sqrt{1 + \left(\dfrac{dy}{dx}\right)^2}$

The positive square root is taken so that s increases as x increases.

Integrating this with respect to x gives an expression for s, the arc length. So, if s is the length of the arc joining $A(x_A, y_A)$ and $B(x_B, y_B)$,

- $s = \displaystyle\int_{x_A}^{x_B} \sqrt{1 + \left(\dfrac{dy}{dx}\right)^2}\, dx$ **(*)**

The curve must be continuous between the end points of the arc. Then, providing the integration is possible, the arc length can be evaluated

Alternatively, dividing **(1)** throughout by $(\delta y)^2$ and proceeding to the limit, gives

$\left(\dfrac{ds}{dy}\right)^2 = 1 + \left(\dfrac{dx}{dy}\right)^2$

Integrating with respect to y gives

- $s = \displaystyle\int_{y_A}^{y_B} \sqrt{1 + \left(\dfrac{dx}{dy}\right)^2}\, dy$

You can use whichever formula is convenient.

If the equation of the curve is given parametrically, i.e. in the form $x = f(t), y = g(t)$, then dividing **(1)** by $(\delta t)^2$ and proceeding to the limit, gives

$\left(\dfrac{ds}{dt}\right)^2 = \left(\dfrac{dx}{dt}\right)^2 + \left(\dfrac{dy}{dt}\right)^2$

Given that the parameters at A and B are t_A and t_B respectively, then integrating with respect to t gives

- $s = \displaystyle\int_{t_A}^{t_B} \sqrt{\left(\dfrac{dx}{dt}\right)^2 + \left(\dfrac{dy}{dt}\right)^2}\, dt$ **(*)**

(*) These formulae are in the formula book.

Example 24 SKILLS PROBLEM-SOLVING

Find the exact length of the arc on the parabola with equation $y = \frac{1}{2}x^2$, from the origin to the point $P(4, 8)$.

$y = \frac{1}{2}x^2 \Rightarrow \frac{dy}{dx} = x$ — First find $\frac{dy}{dx}$

Using $s = \int_{x_A}^{x_B} \sqrt{1 + \left(\frac{dy}{dx}\right)^2}\, dx$, — Choose the appropriate formula.

length of arc $OP = \int_0^4 \sqrt{1 + x^2}\, dx$ — $x_A = 0, x_B = 4$

Using the substitution $x = \sinh u$,
so that $dx = \cosh u\, du$, — Choose the appropriate method of integration.

arc length $= \int_0^{\operatorname{arsinh} 4} \sqrt{1 + \sinh^2 u}\, \cosh u\, du$

$= \int_0^{\operatorname{arsinh} 4} \cosh^2 u\, du$

$= \int_0^{\operatorname{arsinh} 4} \frac{(1 + \cosh 2u)}{2}\, du$

$= \frac{1}{2}\left[u + \frac{1}{2}\sinh 2u\right]_0^{\operatorname{arsinh} 4}$

$= \frac{1}{2}\left[u + \sinh u \cosh u\right]_0^{\operatorname{arsinh} 4}$

$= \frac{1}{2}\operatorname{arsinh} 4 + \frac{1}{2}(4\sqrt{1 + 16})$ — Use $\cosh u = \sqrt{1 + \sinh^2 u}$ and $\sinh u = 4$

$= \frac{1}{2}\operatorname{arsinh} 4 + 2\sqrt{17}$

$= \frac{1}{2}\ln(4 + \sqrt{17}) + 2\sqrt{17}$ — Use $\operatorname{arsinh} x = \ln\{x + \sqrt{1 + x^2}\}$

Example 25 SKILLS PROBLEM-SOLVING

The curve C has parametric equations

$x = t + \frac{1}{t}, y = 2\ln t, t > 0$

Find the length of the arc between points A and B with $t = 1$ and $t = 2$ respectively.

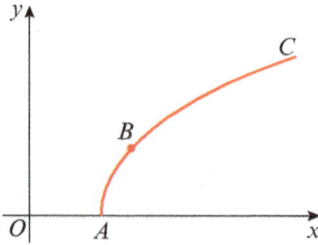

$x = t + \frac{1}{t}, y = 2\ln t, t > 0$

$\frac{dx}{dt} = 1 - \frac{1}{t^2}, \frac{dy}{dt} = \frac{2}{t}$

So $\left(\dfrac{dx}{dt}\right)^2 + \left(\dfrac{dy}{dt}\right)^2 = \left\{\left(1 - \dfrac{2}{t^2} + \dfrac{1}{t^4}\right) + \left(\dfrac{4}{t^2}\right)\right\}$

$= 1 + \dfrac{2}{t^2} + \dfrac{1}{t^4} = \left(1 + \dfrac{1}{t^2}\right)^2$

Arc length $= \displaystyle\int_1^2 \sqrt{\left(1 + \dfrac{1}{t^2}\right)^2}\, dt$

Use $s = \displaystyle\int_{t_A}^{t_B} \sqrt{\left(\dfrac{dx}{dt}\right)^2 + \left(\dfrac{dy}{dt}\right)^2}\, dt$

$= \displaystyle\int_1^2 \left(1 + \dfrac{1}{t^2}\right) dt$

$= \left[t - \dfrac{1}{t}\right]_1^2$

$= 1.5$

Exercise 4G — SKILLS — PROBLEM-SOLVING

1. Find the length of the arc of the curve with equation $y = \dfrac{1}{3}x^{\frac{3}{2}}$, from the origin to the point with x-coordinate 12.

2. The curve C has equation $y = \ln \cos x$. Find the length of the arc of C between the points with x-coordinates 0 and $\dfrac{\pi}{3}$.

3. Find the length of the arc on the curve, with equation $y = 2\cosh\left(\dfrac{x}{2}\right)$, between the points with x-coordinates 0 and $\ln 4$.

4. Find the length of the arc of the curve with equation $y^2 = \dfrac{4}{9}x^3$, from the origin to the point $(3, 2\sqrt{3})$.

5. The curve C has equation $y = \dfrac{1}{2}\sinh^2 2x$. Find the length of the arc on C from the origin to the point whose x-coordinate is 1, giving your answer to 3 significant figures.

(E) 6. The curve C has equation $y = \dfrac{1}{4}(2x^2 - \ln x)$, $x > 0$. The points A and B on C have x-coordinates 1 and 2 respectively. Show that the length of the arc from A to B is $\dfrac{1}{4}(6 + \ln 2)$ **(5 marks)**

(E) 7. Find the length of the arc on the curve $y = 2\operatorname{arcosh}\left(\dfrac{x}{2}\right)$, from the point at which the curve crosses the x-axis to the point with x-coordinate $\dfrac{5}{2}$. Compare your answer with that in Example 25 and explain the relationship. **(6 marks)**

(E) 8. The line $y = 4$ intersects the parabola with equation $y = x^2$ at the points A and B. Find the length of the arc of the parabola from A to B. **(6 marks)**

(E) 9. The circle C has parametric equations $x = r\cos\theta$, $y = r\sin\theta$. Use the formula for arc length on page 79 to show that the length of the circumference is $2\pi r$. **(6 marks)**

10 The diagram shows the asteroid, with parametric equations
$x = 2a \cos^3 t$, $y = 2a \sin^3 t$, $0 \leq t \leq 2\pi$

Find the length of the arc of the curve AB, and hence find
the total length of the curve. **(7 marks)**

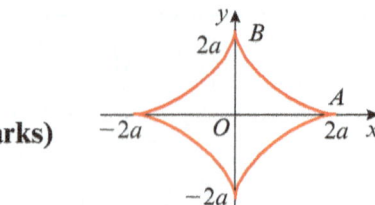

11 Calculate the length of the arc on the curve with parametric equations $x = \tanh u$, $y = \operatorname{sech} u$, between the points with parameters $u = 0$ and $u = 1$ **(7 marks)**

12 The cycloid has parametric equations $x = a(\theta + \sin \theta)$, $y = a(1 - \cos \theta)$. Find the length of the arc from $\theta = 0$ to $\theta = \pi$ **(7 marks)**

13 Show that the length of the arc, between the points with parameters $t = 0$ and $t = \dfrac{\pi}{3}$ on the curve defined by the equations $x = t + \sin t$, $y = 1 - \cos t$, is 2. **(7 marks)**

14 Find the length of the arc of the curve given by the equations $x = e^t \cos t$, $y = e^t \sin t$, between the points with parameters $t = 0$ and $t = \dfrac{\pi}{4}$ **(6 marks)**

15 a Denoting the length of one complete wave of the sine curve with equation $y = \sqrt{3} \sin x$ by L,
show that $L = 4 \displaystyle\int_0^{\frac{\pi}{2}} \sqrt{1 + 3\cos^2 x}\, dx$ **(5 marks)**

b An ellipse has parametric equations $x = \cos t$, $y = 2 \sin t$. Show that the length of its circumference is equal to that of the wave in **a**. **(6 marks)**

Challenge

SKILLS
CREATIVITY

The function f is defined as
$$f(x) = \int_1^x \sqrt{t^3 - 1}\, dt,\ 1 \leq x \leq 4$$
The diagram shows the curve with equation $y = f(x)$.
Find the exact length of the curve.

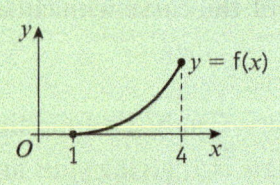

4.8 Finding the area of a surface of revolution

Consider the curve C being rotated completely about the x-axis, and let the surface area generated by an arc between a fixed point on the curve and the point $P(x, y)$ be S.

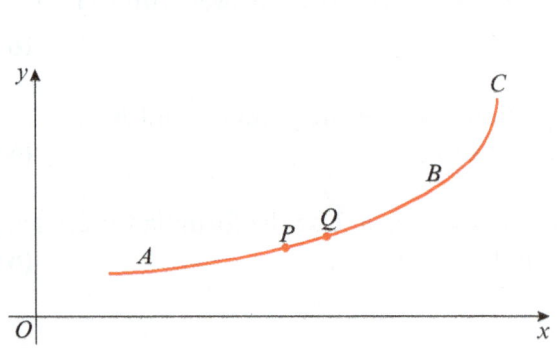
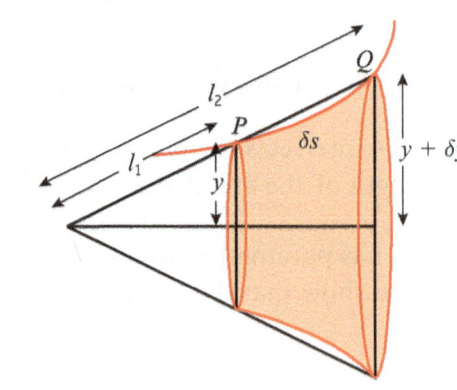

If $Q(x + \delta x, y + \delta y)$ is a neighbouring point on C, then the arc PQ generates a surface of area δS, which is approximately equal to that of a frustum of a cone (see diagram above).

The surface area of a cone is $\pi r l$, where r is the base radius and l the slant height.

So the area of the frustum $= \pi(y + \delta y)l_2 - \pi y l_1$

$\qquad\qquad\qquad\qquad\quad = \pi(2y + \delta y)(l_2 - l_1)$

As P and Q are close together, $\delta s \approx$ chord $PQ = l_2 - l_1$

so $\qquad\qquad\qquad\qquad \delta S \approx \pi(2y + \delta y)\delta s$

and $\qquad\qquad\qquad$ In cosh $\dfrac{\delta S}{\delta x} \approx \pi(2y + \delta y)\dfrac{\delta s}{\delta x}$

> This can be shown using
> $\dfrac{l_2}{l_1} = \dfrac{y + \delta y}{y}$ (similar triangles)
> $\Rightarrow \dfrac{l_2 + l_1}{l_1} = \dfrac{2y + \delta y}{y}$

> Dividing by δx.

As Q approaches P, $\delta x \to 0$, $\delta y \to 0$, $\dfrac{\delta s}{\delta x} \to \dfrac{ds}{dx}$ and $\dfrac{\delta S}{\delta x} \to \dfrac{dS}{dx}$

so in the limit $\dfrac{dS}{dx} = 2\pi y \dfrac{ds}{dx}$

S can be found by integrating with respect to x.

- **The surface area S, generated when an arc is rotated about the x-axis, is**

$$S = \int 2\pi y \dfrac{ds}{dx} dx = \int 2\pi y\, ds \quad (*)$$

Similarly if the curve is rotated about the y-axis, then $S = \int 2\pi x\, ds$

Using results derived in Section 4.7, these formulae for S can be rewritten to apply to the appropriate coordinate system.

For rotation of a curve about the x-axis:

In Cartesian form $\qquad S = 2\pi \displaystyle\int_{x_A}^{x_B} y\sqrt{1 + \left(\dfrac{dy}{dx}\right)^2}\, dx$

> since $S = \int 2\pi y\, ds = \int 2\pi y \dfrac{ds}{dx} dx$

In parametric form $\qquad S = 2\pi \displaystyle\int_{t_A}^{t_B} y\sqrt{\left(\dfrac{dx}{dt}\right)^2 + \left(\dfrac{dy}{dt}\right)^2}\, dt \quad (*)$

> since $S = \int 2\pi y\, ds = \int 2\pi y \dfrac{ds}{dt} dt$

For rotation of a curve about the y-axis:

In Cartesian form $\qquad S = 2\pi \displaystyle\int_{y_A}^{y_B} x\sqrt{1 + \left(\dfrac{dx}{dy}\right)^2}\, dy$

> $S = \int 2\pi x\, ds = \int 2\pi x \dfrac{ds}{dy} dy$

or $\qquad\qquad\qquad\quad S = 2\pi \displaystyle\int_{x_A}^{x_B} x\sqrt{1 + \left(\dfrac{dy}{dx}\right)^2}\, dx$

> $S = \int 2\pi x\, ds = \int 2\pi x \dfrac{ds}{dx} dx$

In parametric form $\qquad S = 2\pi \displaystyle\int_{t_A}^{t_B} x\sqrt{\left(\dfrac{dx}{dt}\right)^2 + \left(\dfrac{dy}{dt}\right)^2}\, dt$

> This form is often the more convenient.

Example 26 — SKILLS: PROBLEM-SOLVING

The curve C has equation $y = \frac{1}{3}\sqrt{x}\,(3 - x)$. The arc of the curve between the points with x-coordinates 1 and 3 is completely rotated about the x-axis. Find the area of the surface generated.

$y = x^{\frac{1}{2}} - \frac{1}{3}x^{\frac{3}{2}} \Rightarrow \frac{dy}{dx} = \frac{1}{2}\left(x^{-\frac{1}{2}} - x^{\frac{1}{2}}\right)$

So $1 + \left(\frac{dy}{dx}\right)^2 = 1 + \frac{1}{4}x^{-1} - \frac{1}{2} + \frac{1}{4}x$

$= \frac{1}{4}x^{-1} + \frac{1}{2} + \frac{1}{4}x$

$= \frac{1}{4}\left(x^{-\frac{1}{2}} + x^{\frac{1}{2}}\right)^2$

Area of the surface generated $= \int_1^3 2\pi y \sqrt{1 + \left(\frac{dy}{dx}\right)^2}\, dx$

$= 2\pi \int_1^3 \left(x^{\frac{1}{2}} - \frac{1}{3}x^{\frac{3}{2}}\right)\frac{1}{2}\left(x^{-\frac{1}{2}} + x^{\frac{1}{2}}\right) dx$

$= \pi \int_1^3 \left(1 + \frac{2}{3}x - \frac{1}{3}x^2\right) dx$

$= \pi \left[x + \frac{1}{3}x^2 - \frac{1}{9}x^3\right]_1^3$

$= 3\pi - \frac{11}{9}\pi$

$= \frac{16}{9}\pi$

Example 27 — SKILLS: PROBLEM-SOLVING

The arc of the curve with equation $y = \cosh x$, from the point $(0, 1)$ to $\left(\ln 2, \frac{5}{4}\right)$ is rotated completely about the y-axis. Find the area of the surface generated.

$y = \cosh x \Rightarrow \frac{dy}{dx} = \sinh x$

Using $\int 2\pi x\, ds$ in the form $\int 2\pi x \frac{ds}{dx}\, dx = \int_{x_2}^{x_1} 2\pi x \sqrt{1 + \left(\frac{dy}{dx}\right)^2}\, dx$

area of the surface generated $= 2\pi \int_0^{\ln 2} x\sqrt{1 + \sinh^2 x}\, dx$

$= 2\pi \int_0^{\ln 2} x \cosh x\, dx$

$= 2\pi \left[x \sinh x - \int \sinh x\, dx\right]_0^{\ln 2}$

$= 2\pi[\ln 2\, \sinh(\ln 2) - \cosh(\ln 2) + 1]$

$= 2\pi \left[\ln 2\left(\frac{3}{4}\right) - \frac{5}{4} + 1\right]$

$= \frac{\pi}{2}[3\ln 2 - 1]$

> Use integration by parts with $u = x$ and $\frac{dv}{dx} = \cosh x$

INTEGRATION　　CHAPTER 4

Example 28

The curve with parametric equations $x = t - \sin t$, $y = 1 - \cos t$, from $t = 0$ to $t = 2\pi$, is rotated through 360° about the x-axis. Find the area of the surface generated.

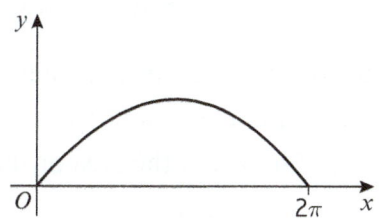

$\dfrac{dx}{dt} = 1 - \cos t,\ \dfrac{dy}{dt} = \sin t$

So $\left(\dfrac{dx}{dt}\right)^2 + \left(\dfrac{dy}{dt}\right)^2 = (1 - \cos t)^2 + \sin^2 t = 2 - 2\cos t = 2(1 - \cos t) = 4\sin^2 \dfrac{t}{2}$

Area of the surface generated $= \displaystyle\int_0^{2\pi} 2\pi y \sqrt{\left(\dfrac{dx}{dt}\right)^2 + \left(\dfrac{dy}{dt}\right)^2}\, dt$

$= 2\pi \displaystyle\int_0^{2\pi} (1 - \cos t) 2\sin\dfrac{t}{2}\, dt$

$= 2\pi \displaystyle\int_0^{2\pi} 4\sin^3\dfrac{t}{2}\, dt$

$= 8\pi \displaystyle\int_0^{2\pi} \sin^2\dfrac{t}{2}\sin\dfrac{t}{2}\, dt$

$= 8\pi \displaystyle\int_0^{2\pi} \left(1 - \cos^2\dfrac{t}{2}\right)\sin\dfrac{t}{2}\, dt$

$= 8\pi \left[-2\cos\dfrac{t}{2} + \dfrac{2}{3}\cos^3\dfrac{t}{2}\right]_0^{2\pi}$

$= 8\pi \left[\left(2 - \dfrac{2}{3}\right) - \left(-2 + \dfrac{2}{3}\right)\right]$

$= \dfrac{64}{3}\pi$

Exercise 4H   SKILLS   PROBLEM-SOLVING

1 a The section of the line $y = \dfrac{3}{4}x$ between points with x-coordinates 4 and 8 is rotated completely about the x-axis. Use integration to find the area of the surface generated.

b The same section of line is rotated completely about the y-axis. Show that the area of the surface generated is 60π.

2 The arc of the curve $y = x^3$, between the origin and the point $(1, 1)$, is rotated through 4 right-angles about the x-axis. Find the area of the surface generated.

3 The arc of the curve $y = \dfrac{1}{2}x^2$, between the origin and the point $(2, 2)$, is rotated through 4 right-angles about the y-axis. Find the area of the surface generated.

4 The points A and B, in the first quadrant, on the curve $y^2 = 16x$ have x-coordinates 5 and 12 respectively. Find, in terms π, the area of the surface generated when the arc AB is rotated completely about the x-axis.

5 The curve C has equation $y = \cosh x$. The arc s on C, has end points $(0, 1)$ and $(1, \cosh 1)$.
   a Find the area of the surface generated when s is rotated completely about the x-axis.
   b Show that the area of the surface generated when s is rotated completely about the y-axis is $2\pi\left(\dfrac{e-1}{e}\right)$

(E/P) 6 The curve C has equation $y = \dfrac{1}{2x} + \dfrac{x^3}{6}$

   a Show that $\sqrt{1 + \left(\dfrac{dy}{dx}\right)^2} = \dfrac{1}{2}\left(x^2 + \dfrac{1}{x^2}\right)$ (3 marks)

   The arc of the curve between points with x-coordinates 1 and 3 is rotated completely about the x-axis.
   b Find the area of the surface generated.   (5 marks)

(E) 7 The diagram shows part of the curve with equation $x^{\frac{2}{3}} + y^{\frac{2}{3}} = 4$
   Find the area of the surface generated when this arc is rotated completely about the y-axis.   (6 marks)

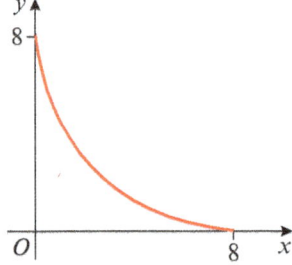

8 a The arc of the circle with equation $x^2 = y^2 = R^2$, between the points $(-R, 0)$ and $(R, 0)$, is rotated through 2π radians about the x-axis. Use integration to find the surface area of the sphere S formed.
   b The axis of a cylinder C of radius R is the x-axis. Show that the areas of the surfaces of S and C, contained between planes with equations $x = a$ and $x = b$, where $a < b < R$, are equal.

(E/P) 9 The finite arc of the parabola with parametric equations $x = at^2$, $y = 2at$, where a is a positive constant, cut off by the line $x = 4a$, is rotated through 180° about the x-axis.
   Show that the area of the surface generated is $\dfrac{8}{3}\pi a^2(5\sqrt{5} - 1)$ (6 marks)

(E/P) 10 The arc, in the first quadrant, of the curve with parametric equations $x = \text{sech } t$, $y = \tanh t$, between the points where $t = 0$ and $t = \ln 2$, is rotated completely about the x-axis. Show that the area of the surface generated is $\dfrac{2\pi}{5}$ (6 marks)

(E/P) 11 The arc of the curve given by $x = 3t^2$, $y = 2t^3$, from $t = 0$ and $t = 2$, is completely rotated about the y-axis.
   a Show that the area of the surface generated can be expressed as
   $36\pi \displaystyle\int_0^2 t^3\sqrt{1 + t^2}\, dt$ (3 marks)
   b Using integration by parts, find the exact value of this area.   (4 marks)

12 The arc of the curve with parametric equations $x = t^2$, $y = t - \frac{1}{3}t^3$, between the points where $t = 0$ and $t = 1$, is rotated through 360° about the x-axis. Calculate the area of the surface generated. **(6 marks)**

13 The asteroid C has parametric equations $x = a\cos^3 t$, $y = a\sin^3 t$, where a is a positive constant. The arc of C, between $t = \frac{\pi}{6}$ and $t = \frac{\pi}{2}$, is rotated through 2π radians about the x-axis. Find the area of the surface of revolution formed.

14 The part of the curve $y = e^x$, between $(0, 1)$ and $(\ln 2, 2)$, is rotated completely about the x-axis. Show that the area of the surface generated is $\pi(\operatorname{arsinh} 2 - \operatorname{arsinh} 1 + 2\sqrt{5} - \sqrt{2})$ **(8 marks)**

Chapter review 4

SKILLS REASONING/ARGUMENTATION; PROBLEM-SOLVING

1 Show that the volume of the solid generated when the finite region enclosed by the curve with equation $y = \tanh x$, the line $x = 1$ and the x-axis is rotated through 2π radians about the x-axis is $\dfrac{2\pi}{1 + e^2}$ **(7 marks)**

2 $4x^2 + 4x + 17 = (ax + b)^2 + c, a > 0$
   a Find the values of a, b and c. **(3 marks)**
   b Find the exact value of $\displaystyle\int_{-0.5}^{1.5} \frac{1}{4x^2 + 4x + 17}\,dx$ **(7 marks)**

3 Find
   a $\displaystyle\int \sinh 4x \cosh 6x\, dx$
   b $\displaystyle\int \frac{\operatorname{sech} x \tanh x}{1 + 2\operatorname{sech} x}\,dx$
   c $\displaystyle\int e^x \sinh x\, dx$

4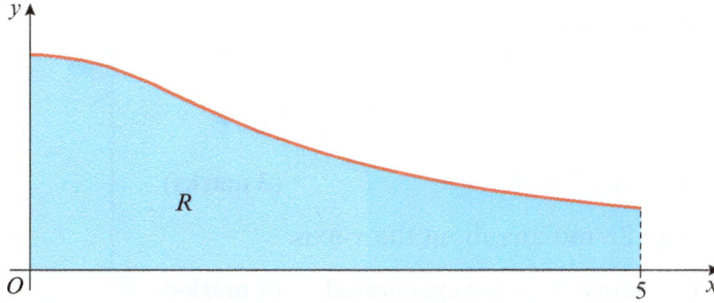

The diagram shows the cross-section R of an artificial ski slope. The slope is modelled by the curve with equation $y = \dfrac{10}{\sqrt{4x^2 + 9}}$, $0 \leq x \leq 5$

Given that 1 unit on each axis represents 10 metres, use integration to calculate the area R. Show your method clearly and give your answer to 2 significant figures. **(6 marks)**

5 a Find $\displaystyle\int \frac{1 + 2x}{1 + 4x^2}\,dx$
   b Find the exact value of $\displaystyle\int_0^{0.5} \frac{1 + 2x}{1 + 4x^2}\,dx$

6 A rope is hung from two points on the same horizontal level. The curve formed by the rope is modelled by the equation

$$y = 4\cosh\left(\frac{x}{4}\right), \quad -20 \leq x \leq 20$$

Find the length of the rope, giving your answer to 3 significant figures.

7 Show that $\int_0^{\frac{1}{2}} \operatorname{artanh} x \, dx = \frac{1}{4}\ln\left(\frac{a}{b}\right)$, where a and b are positive integers to be found.

8 Given that, $I_n = \int_0^{\frac{\pi}{2}} x^n \cos x \, dx$

   a find the values of   i I_0 ii I_1 (6 marks)

   b show, by using integration by parts twice, that $I_n = \left(\frac{\pi}{2}\right)^n - n(n-1)I_{n-2}$, $n \geq 2$ (5 marks)

   c Hence show that $\int_0^{\frac{\pi}{2}} x^3 \cos x \, dx = \frac{1}{8}(\pi^3 - 24\pi + 48)$ (4 marks)

   d Evaluate $\int_0^{\frac{\pi}{2}} x^4 \cos x \, dx$, leaving your answer in terms of π.   (3 marks)

9 a Find $\int \frac{dx}{\sqrt{x^2 - 2x + 10}}$ (4 marks)

   b Find $\int \frac{dx}{x^2 - 2x + 10}$ (4 marks)

   c By using the substitution $x = \sin\theta$, show that $\int_0^{\frac{1}{2}} \frac{x^4}{\sqrt{1-x}} = \frac{4\pi - 7\sqrt{3}}{64}$ (5 marks)

10 Given that $I_n = \int_0^1 x^n(1-x)^{\frac{1}{3}} \, dx$, $n \geq 0$,

   a show that $I_n = \frac{3n}{3n+4} I_{n-1}$, $n \geq n \geq 1$

   b hence find the exact value of $\int_0^1 (1+x)(1-x)^{\frac{4}{3}} \, dx$

11 The curve C has parametric equations

   $x = t - \ln t$

   $y = 4\sqrt{t}$, $1 \leq t \leq 4$

   a Show that the length of C is $3 + \ln 4$ (3 marks)

The curve is rotated through 2π radians about the x-axis.

   b Find the exact area of the curved surface generated.   (4 marks)

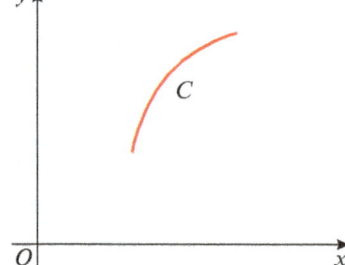

12 The diagram shows part of the curve with equation

   $y = x^2 \operatorname{arsinh} x$

The region R, shown shaded, is bounded by the curve, the x-axis and the line $x = 3$. Show that the area of R is

   $9\ln(3 + \sqrt{10}) - \frac{1}{9}(2 + 7\sqrt{10})$ (9 marks)

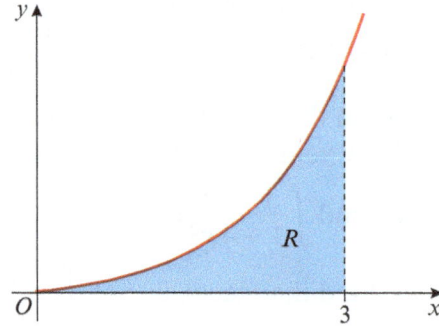

13 a Use the substitution $u = x^2$ to find $\int_0^1 \dfrac{x}{1+x^4}\,dx$

b Find

i $\int \dfrac{1}{\sqrt{4x-x^2}}\,dx$ **ii** $\int \dfrac{4-2x}{\sqrt{4x-x^2}}\,dx$

Hence, or otherwise, evaluate

iii $\int_3^4 \dfrac{5-2x}{\sqrt{4x-x^2}}\,dx$

14 The curve C shown in the diagram has equation $y^2 = 4x$, $0 \le x \le 1$.

The part of the curve in the first quadrant is rotated through 2π radians about the x-axis.

a Show that the surface area of the solid generated is given by

$$4\pi \int_0^1 \sqrt{(1+x)}\,dx$$ (4 marks)

b Find the exact value of this surface area. (3 marks)

c Show also that the length of the curve C, between the points $(1, -2)$ and $(1, 2)$, is given by

$$2\int_0^1 \sqrt{\left(\dfrac{x+1}{x}\right)}\,dx$$ (4 marks)

d Use the substitution $x = \sinh^2\theta$ to show that the exact value of this length is

$2\left[\sqrt{2} + \ln(1+\sqrt{2})\right]$ (5 marks)

15 a Show that $\int x\,\text{arcosh}\,x\,dx = \dfrac{1}{4}(2x^2-1)\text{arcosh}\,x - \dfrac{1}{4}x\sqrt{x^2-1} + C$

b Hence, using the substitution $x = u^2$, find $\int \text{arcosh}(\sqrt{x})\,dx$

16 Given that $I_n = \int \dfrac{\sin(2n+1)x}{\sin x}\,dx$,

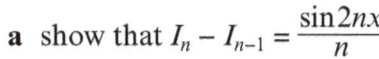

a show that $I_n - I_{n-1} = \dfrac{\sin 2nx}{n}$ (6 marks)

b Hence find I_5 (4 marks)

c Show that, for all positive integers n, $\int_0^{\frac{\pi}{2}} \dfrac{\sin(2n+1)x}{\sin x}\,dx$ always has the same value, which should be found. (5 marks)

17 The diagram shows part of the graph of the curve with equation $y^2 = \dfrac{1}{3}x(x-1)^2$

a Show that the length of the loop is $\dfrac{4\sqrt{3}}{3}$. (6 marks)

The arc OA (in red) is rotated completely about the x-axis.

b Find the area of the surface generated. (7 marks)

18 a Find $\int \dfrac{1}{\sinh x + 2\cosh x}\,dx$ (4 marks)

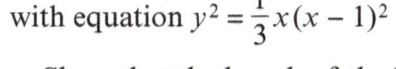

b Show that $\int_1^4 \dfrac{3x-1}{\sqrt{x^2-2x+10}}\,dx = 9(\sqrt{2}-1) + 2\,\text{arsinh}\,1$ (6 marks)

E/P 19 Given that $I_n = \int \sec^n x \, dx$,

    a by writing $\sec^n x = \sec^{n-2} x \sec^2 x$, show that, for $n \geq 2$,
$$(n-1)I_n = \sec^{n-2} x \tan x + (n-2)I_{n-2}$$ (9 marks)

    b find I_5 (3 marks)

    c Hence show that $\int_0^{\frac{\pi}{4}} \sec^5 x \, dx = \frac{1}{8}\{7\sqrt{2} + 3\ln(1+\sqrt{2})\}$ (6 marks)

E/P 20 **a** Show, by using a suitable substitution for x, that
$$\int \sqrt{a^2 - x^2}\, dx = \frac{a^2}{2}\arcsin\left(\frac{x}{a}\right) + \frac{x}{2}\sqrt{a^2 - x^2} + C$$ (8 marks)

    b Hence show that the area of the region enclosed by the ellipse with equation $\frac{x^2}{a^2} + \frac{y^2}{b^2} = 1$ is πab. (5 marks)

> **Challenge**
>
> **SKILLS** **CREATIVITY**
>
> The diagram shows a **Cornu spiral** which has parametric equations
> $$x = \int_0^t \cos\left(\frac{\pi u^2}{2}\right) du, \quad y = \int_0^t \sin\left(\frac{\pi u^2}{2}\right) du, \quad 0 \leq t \leq a$$
>
>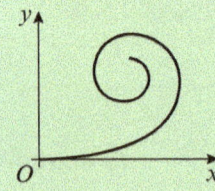
>
> Find the length of the Cornu spiral.

Summary of key points

1 You should be familiar with the following standard forms.
   (Results marked (*) are in the Pearson Edexcel formula booklet)

$\int \sinh x \, dx = \cosh x + C$ (*)

$\int \cosh x \, dx = \sinh x + C$ (*)

$\int \tanh x \, dx = \ln \cosh x$ (*)

$\int \operatorname{sech}^2 x \, dx = \tanh x + C$

$\int \operatorname{cosech}^2 x \, dx = -\coth x + C$

$\int \operatorname{sech} x \tanh x \, dx = -\operatorname{sech} x + C$

$\int \operatorname{cosech} x \coth x \, dx = -\operatorname{cosech} x + C$

$\int \dfrac{1}{\sqrt{(a^2 - x^2)}} \, dx = \arcsin\left(\dfrac{x}{a}\right), |x| < a$ (*) $\int \dfrac{1}{\sqrt{(1 - x^2)}} \, dx = \arcsin x + c, |x| < 1$

$\int \dfrac{1}{a^2 + x^2} \, dx = \dfrac{1}{a} \arctan\left(\dfrac{x}{a}\right)$ (*) $\int \dfrac{1}{1 + x^2} \, dx = \arctan x + C$

$\int \dfrac{1}{\sqrt{(a + x^2)}} \, dx = \operatorname{arsinh}\left(\dfrac{x}{a}\right)$ (*) $\int \dfrac{1}{\sqrt{(1 + x^2)}} \, dx = \operatorname{arsinh} x + C$

$\int \dfrac{1}{\sqrt{(x^2 - a^2)}} \, dx = \operatorname{arcosh}\left(\dfrac{x}{a}\right), x > a$ (*) $\int \dfrac{1}{\sqrt{(x^2 - 1)}} \, dx = \operatorname{arcosh} x + C$

$\int \dfrac{1}{a^2 - x^2} \, dx = \dfrac{1}{2a} \ln\left|\dfrac{a + x}{a - x}\right|, |x| < a$ (*)

$\int \dfrac{1}{x^2 - a^2} \, dx = \dfrac{1}{2a} \ln\left|\dfrac{x - a}{x + a}\right|$ (*)

2 Integration of hyperbolic functions
   a Usually you can use the same method as for the corresponding trigonometric expressions.
   For example: $\int \cosh^2 3x \, dx = \int \dfrac{(1 + \cosh 6x)}{2} \, dx = \dfrac{1}{2}x + \dfrac{1}{12} \sinh 6x + C$
   $\int \tanh^2 x \, dx = \int (1 - \operatorname{sech}^2 x) \, dx = x - \tanh x + C$

   b Sometimes you may need to use, or it may be better to use, the exponential form of the functions.
   For example: $\int e^x \sinh x \, dx = \int e^x \left(\dfrac{e^x - e^{-x}}{2}\right) dx = \dfrac{1}{2} \int (e^{2x} - 1) \, dx = \dfrac{1}{2}\left(\dfrac{e^{2x}}{2} - x\right) + C$
   $\int \operatorname{sech} x \, dx$ (See Example 7)
   $\int \dfrac{1}{\sinh x + \cosh x} \, dx = \int \dfrac{1}{e^x} \, dx = -e^{-x} + C$

3 To integrate expressions of the form $\int \dfrac{1}{px^2 + qx + r} \, dx$ and $\int \dfrac{1}{\sqrt{px^2 + qx + r}} \, dx$, complete the square on $px^2 + qx + r$ and use the appropriate standard form in **1** above.

4 To integrate inverse trigonometric and hyperbolic functions you can use integration by parts in the same way as you integrate $\ln x$. For example, write $\int \arcsin x \, dx$ as $\int (\arcsin x \times 1) \, dx$ and use integration by parts with $u = \arcsin x$ and $\dfrac{dv}{dx} = 1$

5 A reduction formula relates I_n an integral of an expression involving an integer n, usually a power, to an integral of the same form involving a lower value of n.
- This is usually derived using integration by parts.
  For example: if $I_n = \int x^n e^x \, dx$, then $I_n = x^n e^x - n I_{n-1}$, $n > 1$, where $I_{n-1} = \int x^{n-1} e^x \, dx$
  By repeated application of the formula, I_n can be reduced to a form in which it can be found directly.
- Some reduction fomulae can be derived without using integration by parts.
  For example: if $I_n = \int \tan^n x \, dx$, then, by using $\tan^2 x = \sec^2 x - 1$, it can be shown directly that
  $I_n = \dfrac{1}{n-1} \tan^{n-1} x - I_{n-2}$, $n \geq 2$

6 The length of an arc on the curve C, between the points $A(x_A, y_B)$ and $B(x_B, y_B)$ is given by
$$\int_{x_A}^{x_B} \sqrt{1 + \left(\dfrac{dy}{dx}\right)^2} \, dx \quad (*) \quad \text{or} \quad \int_{y_A}^{y_B} \sqrt{1 + \left(\dfrac{dx}{dy}\right)^2} \, dy$$
If the equation of C is given parametrically, with the parameters at A and B being t_A and t_B respectively, then the arc length is given by
$$\int_{t_A}^{t_B} \sqrt{\left(\dfrac{dx}{dt}\right)^2 + \left(\dfrac{dy}{dt}\right)^2} \, dt \quad (*)$$

7 The area of the surface generated when the arc AB on the curve C, is rotated completely about the x-axis is $2\pi \int y \, ds$ (*) or about the y-axis is $2\pi \int x \, ds$.
These can be used to give the following results:

i about the x-axis $S = 2\pi \displaystyle\int_{x_A}^{x_B} y \sqrt{1 + \left(\dfrac{dy}{dx}\right)^2} \, dx$

ii about the y-axis $S = 2\pi \displaystyle\int_{y_A}^{y_B} x \sqrt{1 + \left(\dfrac{dx}{dy}\right)^2} \, dy \quad \text{or} \quad S = 2\pi \displaystyle\int_{x_A}^{x_B} x \sqrt{1 + \left(\dfrac{dy}{dx}\right)^2} \, dx$ — This is often the more convenient.

If C is given in parametric form, the results are

i about the x-axis $S = 2\pi \displaystyle\int_{t_A}^{t_B} y \sqrt{\left(\dfrac{dx}{dt}\right)^2 + \left(\dfrac{dy}{dt}\right)^2} \, dt \quad (*)$

ii about the y-axis $S = 2\pi \displaystyle\int_{t_A}^{t_B} x \sqrt{\left(\dfrac{dx}{dt}\right)^2 + \left(\dfrac{dy}{dt}\right)^2} \, dt$

Review exercise

1 Find the value of x for which
$$2\tanh x - 1 = 0$$
giving your answer in terms of a natural logarithm. **(4)**
← Further Pure 3 Section 1.1

2 Starting from the definition of $\cosh x$ in terms of exponentials, find, in terms of natural logarithms, the values of x for which $5 = 3\cosh x$ **(4)**
← Further Pure 3 Section 1.1

3 The curves with equations $y = 5\sinh x$ and $y = 4\cosh x$ meet at the point $A(\ln p, q)$. Find the exact values of p and q. **(4)**
← Further Pure 3 Section 1.3

4 Find the values of x for which
$$5\cosh x - 2\sinh x = 11$$
giving your answers as natural logarithms. **(5)**
← Further Pure 3 Section 1.3

5 By expressing $\sinh 2x$ and $\cosh 2x$ in terms of exponentials, find the exact values of x for which
$$6\sinh 2x + 9\cosh 2x = 7$$
giving each answer in the form $\frac{1}{2}\ln p$, where p is a rational number. **(5)**
← Further Pure 3 Section 1.3

6 Given that
$$\sinh x + 2\cosh x = k$$
where k is a positive constant,
a find the set of values of k for which at least one real solution of this equation exists **(4)**
b solve the equation when $k = 2$ **(3)**
← Further Pure 3 Section 1.3

7 Using the definitions of $\cosh x$ and $\sinh x$ in terms of exponentials,
a prove that $\cosh^2 x - \sinh^2 x \equiv 1$ **(3)**

b solve the equation $\dfrac{1}{\sinh x} - \dfrac{2}{\tanh x} = 2$
giving your answer in the form $k\ln a$, where k and a are integers. **(5)**
← Further Pure 3 Section 1.3

8 a From the definition of $\cosh x$ in terms of exponentials, show that
$$\cosh 2x \equiv 2\cosh^2 x - 1$$ **(3)**
b Solve the equation
$$\cosh 2x - 5\cosh x = 2$$
giving the answers in terms of natural logarithms. **(5)**
← Further Pure 3 Section 1.3

9 a Using the definition of $\cosh x$ in terms of exponentials, prove that
$$4\cosh^3 x - 3\cosh x \equiv \cosh 3x$$ **(4)**
b Hence, or otherwise, solve the equation
$$\cosh 3x = 5\cosh x$$
giving your answer as natural logarithms. **(4)**
← Further Pure 3 Section 1.3

10 a Starting from the definitions of $\cosh x$ and $\sinh x$ in terms of exponentials, prove that
$$\cosh(A - B) \equiv \cosh A \cosh B - \sinh A \sinh B$$ **(4)**
b Hence, or otherwise, given that $\cosh(x - 1) = \sinh x$, show that
$$\tanh x = \frac{e^2 + 1}{e^2 + 2e - 1}$$ **(5)**
← Further Pure 3 Section 1.3

11 a Starting from the definition
$$\sinh y = \frac{e^y - e^{-y}}{2}$$
prove that, for all real values of x,
$$\operatorname{arsinh} x = \ln(x + \sqrt{(1 + x^2)})$$ **(4)**

b Hence, or otherwise, prove that, for $0 < \theta < \pi$,

$$\text{arsinh}(\cot \theta) = \ln\left(\cot\frac{\theta}{2}\right) \quad (5)$$

← Further Pure 3 Sections 1.2, 1.3

E/P 12 a Starting from the definition of $\tanh x$ in terms of e^x, show that

$$\text{artanh } x = \frac{1}{2}\ln\left(\frac{1+x}{1-x}\right), |x| < 1 \quad (5)$$

b Sketch the graph of $y = \text{artanh } x$. (2)

c Solve the equation $x = \tanh(\ln\sqrt{6x})$ for $0 < x < 1$. (5)

← Further Pure 3 Sections 1.2, 1.3

E 13 a Show that, for $0 < x \leq 1$,

$$\ln\left(\frac{1-\sqrt{(1-x^2)}}{x}\right) = -\ln\left(\frac{1+\sqrt{(1-x^2)}}{x}\right) \quad (2)$$

b Using the definitions of $\cosh x$ and $\sinh x$ in terms of exponentials, show that, for $0 < x \leq 1$

$$\text{arcosh}\left(\frac{1}{x}\right) = \ln\left(\frac{1+\sqrt{1-x^2}}{x}\right) \quad (3)$$

c Solve the equation

$$3\tanh^2 x - 4\,\text{sech } x + 1 = 0$$

giving exact answers in terms of natural logarithms. (4)

← Further Pure 3 Sections 1.2, 1.3

E/P 14 a Express $\cosh 3\theta$ and $\cosh 5\theta$ in terms of $\cosh \theta$. (4)

b Hence determine the real roots of the equation

$$2\cosh 5x + 10\cosh 3x + 20\cosh x = 243$$

giving your answers to 2 decimal places. (6)

← Further Pure 3 Section 1.3

E 15 An ellipse has equation $\frac{x^2}{16} + \frac{y^2}{9} = 1$

a Sketch the ellipse. (2)

b Find the value of the eccentricity e. (2)

c State the coordinates of the foci of the ellipse. (2)

← Further Pure 3 Sections 2.1, 2.3

E 16 The hyperbola H has equation $\frac{x^2}{16} - \frac{y^2}{4} = 1$

Find

a the value of the eccentricity of H (2)

b the distance between the foci of H. (2)

The ellipse E has equation $\frac{x^2}{16} + \frac{y^2}{4} = 1$

c Sketch H and E on the same diagram, showing the coordinates of the points where each curve crosses the axes. (4)

← Further Pure 3 Sections 2.1, 2.2, 2.3

E/P 17 An ellipse, with equation $\frac{x^2}{9} + \frac{y^2}{4} = 1$, has foci S and S'.

a Find the coordinates of the foci of the ellipse. (2)

b Using the focus–directrix property of the ellipse, prove that, for any point P on the ellipse, $SP + S'P = 6$ (5)

← Further Pure 3 Sections 2.1, 2.3

E 18 a Find the eccentricity of the ellipse with equation $3x^2 + 4y^2 = 12$ (3)

b Find an equation of the tangent to the ellipse with equation $3x^2 + 4y^2 = 12$ at the point with coordinates $\left(1, \frac{3}{2}\right)$. (4)

This tangent meets the y-axis at G. Given that S and S' are the foci of the ellipse,

c find the area of triangle $SS'G$. (5)

← Further Pure 3 Sections 2.3, 2.4

E 19 The points S_1 and S_2 have Cartesian coordinates $\left(-\frac{a}{2}\sqrt{3}, 0\right)$ and $\left(\frac{a}{2}\sqrt{3}, 0\right)$ respectively.

a Find a Cartesian equation of the ellipse which has S_1 and S_2 as its two foci, and a major axis of length $2a$. (4)

b Write down the equations of the directrices of this ellipse. (1)

Given that parametric equations of this ellipse are

$$x = a\cos\phi, \; y = b\sin\phi$$

c express b in terms of a. (4)

The point P is such that $\phi = \frac{\pi}{4}$ and the point Q such that $\phi = \frac{\pi}{2}$

d Show that an equation of the chord PQ is $(\sqrt{2} - 1)x + 2y - a = 0$ (3)

← Further Pure 3 Section 2.3

E/P 20 a Find the eccentricity of the ellipse
$$\frac{x^2}{9} + \frac{y^2}{4} = 1 \quad (2)$$

b Find also the coordinates of both foci and equations of both directrices of this ellipse. (2)

c Show that an equation for the tangent to this ellipse at the point $P(3\cos\theta, 2\sin\theta)$ is
$$\frac{x\cos\theta}{3} + \frac{y\sin\theta}{2} = 1 \quad (4)$$

d Show that, as θ varies, the foot of the perpendicular from the origin to the tangent at P lies on the curve
$$(x^2 + y^2)^2 = 9x^2 + 4y^2 \quad (6)$$

← Further Pure 3 Sections 2.3, 2.4

E/P 21 a Show that an equation of the normal to the ellipse $\frac{x^2}{a^2} + \frac{y^2}{b^2} = 1$ at the point $P(a\cos\theta, b\sin\theta)$ is
$$ax\sec\theta - by\cosec\theta = a^2 - b^2 \quad (3)$$

The normal at P cuts the x-axis at G.

b Show that the coordinates of M, the midpoint of PG are
$$\left(\frac{2a^2 - b^2}{2a}\cos\theta, \frac{b}{2}\sin\theta\right) \quad (3)$$

c Prove that, as θ varies, the locus of M is an ellipse and determine the equation of this ellipse. (4)

Given that the normal at P meets the y-axis at H and that O is the origin,

d prove that, if $a > b$, then the ratio of the area of triangle OMG to the area of triangle OGH is $b^2 : 2(a^2 - b^2)$. (4)

← Further Pure 3 Sections 2.4, 2.5

E/P 22 The diagram shows the ellipse with equation $\frac{x^2}{8^2} + \frac{y^2}{4^2} = 1$

The point P has coordinates $(4, 2\sqrt{3})$

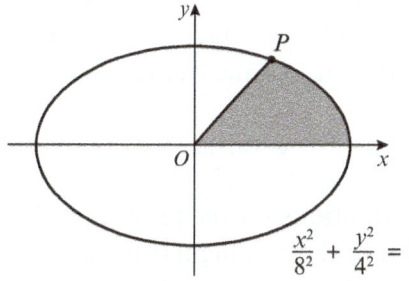

Show that the exact value for the area of the shaded region is $a\pi$, where a is a rational number to be found. (6)

← Further Pure 3 Section 2.2

E/P 23 The line with equation $y = mx + c$ is a tangent to the ellipse with equation
$$\frac{x^2}{a^2} + \frac{y^2}{b^2} = 1$$

a Show that $c^2 = a^2m^2 + b^2$ (4)

b Hence, or otherwise, find the equations of the tangents from the point $(3, 4)$ to the ellipse with equation $\frac{x^2}{16} + \frac{y^2}{25} = 1$ (4)

← Further Pure 3 Section 2.4

E/P 24 The ellipse E has equation $\frac{x^2}{a^2} + \frac{y^2}{b^2} = 1$ and the line L has equation $y = mx + c$, where $m > 0$ and $c > 0$

a Show that, if L and E have any points of intersection, the x-coordinates of these points are the roots of the equation
$$(b^2 + a^2m^2)x^2 + 2a^2mcx + a^2(c^2 - b^2) = 0 \quad (4)$$

Hence, given that L is a tangent to E,

b show that $c^2 = b^2 + a^2m^2$ (2)

The tangent L meets the negative x-axis at the point A and the positive y-axis at the point B, and O is the origin.

c Find, in terms of m, a and b, the area of the triangle OAB. (3)

d Prove that, as m varies, the minimum area of the triangle OAB is ab. (3)

e Find, in terms of a, the x-coordinate of the point of contact of L and E when the area of the triangle is a minimum. (2)

← Further Pure 3 Section 2.4

(E/P) 25 a Find equations for the tangent and normal to the rectangular hyperbola $x^2 - y^2 = 1$, at the point P with coordinates $(\cosh t, \sinh t)$, $t > 0$ (5)

The tangent and normal cut the x-axis at T and G respectively. The perpendicular from P to the x-axis meets an asymptote in the first quadrant at Q.

b Show that GQ is perpendicular to this asymptote. (4)

The normal cuts the y-axis at R.

c Show that R lies on the circle with centre at T and radius TG. (4)

← Further Pure 3 Section 2.5

(E/P) 26 The point P lies on the hyperbola $\dfrac{x^2}{a^2} - \dfrac{y^2}{b^2} = 1$, and N is the foot of the perpendicular from P onto the x-axis. The tangent to the hyperbola at P meets the x-axis at T.

Show that $OT \times ON = a^2$, where O is the origin. (6)

← Further Pure 3 Section 2.5

(E/P) 27 The hyperbola C has equation $\dfrac{x^2}{a^2} - \dfrac{y^2}{b^2} = 1$

a Show that an equation of the normal to C at the point $P(a \sec t, b \tan t)$ is

$$ax \sin t + by = (a^2 + b^2) \tan t$$ (6)

The normal to C at P cuts the x-axis at the point A and S is a focus of C. Given that the eccentricity of C is $\frac{3}{2}$, and that $OA = 3OS$, where O is the origin,

b determine the possible values of t, for $0 \leqslant t \leqslant 2\pi$ (3)

← Further Pure 3 Section 2.5

(E) 28 a Show that the hyperbola $x^2 - y^2 = a^2$ $a > 0$, has eccentricity equal to $\sqrt{2}$. (3)

b Hence state the coordinates of the focus S and an equation of the corresponding directrix L, where both S and L lie in the region $x > 0$ (2)

The perpendicular from S to the line $y = x$ meets the line $y = x$ at P and the perpendicular from S to the line $y = -x$ meets the line $y = -x$ at Q.

c Show that both P and Q lie on the directrix L and give the coordinates of P and Q. (3)

Given that the line SP meets the hyperbola at the point R,

d prove that the tangent at R passes through the point Q. (4)

← Further Pure 3 Sections 2.3, 2.5

(E/P) 29 Show that the equations of the tangents with gradient m to the hyperbola with equation $x^2 - 4y^2 = 4$ are

$$y = mx \pm \sqrt{4m^2 - 1}, \text{ where } |m| > \tfrac{1}{2}$$ (6)

← Further Pure 3 Section 2.5

(E) 30 An ellipse has equation $\dfrac{x^2}{a^2} + \dfrac{y^2}{b^2} = 1$ where a and b are constants and $a > b$.

a Find an equation of the tangent at the point $P(a \cos t, b \sin t)$. (3)

b Find an equation of the normal at the point $P(a \cos t, b \sin t)$. (3)

The normal at P meets the x-axis at the point Q. The tangent at P meets the y-axis at the point R.

c Find, in terms of a, b and t, the coordinates of M, the midpoint of QR. (4)

Given that $0 < t < \dfrac{\pi}{2}$

REVIEW EXERCISE 1

d prove that, as t varies, the locus of M has equation $\left(\dfrac{2ax}{a^2-b^2}\right)^2 + \left(\dfrac{b}{2y}\right)^2 = 1$ **(4)**

← Further Pure 3 Sections 2.5, 2.6

(E/P) 31 a Find the equations for the tangent and normal to the hyperbola
$$\dfrac{x^2}{a^2} - \dfrac{y^2}{b^2} = 1$$
at the point $(a\sec\theta, b\tan\theta)$. **(6)**

b If these lines meet the y-axis at P and Q respectively, prove that the circle with PQ as a diameter passes through the foci of the hyperbola. **(5)**

← Further Pure 3 Sections 2.5, 2.6

(E/P) 32 a Show that, for $x = \ln k$, where k is a positive constant,
$$\cosh 2x = \dfrac{k^4+1}{2k^2}$$ **(4)**

b Given that $f(x) = px - \tanh 2x$, where p is a constant, find the value of p for which $f(x)$ has a stationary value at $x = \ln 2$, giving your answer as an exact fraction. **(6)**

← Further Pure 3 Sections 1.1, 3.1

(E) 33 The curve with equation
$$y = -x + \tanh 4x, \; x \geqslant 0$$
has a maximum turning point A.

a Find, in exact logarithmic form, the x-coordinate of A. **(5)**

b Show that the y-coordinate of A is
$$\dfrac{1}{4}(2\sqrt{3} - \ln(2+\sqrt{3}))$$ **(3)**

← Further Pure 3 Section 3.1

(E/P) 34 Use the substitution $x = \dfrac{a}{\sinh\theta}$, where a is a constant, to show that, for $x > 0$, $a > 0$
$$\int \dfrac{1}{x\sqrt{x^2+a^2}}\,dx = -\dfrac{1}{a}\operatorname{arsinh}\left(\dfrac{a}{x}\right) + \text{constant.}$$ **(6)**

← Further Pure 3 Section 4.3

(E/P) 35 a Prove that the derivative of $\operatorname{artanh} x$, $-1 < x < 1$, is $\dfrac{1}{1-x^2}$ **(4)**

b Find $\int \operatorname{artanh} x \, dx$ **(4)**

← Further Pure 3 Section 4.2

(E/P) 36 a Starting from the definition of $\sinh x$ in terms of e^x, prove that
$$\operatorname{arsinh} x = \ln(x + \sqrt{x^2+1})$$ **(2)**

b Prove that the derivative of $\operatorname{arsinh} x$ is $(1+x^2)^{-\frac{1}{2}}$ **(4)**

c Show that the equation
$$(1+x^2)\dfrac{d^2y}{dx^2} + x\dfrac{dy}{dx} - 2 = 0$$
is satisfied when $y = (\operatorname{arsinh} x)^2$ **(4)**

d Use integration by parts to find
$$\int_0^1 \operatorname{arsinh} x \, dx,$$
giving your answer in terms of a natural logarithm. **(5)**

← Further Pure 3 Sections 1.2, 3.2, 4.2

(E) 37 $4x^2 + 4x + 5 \equiv (px+q)^2 + r$

a Find the values of the constants p, q and r. **(2)**

b Hence, or otherwise, find
$$\int \dfrac{1}{4x^2+4x+5}\,dx$$ **(4)**

c Show that
$$\int \dfrac{2}{\sqrt{4x^2+4x+5}}\,dx$$
$$= \ln((2x+1) + \sqrt{4x^2+4x+5}) + k$$
where k is an arbitrary constant. **(5)**

← Further Pure 3 Section 4.1

(E) 38 Find $\displaystyle\int \dfrac{x+2}{\sqrt{4x^2+9}}\,dx$ **(7)**

← Further Pure 3 Section 4.3

(E) 39 Show that $\displaystyle\int_2^5 \dfrac{1}{\sqrt{x^2-4x+8}}\,dx = \operatorname{arsinh} k$ where k is a rational constant to be found. **(6)**

← Further Pure 3 Section 4.4

40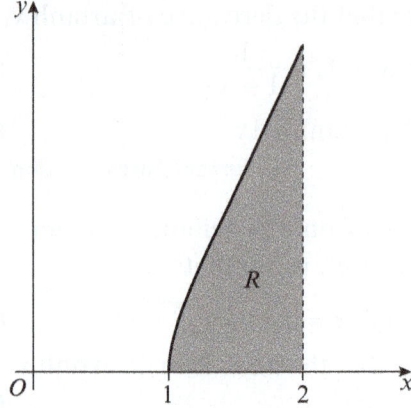

The figure above shows a sketch of the curve with equation

$y = x \operatorname{arcosh} x$, $1 \leq x \leq 2$

The region R, shaded in the figure, is bounded by the curve, the x-axis and the line $x = 2$

Show that the area of R is

$$\frac{7}{4}\ln(2 + \sqrt{3}) - \frac{\sqrt{3}}{2} \qquad (8)$$

← Further Pure 3 Section 4.4

41 $I_n = \int \sec^n x \, dx$, $n \geq 0$

a Prove that, for $n \geq 2$

$(n-1)I_n = \sec^{n-2} x \tan x + (n-2)I_{n-2}$ (6)

b Using your reduction formula, find

$\int \sec^4 x \, dx$ (2)

← Further Pure 3 Section 4.6

42 $I_n = \int_0^{\frac{\pi}{4}} x^n \cos x \, dx$, $n \geq 0$

a Prove that, for $n \geq 2$

$I_n = \frac{1}{\sqrt{2}}\left(\frac{\pi}{4}\right)^{n-1}\left(\frac{\pi}{4} + n\right) - n(n-1)I_{n-2}$ (6)

b Find the value of I_4 to four decimal places. (4)

← Further Pure 3 Section 4.6

43 $I_n = \int_0^a x^n (a-x)^{\frac{1}{3}} dx$, $n \geq 0$, $a > 0$

a Prove that $I_n = \frac{3an}{3n+4} I_{n-1}$, $n \geq 1$ (6)

Given that $I_2 = \frac{27}{49} a^{\frac{4}{3}}$

b find the value of a. (6)

← Further Pure 3 Section 4.6

44 An arc is defined by the curve $y = (ax^3)^{\frac{1}{2}}$ between $x = 0$ and $x = 4$ where $a > 0$

a Find the length of the arc in terms of a. (5)

Given that the arc has length 16,

b calculate the value of a, giving your answer to four decimal places. (4)

← Further Pure 3 Section 4.7

45 A parabola has equation $y^2 = 2x + 16$

An arc of length L forms a section of this curve between $y = 0$ and $y = 3$

a Show that L is given by

$$L = \int_0^3 \sqrt{1 + y^2} \, dy \qquad (3)$$

b Hence find the exact value of L. (5)

← Further Pure 3 Section 4.7

46 An arc lies on a curve with parametric equations $x = t^2 - 1$ and $y = \frac{1}{3}t^3 - 2$ with $0 \leq t \leq 2$

Show that this arc has length $\frac{8^{\frac{3}{2}} - 8}{3}$ (6)

← Further Pure 3 Section 4.7

47 A length of wire is bent into a flat spiral such that its shape is defined by the polar equation $r = \theta$ for $0 \leq \theta \leq 4\pi$

a Using a substitution of the form $\theta = f(x)$, or otherwise, show that the length of wire, W, needed to make this spiral is given by

$$W = \int_0^{\arctan(4\pi)} \sec^3 x \, dx \qquad (6)$$

b Hence calculate W to two decimal places. (3)

← Further Pure 3 Section 4.7

48 A parametric curve is defined by $x = 2\sqrt{t}$ and $y = 1 - t$, $t \geq 0$. A surface is created by rotating an arc of this curve, defined by $1 \leq t \leq 4$, around the y-axis. Find an exact expression for the area of this surface of revolution. (5)

← Further Pure 3 Section 4.8

REVIEW EXERCISE 1

E/P 49 An arc is defined by the curve
$y = \sqrt{a - x^2}, -1 \leq x \leq 1$
The area of the surface obtained by rotating this arc around the x-axis is 24π.

a Find a. (6)

b Using a sketch or otherwise describe the nature of this surface. (3)

← Further Pure 3 Section 4.8

E/P 50 Consider a curve with polar equation
$r = \sqrt{\cos 2\theta}$

An arc is formed by a section of this curve when $0 \leq \theta \leq \dfrac{\pi}{4}$

This arc is rotated around the initial line, giving rise to a curved surface.

Find the exact area of this curved surface. (6)

← Further Pure 3 Section 4.8

E/P 51 An arc is defined by the function $f(x) = e^x$ when $0 \leq x \leq 1$

Find the area of the surface when this arc is rotated about the x-axis, giving your answer correct to three decimal places. (8)

← Further Pure 3 Section 4.8

Challenge

1 **a** Prove that for two lines $y = m_1 x + c_1$ and $y = m_2 x + c_2$, $m_1 \neq m_2$, the acute angle α between the two lines satisfies
$$\tan \alpha = \frac{m_2 - m_1}{1 + m_1 m_2}$$

b Hence, or otherwise, prove that the normal to an ellipse at any point P bisects the angle SPS', where S and S' are the foci of the ellipse.

← Further Pure 3 Section 2.4

2 The solid formed by rotating the curve $y = \dfrac{1}{x}$, $x > 1$ by 2π about the x-axis is sometimes called Torricelli's Trumpet.

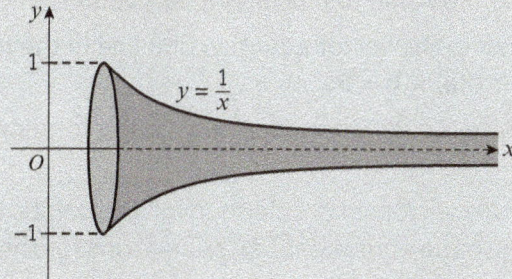

a Find the volume generated by the solid.

b Show that the area of the surface of revolution generated is $2\pi \displaystyle\int_1^\infty \dfrac{1}{x}\sqrt{1 + \dfrac{1}{x^4}}\,dx$

c Explain why
$$2\pi \int_1^a \frac{1}{x}\sqrt{1 + \frac{1}{x^4}}\,dx > 2\pi \int_1^a \frac{1}{x}\,dx$$
for all $x > 0$

d Hence explain why Torricelli's trumpet has infinite surface area.

← Further Pure 3 Section 4.8

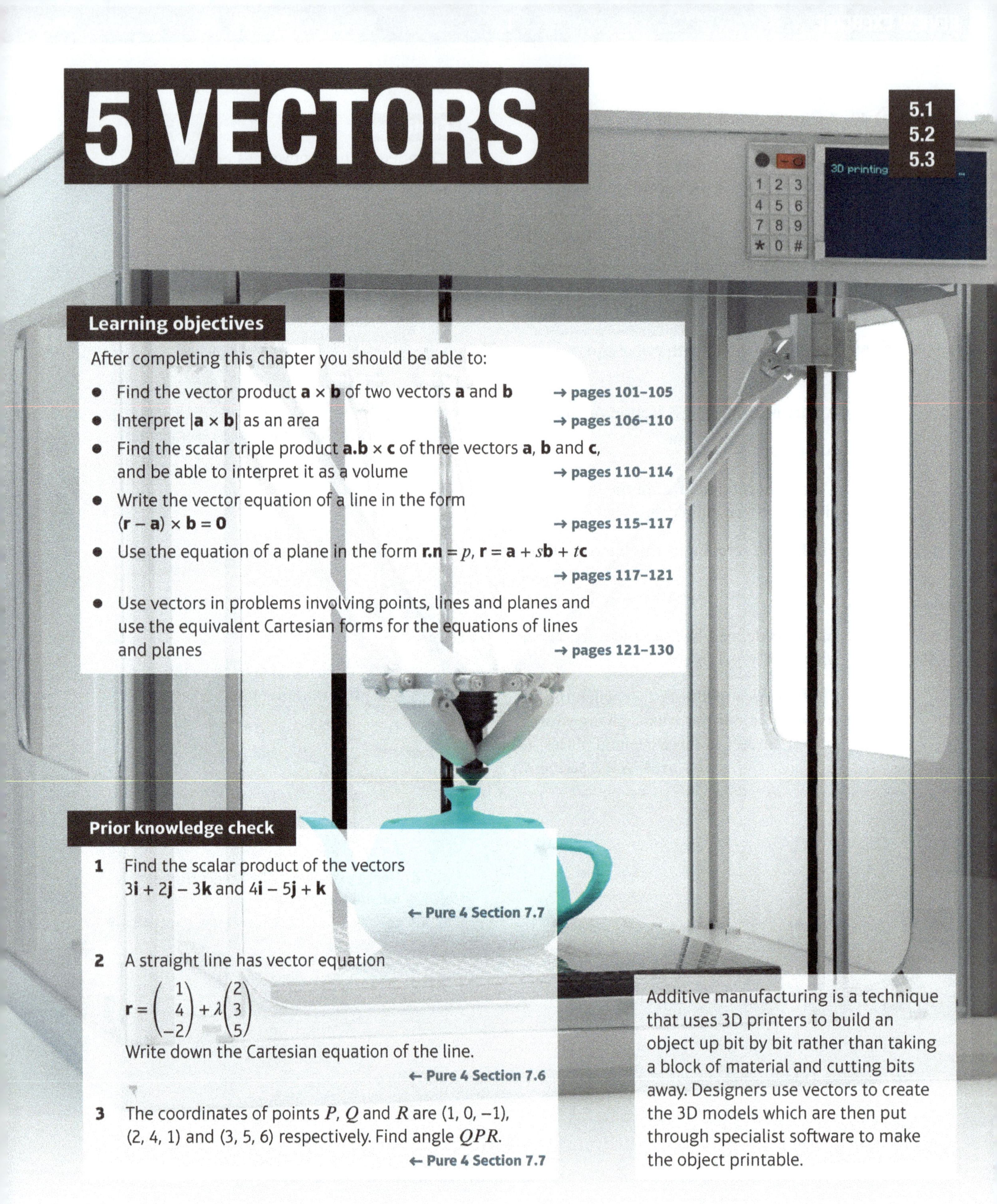

5 VECTORS

5.1
5.2
5.3

Learning objectives

After completing this chapter you should be able to:

- Find the vector product **a** × **b** of two vectors **a** and **b** → pages 101–105
- Interpret |**a** × **b**| as an area → pages 106–110
- Find the scalar triple product **a.b** × **c** of three vectors **a**, **b** and **c**, and be able to interpret it as a volume → pages 110–114
- Write the vector equation of a line in the form (**r** − **a**) × **b** = **0** → pages 115–117
- Use the equation of a plane in the form **r.n** = p, **r** = **a** + s**b** + t**c** → pages 117–121
- Use vectors in problems involving points, lines and planes and use the equivalent Cartesian forms for the equations of lines and planes → pages 121–130

Prior knowledge check

1. Find the scalar product of the vectors $3\mathbf{i} + 2\mathbf{j} - 3\mathbf{k}$ and $4\mathbf{i} - 5\mathbf{j} + \mathbf{k}$
   ← Pure 4 Section 7.7

2. A straight line has vector equation
$$\mathbf{r} = \begin{pmatrix} 1 \\ 4 \\ -2 \end{pmatrix} + \lambda \begin{pmatrix} 2 \\ 3 \\ 5 \end{pmatrix}$$
   Write down the Cartesian equation of the line.
   ← Pure 4 Section 7.6

3. The coordinates of points P, Q and R are $(1, 0, -1)$, $(2, 4, 1)$ and $(3, 5, 6)$ respectively. Find angle QPR.
   ← Pure 4 Section 7.7

Additive manufacturing is a technique that uses 3D printers to build an object up bit by bit rather than taking a block of material and cutting bits away. Designers use vectors to create the 3D models which are then put through specialist software to make the object printable.

5.1 Vector product

You have already encountered the **scalar** (or **dot**) **product** of two vectors.

The scalar (or dot) product of two vectors **a** and **b** is written as **a.b**, and defined as

$$\mathbf{a.b} = |\mathbf{a}||\mathbf{b}|\cos\theta$$

where θ is the angle between **a** and **b**.

The scalar product produces a number (or scalar) as an answer. It is useful to define a second type of product that gives an answer as a vector.

Links If $\mathbf{a} = \begin{pmatrix} x_1 \\ y_1 \\ z_1 \end{pmatrix}$ and $\mathbf{b} = \begin{pmatrix} x_2 \\ y_2 \\ z_2 \end{pmatrix}$

then $\mathbf{a.b} = x_1x_2 + y_1y_2 + z_1z_2$

← Pure 4 Section 7.11

- **The vector (or cross) product of the vectors a and b is defined as**

$$\mathbf{a} \times \mathbf{b} = |\mathbf{a}||\mathbf{b}|\sin\theta\,\hat{\mathbf{n}}$$

where θ is the angle between **a** and **b**.

Online Use GeoGebra to explore the cross product of two vectors.

Notation $\hat{\mathbf{n}}$ is the unit vector that is perpendicular to both **a** and **b**.

Since $0 \leq \theta \leq 180°$, $|\mathbf{a}||\mathbf{b}|\sin\theta$ is a positive scalar quantity. This means that $\mathbf{a} \times \mathbf{b}$ is a vector quantity with magnitude $|\mathbf{a}||\mathbf{b}|\sin\theta$ that acts in the direction of $\hat{\mathbf{n}}$.

The direction of $\hat{\mathbf{n}}$ is that in which a right-handed screw would move when turned from **a** to **b**.

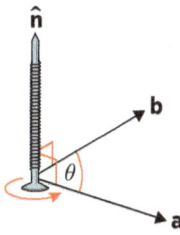

Problem-solving

You can also use a 'right-hand rule' to determine the direction of $\hat{\mathbf{n}}$, and hence the direction of $\mathbf{a} \times \mathbf{b}$. If **a** is your first finger, and **b** is your second finger, then $\mathbf{a} \times \mathbf{b}$ acts in the direction of your thumb:

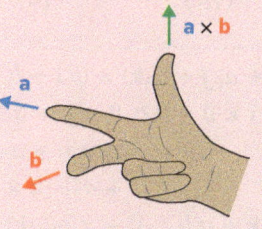

If the turn is in the opposite sense, i.e. from **b** to **a**, then the movement of the screw is in the opposite direction to $\hat{\mathbf{n}}$, i.e. in the direction of $-\hat{\mathbf{n}}$.

So $\mathbf{b} \times \mathbf{a} = |\mathbf{b}||\mathbf{a}|\sin\theta\,(-\hat{\mathbf{n}})$
$= -|\mathbf{a}||\mathbf{b}|\sin\theta\,\hat{\mathbf{n}}$
$= -\mathbf{a} \times \mathbf{b}$

- $\mathbf{b} \times \mathbf{a} = -\mathbf{a} \times \mathbf{b}$

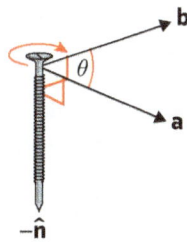

Watch out The **vector product** is not commutative: the order of multiplication matters.

Example 1

Find the values of

a $\mathbf{i} \times \mathbf{i}$ **b** $\mathbf{j} \times \mathbf{k}$ **c** $\mathbf{i} \times \mathbf{k}$

a $\mathbf{i} \times \mathbf{i} = \mathbf{0}$ — $\sin\theta = 0$, as the angle between $\mathbf{i}$ and itself is zero.

b $\mathbf{j} \times \mathbf{k} = 1 \times 1 \times \sin 90° \,\mathbf{i} = \mathbf{i}$ — The angle between $\mathbf{j}$ and $\mathbf{k}$ is 90° and, as $\mathbf{j}$ and $\mathbf{k}$ are unit vectors, each has magnitude 1 unit.

c $\mathbf{i} \times \mathbf{k} = -\mathbf{k} \times \mathbf{i} = -1 \times 1 \times \sin 90° \,\mathbf{j} = -\mathbf{j}$ — Use the right-hand rule. If $\mathbf{i}$ is your first finger and $\mathbf{k}$ is your second finger, your thumb will point **away** from $\mathbf{j}$, so $\mathbf{i} \times \mathbf{k} = -\mathbf{j}$

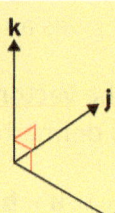

- $\mathbf{i} \times \mathbf{i} = \mathbf{0}$
- $\mathbf{j} \times \mathbf{j} = \mathbf{0}$
- $\mathbf{k} \times \mathbf{k} = \mathbf{0}$
- $\mathbf{i} \times \mathbf{j} = \mathbf{k}$ and $\mathbf{j} \times \mathbf{i} = -\mathbf{k}$
- $\mathbf{j} \times \mathbf{k} = \mathbf{i}$ and $\mathbf{k} \times \mathbf{j} = -\mathbf{i}$
- $\mathbf{k} \times \mathbf{i} = \mathbf{j}$ and $\mathbf{i} \times \mathbf{k} = -\mathbf{j}$

As $\mathbf{a} \times \mathbf{b} = |\mathbf{a}||\mathbf{b}|\sin\theta\,\hat{\mathbf{n}}$, $\mathbf{a} \times \mathbf{b} = \mathbf{0}$ implies that $\mathbf{a} = \mathbf{0}$, $\mathbf{b} = \mathbf{0}$ or $\sin\theta = 0$
$\sin\theta = 0$ implies that $\theta = 0°$ or $180°$, so $\mathbf{a}$ and $\mathbf{b}$ must be parallel.

- If $\mathbf{a} \times \mathbf{b} = \mathbf{0}$ then either $\mathbf{a} = \mathbf{0}$, $\mathbf{b} = \mathbf{0}$ or $\mathbf{a}$ and $\mathbf{b}$ are parallel.

Example 2 SKILLS ANALYSIS

Given that $\mathbf{a} = \begin{pmatrix} a_1 \\ a_2 \\ a_3 \end{pmatrix}$ and $\mathbf{b} = \begin{pmatrix} b_1 \\ b_2 \\ b_3 \end{pmatrix}$ find $\mathbf{a} \times \mathbf{b}$

Notation You may assume the vector product is **distributive** over vector addition. This means that

$\mathbf{a} \times (\mathbf{b} + \mathbf{c}) = (\mathbf{a} \times \mathbf{b}) + (\mathbf{a} \times \mathbf{c})$

$\begin{aligned}
\mathbf{a} \times \mathbf{b} &= (a_1\mathbf{i} + a_2\mathbf{j} + a_3\mathbf{k}) \times (b_1\mathbf{i} + b_2\mathbf{j} + b_3\mathbf{k}) \\
&= a_1b_1(\mathbf{i} \times \mathbf{i}) + a_1b_2(\mathbf{i} \times \mathbf{j}) + a_1b_3(\mathbf{i} \times \mathbf{k}) \\
&\quad + a_2b_1(\mathbf{j} \times \mathbf{i}) + a_2b_2(\mathbf{j} \times \mathbf{j}) + a_2b_3(\mathbf{j} \times \mathbf{k}) \\
&\quad + a_3b_1(\mathbf{k} \times \mathbf{i}) + a_3b_2(\mathbf{k} \times \mathbf{j}) + a_3b_3(\mathbf{k} \times \mathbf{k}) \\
&= a_1b_2\mathbf{k} + a_1b_3(-\mathbf{j}) + a_2b_1(-\mathbf{k}) + a_2b_3(\mathbf{i}) + a_3b_1(\mathbf{j}) + a_3b_2(-\mathbf{i}) \\
&= (a_2b_3 - a_3b_2)\mathbf{i} + (a_3b_1 - a_1b_3)\mathbf{j} + (a_1b_2 - a_2b_1)\mathbf{k}
\end{aligned}$

Simplify the cross product and collect like terms.

In determinant form,

$\mathbf{a} \times \mathbf{b} = \begin{vmatrix} \mathbf{i} & \mathbf{j} & \mathbf{k} \\ a_1 & a_2 & a_3 \\ b_1 & b_2 & b_3 \end{vmatrix} = \mathbf{i}\begin{vmatrix} a_2 & a_3 \\ b_2 & b_3 \end{vmatrix} - \mathbf{j}\begin{vmatrix} a_1 & a_3 \\ b_1 & b_3 \end{vmatrix} + \mathbf{k}\begin{vmatrix} a_1 & a_2 \\ b_1 & b_2 \end{vmatrix}$

$= (a_2b_3 - a_3b_2)\mathbf{i} + (a_3b_1 - a_1b_3)\mathbf{j} + (a_1b_2 - a_2b_1)\mathbf{k}$

You can write each component as the **determinant** of a 2 × 2 **matrix**, or the whole vector product as a determinant of a 3 × 3 matrix.

← Further Pure 3 Section 6.2

- $\mathbf{a} \times \mathbf{b} = (a_2b_3 - a_3b_2)\mathbf{i} + (a_3b_1 - a_1b_3)\mathbf{j} + (a_1b_2 - a_2b_1)\mathbf{k}$

- $\mathbf{a} \times \mathbf{b} = \begin{vmatrix} \mathbf{i} & \mathbf{j} & \mathbf{k} \\ a_1 & a_2 & a_3 \\ b_1 & b_2 & b_3 \end{vmatrix} = \mathbf{i}\begin{vmatrix} a_2 & a_3 \\ b_2 & b_3 \end{vmatrix} - \mathbf{j}\begin{vmatrix} a_1 & a_3 \\ b_1 & b_3 \end{vmatrix} + \mathbf{k}\begin{vmatrix} a_1 & a_2 \\ b_1 & b_2 \end{vmatrix}$

Example 3

Given that **a** = 2**i** − 3**j** and **b** = 4**i** + **j** − **k**, find **a** × **b**:

a directly

b by a method involving a determinant.

c Verify that **a** × **b** is perpendicular to both **a** and **b**.

a (2**i** − 3**j**) × (4**i** + **j** − **k**)
= 8(**i** × **i**) + 2(**i** × **j**) − 2(**i** × **k**) − 12(**j** × **i**) − 3(**j** × **j**) + 3(**j** × **k**)
= **0** + 2**k** + 2**j** + 12**k** − **0** + 3**i**
= 3**i** + 2**j** + 14**k**

Use the distributive property to multiply out the brackets.

Simplify the cross products of unit vectors.

b $\begin{vmatrix} \mathbf{i} & \mathbf{j} & \mathbf{k} \\ 2 & -3 & 0 \\ 4 & 1 & -1 \end{vmatrix} = \mathbf{i}\begin{vmatrix} -3 & 0 \\ 1 & -1 \end{vmatrix} - \mathbf{j}\begin{vmatrix} 2 & 0 \\ 4 & -1 \end{vmatrix} + \mathbf{k}\begin{vmatrix} 2 & -3 \\ 4 & 1 \end{vmatrix}$
= **i**(3 − 0) − **j**(−2 − 0) + **k**(2 + 12)
= 3**i** + 2**j** + 14**k**

Problem-solving

Using the determinant is usually a quicker way to evaluate the cross product.

c (3**i** + 2**j** + 14**k**).(2**i** − 3**j**) = (3 × 2) + (2 × (−3)) + (14 × 0) = 0
(3**i** + 2**j** + 14**k**).(4**i** + **j** − **k**) = (3 × 4) + (2 × 1) + (14 × (−1)) = 0

*Work out (**a** × **b**).**a** and (**a** × **b**).**b**. If both answers are 0 then **a** × **b** is perpendicular to both **a** and **b**.*

Example 4

SKILLS PROBLEM-SOLVING

Find a unit vector perpendicular to both (4**i** + 3**j** + 2**k**) and (8**i** + 3**j** + 3**k**)

The vector product will give a perpendicular vector.

$\begin{vmatrix} \mathbf{i} & \mathbf{j} & \mathbf{k} \\ 4 & 3 & 2 \\ 8 & 3 & 3 \end{vmatrix} = \mathbf{i}\begin{vmatrix} 3 & 2 \\ 3 & 3 \end{vmatrix} - \mathbf{j}\begin{vmatrix} 4 & 2 \\ 8 & 3 \end{vmatrix} + \mathbf{k}\begin{vmatrix} 4 & 3 \\ 8 & 3 \end{vmatrix}$
= **i**(9 − 6) − **j**(12 − 16) + **k**(12 − 24)
= 3**i** + 4**j** − 12**k**

Since |3**i** + 4**j** − 12**k**| = $\sqrt{3^2 + 4^2 + (-12)^2}$ = 13

a suitable unit vector is $\frac{1}{13}$(3**i** + 4**j** − 12**k**)

Watch out You can find vector products using your calculator. But you might encounter a vector with an unknown in it, so it is important that you know how to find the vector product manually.

Find the magnitude of your product vector.

Divide the vector by its magnitude to obtain a unit vector.

Example 5

Find the sine of the acute angle between the vectors $\mathbf{a} = 2\mathbf{i} + \mathbf{j} + 2\mathbf{k}$ and $\mathbf{b} = -3\mathbf{j} + 4\mathbf{k}$

$\mathbf{a} \times \mathbf{b} = |\mathbf{a}||\mathbf{b}|\sin\theta\, \hat{\mathbf{n}}$

So $\dfrac{|\mathbf{a} \times \mathbf{b}|}{|\mathbf{a}||\mathbf{b}|} = \sin\theta$ — Rearrange the formula to make $\sin\theta$ the subject. $|\hat{\mathbf{n}}| = 1$ so $|\mathbf{a} \times \mathbf{b}| = |\mathbf{a}||\mathbf{b}|\sin\theta$

$\mathbf{a} \times \mathbf{b} = \begin{vmatrix} \mathbf{i} & \mathbf{j} & \mathbf{k} \\ 2 & 1 & 2 \\ 0 & -3 & 4 \end{vmatrix}$

$= \mathbf{i}(4 + 6) - \mathbf{j}(8 - 0) + \mathbf{k}(-6 - 0)$

$= 10\mathbf{i} - 8\mathbf{j} - 6\mathbf{k}$ — Calculate the vector product.

and $|10\mathbf{i} - 8\mathbf{j} - 6\mathbf{k}| = \sqrt{100 + 64 + 36}$ — Find the magnitude of $\mathbf{a} \times \mathbf{b}$

So $\sin\theta = \dfrac{\sqrt{200}}{\sqrt{2^2 + 1^2 + 2^2}\,\sqrt{(-3)^2 + 4^2}}$ — Also find the magnitude of $\mathbf{a}$ and of $\mathbf{b}$ and substitute the three surds into the formula for $\sin\theta$.

$= \dfrac{\sqrt{200}}{\sqrt{9}\,\sqrt{25}}$

$= \dfrac{10\sqrt{2}}{3 \times 5}$ — Simplify your answer.

$= \dfrac{2\sqrt{2}}{3}$

Watch out In general, to find the angle between two vectors use the scalar product. This gives the cosine of the angle. Immediately we know whether the angle is acute or obtuse. In this example it is not clear whether the angle θ is acute or obtuse. This is similar to the ambiguous case when using the sine rule.

Exercise 5A SKILLS ANALYSIS

1 Simplify

  a $5\mathbf{j} \times \mathbf{k}$ **b** $3\mathbf{i} \times \mathbf{k}$ **c** $\mathbf{k} \times 3\mathbf{i}$

  d $3\mathbf{i} \times (9\mathbf{i} - \mathbf{j} + \mathbf{k})$ **e** $2\mathbf{j} \times (3\mathbf{i} + \mathbf{j} - \mathbf{k})$ **f** $(3\mathbf{i} + \mathbf{j} - \mathbf{k}) \times 2\mathbf{j}$

  g $\begin{pmatrix} 5 \\ 2 \\ -1 \end{pmatrix} \times \begin{pmatrix} 1 \\ -1 \\ 3 \end{pmatrix}$ **h** $\begin{pmatrix} 2 \\ -1 \\ 6 \end{pmatrix} \times \begin{pmatrix} 1 \\ -2 \\ 3 \end{pmatrix}$ **i** $\begin{pmatrix} 1 \\ 5 \\ -4 \end{pmatrix} \times \begin{pmatrix} 2 \\ -1 \\ -1 \end{pmatrix}$ **j** $\begin{pmatrix} 3 \\ 0 \\ 2 \end{pmatrix} \times \begin{pmatrix} 1 \\ -1 \\ 2 \end{pmatrix}$

2 Find the vector product of the vectors $\mathbf{a}$ and $\mathbf{b}$, leaving your answers in terms of λ in each case.

  a $\mathbf{a} = \lambda\mathbf{i} + 2\mathbf{j} + \mathbf{k}$ $\mathbf{b} = \mathbf{i} - 3\mathbf{k}$

  b $\mathbf{a} = 2\mathbf{i} - \mathbf{j} + 7\mathbf{k}$ $\mathbf{b} = \mathbf{i} - \lambda\mathbf{j} + 3\mathbf{k}$

3 Find a unit vector that is perpendicular to both $2\mathbf{i} - \mathbf{j}$ and to $4\mathbf{i} + \mathbf{j} + 3\mathbf{k}$

4 Find a unit vector that is perpendicular to both $4\mathbf{i} + \mathbf{k}$ and $\mathbf{j} - \sqrt{2}\mathbf{k}$

5 Find a unit vector that is perpendicular to both $\mathbf{i} - \mathbf{j}$ and $3\mathbf{i} + 4\mathbf{j} - 6\mathbf{k}$

6 Find a unit vector that is perpendicular to both $\begin{pmatrix} 1 \\ 6 \\ 4 \end{pmatrix}$ and to $\begin{pmatrix} 5 \\ 9 \\ 8 \end{pmatrix}$

7 Find a vector of magnitude 5 which is perpendicular to both $\begin{pmatrix} 4 \\ 0 \\ 1 \end{pmatrix}$ and $\begin{pmatrix} 0 \\ \sqrt{2} \\ 1 \end{pmatrix}$

8 Find the magnitude of $(\mathbf{i} + \mathbf{j} - \mathbf{k}) \times (\mathbf{i} - \mathbf{j} + \mathbf{k})$

9 Given that $\mathbf{a} = -\mathbf{i} + 2\mathbf{j} - 5\mathbf{k}$ and $\mathbf{b} = 5\mathbf{i} - 2\mathbf{j} + \mathbf{k}$, find:
  a $\mathbf{a}.\mathbf{b}$
  b $\mathbf{a} \times \mathbf{b}$
  c the unit vector in the direction $\mathbf{a} \times \mathbf{b}$

10 Find the sine of the angle between each of the following pairs of vectors $\mathbf{a}$ and $\mathbf{b}$. You may leave your answers as surds, in their simplest form.
  a $\mathbf{a} = 3\mathbf{i} - 4\mathbf{j}$, $\mathbf{b} = 2\mathbf{i} + 2\mathbf{j} + \mathbf{k}$
  b $\mathbf{a} = \mathbf{j} + 2\mathbf{k}$, $\mathbf{b} = 5\mathbf{i} + 4\mathbf{j} - 2\mathbf{k}$
  c $\mathbf{a} = 5\mathbf{i} + 2\mathbf{j} + 2\mathbf{k}$, $\mathbf{b} = 4\mathbf{i} + 4\mathbf{j} + \mathbf{k}$

11 The line l_1 has equation $\mathbf{r} = \mathbf{i} - \mathbf{j} + \lambda(\mathbf{i} + 2\mathbf{j} + 3\mathbf{k})$ and the line l_2 has equation $\mathbf{r} = 2\mathbf{i} + \mathbf{j} + \mathbf{k} + \mu(2\mathbf{i} - \mathbf{j} + \mathbf{k})$. Find a vector that is perpendicular to both l_1 and l_2.

(P) 12 It is given that $\mathbf{a} = \begin{pmatrix} 1 \\ 3 \\ -1 \end{pmatrix}$ and $\mathbf{b} = \begin{pmatrix} 2 \\ u \\ v \end{pmatrix}$ and that $\mathbf{a} \times \mathbf{b} = \begin{pmatrix} w \\ -6 \\ -7 \end{pmatrix}$, where u, v and w are scalar constants. Find the values of u, v and w.

(P) 13 Given that $\mathbf{p} = a\mathbf{i} - \mathbf{j} + 4\mathbf{k}$, that $\mathbf{q} = \mathbf{j} - \mathbf{k}$ and that their vector product $\mathbf{q} \times \mathbf{p} = 3\mathbf{i} - \mathbf{j} + b\mathbf{k}$ where a and b are scalar constants,
  a find the values of a and b
  b find the value of the cosine of the angle between $\mathbf{p}$ and $\mathbf{q}$.

(P) 14 If $\mathbf{a} \times \mathbf{b} = \mathbf{0}$, $\mathbf{a} = 2\mathbf{i} + \mathbf{j} - \mathbf{k}$ and $\mathbf{b} = 3\mathbf{i} + \lambda\mathbf{j} + \mu\mathbf{k}$, where λ and μ are scalar constants, find the values of λ and μ.

(P) 15 If three vectors $\mathbf{a}$, $\mathbf{b}$ and $\mathbf{c}$ satisfy $\mathbf{a} + \mathbf{b} + \mathbf{c} = \mathbf{0}$, show that
$$\mathbf{a} \times \mathbf{b} = \mathbf{b} \times \mathbf{c} = \mathbf{c} \times \mathbf{a}$$

> **Challenge**
>
> **SKILLS CREATIVITY**
>
> $\mathbf{a}$ is a non-zero vector and $\mathbf{b}$ and $\mathbf{c}$ are non-parallel vectors.
> Given that $\mathbf{a} \times \mathbf{b} = \mathbf{c} \times \mathbf{a}$, show that $\mathbf{a}$ is parallel to $\mathbf{b} + \mathbf{c}$

5.2 Finding areas

You can use the vector product to solve problems involving areas of triangles and parallelograms.

Example 6

Find the area of triangle OAB, where O is the origin, A is the point with position vector $\mathbf{a}$ and B is the point with position vector $\mathbf{b}$.

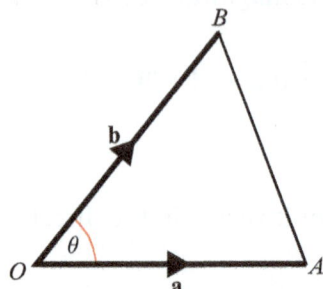

$$\text{Area of triangle } OAB = \tfrac{1}{2}(OA)(OB)\sin\theta$$
$$= \tfrac{1}{2}|\mathbf{a}||\mathbf{b}|\sin\theta$$
$$= \tfrac{1}{2}|\mathbf{a}\times\mathbf{b}|$$

Use the formula for area of triangle, Area $= \tfrac{1}{2}ab\sin C$, and let the angle $AOB = \theta$.

Use the definition of vector product to obtain this result.

- If A and B have position vectors $\mathbf{a}$ and $\mathbf{b}$ respectively, then
$$\text{Area of triangle } OAB = \tfrac{1}{2}|\mathbf{a}\times\mathbf{b}|$$

Example 7

Find the area of triangle ABC, where the position vectors of A, B and C are $\mathbf{a}$, $\mathbf{b}$ and $\mathbf{c}$ respectively.

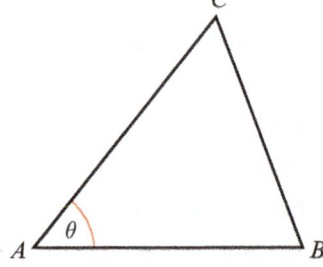

$$\text{Area of triangle } ABC = \tfrac{1}{2}(AB)(AC)\sin\theta$$
$$= \tfrac{1}{2}|\mathbf{b}-\mathbf{a}||\mathbf{c}-\mathbf{a}|\sin\theta$$
$$= \tfrac{1}{2}|(\mathbf{b}-\mathbf{a})\times(\mathbf{c}-\mathbf{a})|$$
$$= \tfrac{1}{2}|(\mathbf{b}\times\mathbf{c})-(\mathbf{b}\times\mathbf{a})-(\mathbf{a}\times\mathbf{c})+(\mathbf{a}\times\mathbf{a})|$$
$$= \tfrac{1}{2}|(\mathbf{b}\times\mathbf{c})+(\mathbf{c}\times\mathbf{a})+(\mathbf{a}\times\mathbf{b})|$$
$$= \tfrac{1}{2}|(\mathbf{a}\times\mathbf{b})+(\mathbf{b}\times\mathbf{c})+(\mathbf{c}\times\mathbf{a})|$$

Let the angle $BAC = \theta$.

Use the definition of the vector product.

Expand using the distributive law.

Use $\mathbf{a}\times\mathbf{a} = \mathbf{0}$, $\mathbf{a}\times\mathbf{b} = -\mathbf{b}\times\mathbf{a}$ and $\mathbf{c}\times\mathbf{a} = -\mathbf{a}\times\mathbf{c}$

- If A, B and C have position vectors $\mathbf{a}$, $\mathbf{b}$ and $\mathbf{c}$ respectively, then
$$\text{Area of triangle } ABC = \tfrac{1}{2}|\overrightarrow{AB}\times\overrightarrow{AC}|$$
$$= \tfrac{1}{2}|(\mathbf{b}-\mathbf{a})\times(\mathbf{c}-\mathbf{a})|$$
$$= \tfrac{1}{2}|(\mathbf{a}\times\mathbf{b})+(\mathbf{b}\times\mathbf{c})+(\mathbf{c}\times\mathbf{a})|$$

Example 8 — SKILLS: PROBLEM-SOLVING

Find the area of the parallelogram $ABCD$, where the position vectors of A, B and D are **a**, **b** and **d** respectively.

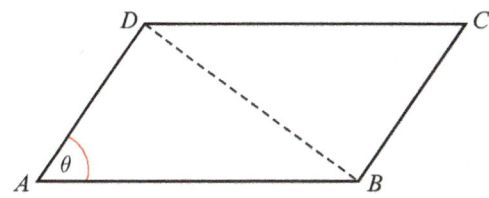

Area of parallelogram $ABCD$
= area of triangle ABD + area of triangle BCD
= 2 × area of triangle ABD ← The two triangles are congruent so have equal area.
= $(AB)(AD)\sin\theta$ ← θ is the angle BAD.
= $|(\mathbf{b} - \mathbf{a}) \times (\mathbf{d} - \mathbf{a})|$
= $|(\mathbf{a} \times \mathbf{b}) + (\mathbf{b} \times \mathbf{d}) + (\mathbf{d} \times \mathbf{a})|$

- If A and B have position vectors **a** and **b** respectively, then

    Area of parallelogram $OABC = |\mathbf{a} \times \mathbf{b}|$

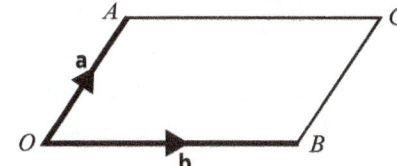

- If A, B, C and D have position vectors **a**, **b**, **c** and **d** respectively, then

    Area of parallelogram $ABCD = |\overrightarrow{AB} \times \overrightarrow{AD}|$

    = $|(\mathbf{b} - \mathbf{a}) \times (\mathbf{d} - \mathbf{a})|$

    = $|(\mathbf{a} \times \mathbf{b}) + (\mathbf{b} \times \mathbf{d}) + (\mathbf{d} \times \mathbf{a})|$

Online Use GeoGebra to explore this relationship.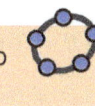

Example 9

Find the area of triangle OAB, where O is the origin, A is the point with position vector $\mathbf{i} - \mathbf{j}$ and B is the point with position vector $3\mathbf{i} + 4\mathbf{j} - 6\mathbf{k}$

Area of triangle $OAB = \frac{1}{2}|(\mathbf{i} - \mathbf{j}) \times (3\mathbf{i} + 4\mathbf{j} - 6\mathbf{k})|$

$(\mathbf{i} - \mathbf{j}) \times (3\mathbf{i} + 4\mathbf{j} - 6\mathbf{k}) = \begin{vmatrix} \mathbf{i} & \mathbf{j} & \mathbf{k} \\ 1 & -1 & 0 \\ 3 & 4 & -6 \end{vmatrix}$ ← First find the vector product using the determinant method.

$= 6\mathbf{i} + 6\mathbf{j} + 7\mathbf{k}$

So area of triangle $= \frac{1}{2}|6\mathbf{i} + 6\mathbf{j} + 7\mathbf{k}| = \frac{1}{2}\sqrt{6^2 + 6^2 + 7^2}$ ← Then use this to find the area of the triangle.

$= \frac{\sqrt{121}}{2} = 5.5$

Example 10

Find the area of triangle ABC, where the position vectors of A, B and C are
$4\mathbf{i} - 2\mathbf{j} + \mathbf{k}$, $-12\mathbf{i} + 14\mathbf{j} + \mathbf{k}$ and $-4\mathbf{i} - 2\mathbf{j} + \mathbf{k}$ respectively.

$\overrightarrow{AB} = (-12\mathbf{i} + 14\mathbf{j} + \mathbf{k}) - (4\mathbf{i} - 2\mathbf{j} + \mathbf{k}) = -16\mathbf{i} + 16\mathbf{j}$

$\overrightarrow{AC} = (-4\mathbf{i} - 2\mathbf{j} + \mathbf{k}) - (4\mathbf{i} - 2\mathbf{j} + \mathbf{k}) = -8\mathbf{i}$

— Find vectors representing two of the sides of the triangle.

$\overrightarrow{AB} \times \overrightarrow{AC} = \begin{vmatrix} \mathbf{i} & \mathbf{j} & \mathbf{k} \\ -16 & 16 & 0 \\ -8 & 0 & 0 \end{vmatrix} = 128\mathbf{k}$

So area of triangle $ABC = \frac{1}{2}|128\mathbf{k}| = 64$

— Area of triangle $= \frac{1}{2}|\overrightarrow{AB} \times \overrightarrow{AC}|$. Find $\overrightarrow{AB} \times \overrightarrow{AC}$ using the determinant method, then find half its modulus. Remember that $|p\mathbf{k}| = p$ for any scalar p.

Example 11

Find the area of the parallelogram $ABCD$, where the position vectors of A, B and D are
$2\mathbf{i} + \mathbf{j} - \mathbf{k}$, $6\mathbf{i} + 4\mathbf{j} - 3\mathbf{k}$ and $14\mathbf{i} + 7\mathbf{j} - 6\mathbf{k}$ respectively.

Area of parallelogram $ABCD = |\overrightarrow{AB} \times \overrightarrow{AD}|$

$\overrightarrow{AB} = (6\mathbf{i} + 4\mathbf{j} - 3\mathbf{k}) - (2\mathbf{i} + \mathbf{j} - \mathbf{k}) = 4\mathbf{i} + 3\mathbf{j} - 2\mathbf{k}$

$\overrightarrow{AD} = (14\mathbf{i} + 7\mathbf{j} - 6\mathbf{k}) - (2\mathbf{i} + \mathbf{j} - \mathbf{k}) = 12\mathbf{i} + 6\mathbf{j} - 5\mathbf{k}$

— Find vectors representing two adjacent sides of the parallelogram.

$\overrightarrow{AB} \times \overrightarrow{AD} = \begin{vmatrix} \mathbf{i} & \mathbf{j} & \mathbf{k} \\ 4 & 3 & -2 \\ 12 & 6 & -5 \end{vmatrix} = -3\mathbf{i} - 4\mathbf{j} - 12\mathbf{k}$

So area of parallelogram $= |-3\mathbf{i} - 4\mathbf{j} - 12\mathbf{k}| = 13$

— Area of parallelogram $= |\overrightarrow{AB} \times \overrightarrow{AD}|$

Exercise 5B   SKILLS   PROBLEM-SOLVING

1 Find the area of triangle OAB, where O is the origin, A is the point with position vector $\mathbf{a}$ and B is the point with position vector $\mathbf{b}$ in the following cases.

  a $\mathbf{a} = \mathbf{i} + \mathbf{j} - 4\mathbf{k}$ $\mathbf{b} = 2\mathbf{i} - \mathbf{j} - 2\mathbf{k}$

  b $\mathbf{a} = 3\mathbf{i} + 4\mathbf{j} - 5\mathbf{k}$ $\mathbf{b} = 2\mathbf{i} + \mathbf{j} - 2\mathbf{k}$

  c $\mathbf{a} = \begin{pmatrix} 2 \\ 3 \\ 0 \end{pmatrix}$ $\mathbf{b} = \begin{pmatrix} 2 \\ 6 \\ -9 \end{pmatrix}$

2 Find the area of triangle ABC, where the position vectors of A, B and C are $\mathbf{a}$, $\mathbf{b}$ and $\mathbf{c}$ respectively, in the following cases:

  a $\mathbf{a} = \mathbf{i} - \mathbf{j} - \mathbf{k}$ $\mathbf{b} = 4\mathbf{i} + \mathbf{j} + \mathbf{k}$ $\mathbf{c} = 4\mathbf{i} - 3\mathbf{j} + \mathbf{k}$

  b $\mathbf{a} = \begin{pmatrix} 0 \\ 1 \\ 2 \end{pmatrix}$ $\mathbf{b} = \begin{pmatrix} 1 \\ 0 \\ 2 \end{pmatrix}$ $\mathbf{c} = \begin{pmatrix} 2 \\ 0 \\ -10 \end{pmatrix}$

3 Find the area of the triangle with vertices $A(1, 0, 2)$, $B(2, -2, 0)$ and $C(3, -1, 1)$.

4 Find the area of the triangle with vertices $A(-1, 1, 1)$, $B(1, 0, 2)$ and $C(0, 3, 4)$.

5 Find the area of the parallelogram $ABCD$, shown in the diagram, where the position vectors of A, B and D are $\mathbf{i} + \mathbf{j} + \mathbf{k}$, $-3\mathbf{i} + 4\mathbf{j} + \mathbf{k}$ and $2\mathbf{i} - \mathbf{j}$ respectively.

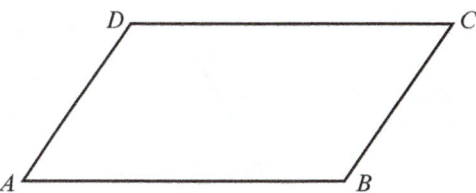

6 Find the area of the parallelogram $ABCD$, shown in the diagram, in which the vertices A, B and D have coordinates $(0, 5, 3)$, $(2, 1, -1)$ and $(1, 6, 6)$ respectively.

7 Find the area of the parallelogram $ABCD$, shown in the diagram, where the position vectors of A, B and D are $\mathbf{j}$, $\mathbf{i} + 4\mathbf{j} + \mathbf{k}$ and $2\mathbf{i} + 6\mathbf{j} + 3\mathbf{k}$ respectively.

(P) **8** Relative to an origin O, the points P and Q have position vectors $\mathbf{p}$ and $\mathbf{q}$ respectively, where $\mathbf{p} = a(\mathbf{i} + \mathbf{j} + 2\mathbf{k})$, $\mathbf{q} = a(2\mathbf{i} + \mathbf{j} + 3\mathbf{k})$ and $a > 0$

Find the area of triangle OPQ, giving your answer in terms of a.

(P) **9 a** Prove that the area of the parallelogram $ABCD$ is $|(\mathbf{b} - \mathbf{a}) \times (\mathbf{c} - \mathbf{a})|$

b Show that $(\mathbf{b} - \mathbf{a}) \times (\mathbf{c} - \mathbf{a}) = (\mathbf{b} - \mathbf{a}) \times (\mathbf{d} - \mathbf{a})$ implies that $(\mathbf{b} - \mathbf{a}) \times (\mathbf{c} - \mathbf{d}) = \mathbf{0}$, and explain the geometrical significance of this vector product.

(E) **10** The position vectors of the points A, B and C relative to an origin O are $2\mathbf{i} - \mathbf{j} - \mathbf{k}$, $6\mathbf{i} - 2\mathbf{k}$ and $3\mathbf{i} + 3\mathbf{j}$ respectively.

Find

a $\overrightarrow{AC} \times \overrightarrow{BC}$ **(3 marks)**

b the exact area of triangle ABC. **(2 marks)**

(E) **11** The sail of a yacht is modelled as a triangle with vertices at $A(-3, 2, -4)$, $B(-2, -3, 1)$ and $C(1, 2, -1)$, where the dimensions are in metres.

a Find $\overrightarrow{AB} \times \overrightarrow{AC}$ **(3 marks)**

b Hence find the area of fabric needed to construct the sail according to this model. **(2 marks)**

c Suggest, with a reason, whether the actual area of fabric needed to construct the sail will be larger or smaller than this value. **(1 mark)**

(E) **12** A jeweller makes gold pendants in the shape of a parallelogram $ABCD$ where sides AB and DC are equal and parallel. She designs the pendants in 3D space and models the pendants as having vertices $A(-1, 2, 0)$, $B(3, -3, -2)$ and $D(-2, 0, 3)$ where each unit represents 1 cm.

a Find the coordinates of point C. **(2 marks)**

Given that gold costs £595 per cm³, and that the pendants will be 3 mm thick,

b find, correct to the nearest pound, the cost of making one pendant. **(4 marks)**

> **Challenge**
>
> **SKILLS INNOVATION**
>
> In the diagram below, $ABCD$ and $CDEF$ are parallelograms which lie in the same plane.
>
>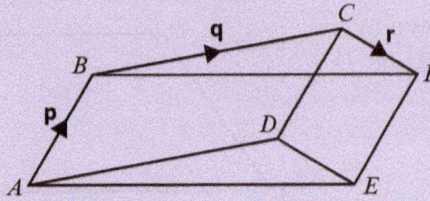
>
> $\overrightarrow{AB} = \mathbf{p}$, $\overrightarrow{BC} = \mathbf{q}$ and $\overrightarrow{CF} = \mathbf{r}$
>
> By considering area, show that $|\mathbf{p} \times (\mathbf{q} + \mathbf{r})| = |\mathbf{p} \times \mathbf{q}| + |\mathbf{p} \times \mathbf{r}|$

5.3 Scalar triple product

You can find the **scalar triple product** of three vectors **a**, **b** and **c**, and use it to find the volume of a **parallelepiped** and of a tetrahedron.

Online Use GeoGebra to explore the scalar triple product.

Notation A **parallelepiped** is a three-dimensional solid with six parallelogram-shaped faces.

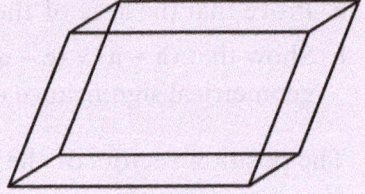

You know that $\mathbf{b} \times \mathbf{c} = (b_2c_3 - b_3c_2)\mathbf{i} + (b_3c_1 - b_1c_3)\mathbf{j} + (b_1c_2 - b_2c_1)\mathbf{k}$, where $\mathbf{b} = b_1\mathbf{i} + b_2\mathbf{j} + b_3\mathbf{k}$ and $\mathbf{c} = c_1\mathbf{i} + c_2\mathbf{j} + c_3\mathbf{k}$

So if $\mathbf{a} = a_1\mathbf{i} + a_2\mathbf{j} + a_3\mathbf{k}$, then

■ $\mathbf{a}.(\mathbf{b} \times \mathbf{c}) = a_1(b_2c_3 - b_3c_2) + a_2(b_3c_1 - b_1c_3) + a_3(b_1c_2 - b_2c_1)$

This can also be written as

■ $\mathbf{a}.(\mathbf{b} \times \mathbf{c}) = \begin{vmatrix} a_1 & a_2 & a_3 \\ b_1 & b_2 & b_3 \\ c_1 & c_2 & c_3 \end{vmatrix}$, and $\mathbf{a}.(\mathbf{b} \times \mathbf{c})$ is known as the scalar triple product.

Example 12 SKILLS ANALYSIS

Given that $\mathbf{a} = 3\mathbf{i} - \mathbf{j} + 4\mathbf{k}$, $\mathbf{b} = \mathbf{i} + \mathbf{j} - \mathbf{k}$ and $\mathbf{c} = 2\mathbf{i} + 3\mathbf{j} + 5\mathbf{k}$, find

a $\mathbf{a}.(\mathbf{b} \times \mathbf{c})$ b $\mathbf{b}.(\mathbf{c} \times \mathbf{a})$ c $\mathbf{a}.(\mathbf{a} \times \mathbf{c})$

$$\mathbf{a} \cdot \mathbf{b} \times \mathbf{c} = \begin{vmatrix} \mathbf{i} & \mathbf{j} & \mathbf{k} \\ 1 & 1 & -1 \\ 2 & 3 & 5 \end{vmatrix} = 8\mathbf{i} - 7\mathbf{j} + \mathbf{k}$$

So $\mathbf{a} \cdot (\mathbf{b} \times \mathbf{c}) = (3\mathbf{i} - \mathbf{j} + 4\mathbf{k}) \cdot (8\mathbf{i} - 7\mathbf{j} + \mathbf{k})$
$= 24 + 7 + 4$
$= 35$

You could calculate $\mathbf{a} \cdot (\mathbf{b} \times \mathbf{c})$ directly as a determinant:

$$\begin{vmatrix} 3 & -1 & 4 \\ 1 & 1 & -1 \\ 2 & 3 & 5 \end{vmatrix} = 3\begin{vmatrix} 1 & -1 \\ 3 & 5 \end{vmatrix} - (-1)\begin{vmatrix} 1 & -1 \\ 2 & 5 \end{vmatrix} + 4\begin{vmatrix} 1 & 1 \\ 2 & 3 \end{vmatrix}$$

$= 24 + 7 + 4 = 35$

$$\mathbf{b} \ \mathbf{c} \times \mathbf{a} = \begin{vmatrix} \mathbf{i} & \mathbf{j} & \mathbf{k} \\ 2 & 3 & 5 \\ 3 & -1 & 4 \end{vmatrix} = 17\mathbf{i} + 7\mathbf{j} - 11\mathbf{k}$$

So $\mathbf{b} \cdot (\mathbf{c} \times \mathbf{a}) = (\mathbf{i} + \mathbf{j} - \mathbf{k}) \cdot (17\mathbf{i} + 7\mathbf{j} - 11\mathbf{k})$
$= 17 + 7 + 11$
$= 35$

Notice that $\mathbf{a} \cdot (\mathbf{b} \times \mathbf{c}) = \mathbf{b} \cdot (\mathbf{c} \times \mathbf{a})$

$\mathbf{c} \ \mathbf{a} \times \mathbf{c} = -\mathbf{c} \times \mathbf{a} = -17\mathbf{i} - 7\mathbf{j} + 11\mathbf{k}$

Use the result that $\mathbf{a} \times \mathbf{c} = -\mathbf{c} \times \mathbf{a}$

So $\mathbf{a} \cdot (\mathbf{a} \times \mathbf{c}) = (3\mathbf{i} - \mathbf{j} + 4\mathbf{k}) \cdot (-17\mathbf{i} - 7\mathbf{j} + 11\mathbf{k})$
$= -51 + 7 + 44$
$= 0$

This scalar product is zero since $\mathbf{a} \times \mathbf{c}$ is perpendicular to $\mathbf{a}$.

The above worked example illustrates two important points.

- The scalar triple product is cyclic:
  $\mathbf{a} \cdot (\mathbf{b} \times \mathbf{c}) = \mathbf{b} \cdot (\mathbf{c} \times \mathbf{a}) = \mathbf{c} \cdot (\mathbf{a} \times \mathbf{b})$

- If a vector is repeated then the scalar triple product is equal to zero:
  $\mathbf{a} \cdot (\mathbf{a} \times \mathbf{p}) = \mathbf{a} \cdot (\mathbf{p} \times \mathbf{a}) = 0$ for any vector $\mathbf{p}$.

Hint You can use the first of these to prove the second:
$\mathbf{a} \cdot (\mathbf{a} \times \mathbf{p}) = \mathbf{p} \cdot (\mathbf{a} \times \mathbf{a}) = \mathbf{p} \cdot \mathbf{0} = 0$

Example 13

Find the volume of the parallelepiped shown in the figure, given that O is the origin and A, B and C have position vectors $\mathbf{a}$, $\mathbf{b}$ and $\mathbf{c}$ respectively. The angle between $\mathbf{b}$ and $\mathbf{c}$ is θ and the angle between the perpendicular height and $\mathbf{a}$ is ϕ.

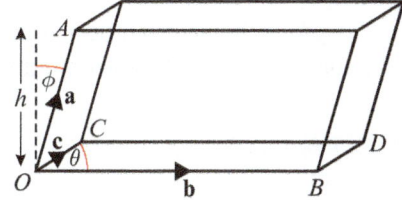

The volume of the parallelepiped is given by (area of base) $\times h$ where h is the perpendicular distance between the base and the top face.

The base, $OBDC$ is a parallelogram and its area is $|\mathbf{b} \times \mathbf{c}|$

So the volume of the parallelepiped is $|\mathbf{b} \times \mathbf{c}|h$

But $h = OA \cos \phi$

So volume is $|\mathbf{b} \times \mathbf{c}|OA \cos \phi$
$= |\mathbf{b} \times \mathbf{c}||\mathbf{a}|\cos \phi$
$= \mathbf{a} \cdot (\mathbf{b} \times \mathbf{c})$

As $\cos \phi = \dfrac{h}{OA}$

Since $\mathbf{b} \times \mathbf{c}$ is in the direction of the perpendicular height, ϕ is the angle between vector $\mathbf{a}$ and vector $\mathbf{b} \times \mathbf{c}$

From the definition of scalar product.

- If three sides of a parallelepiped are given by vectors **a**, **b** and **c** as shown in the diagram, then the volume of the parallelepiped is given by |**a**.(**b** × **c**)|

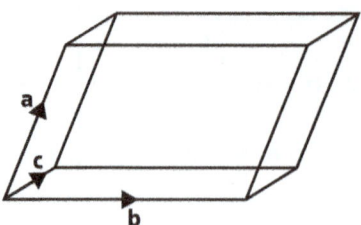

Notation **a**, **b** and **c** can be any three non-parallel sides of the parallelepiped.

Example 14   SKILLS   PROBLEM-SOLVING

Find the volume of the tetrahedron shown in the figure, given that O is the origin and A, B and C have position vectors **a**, **b** and **c** respectively. The angle between **b** and **c** is θ and the angle between the perpendicular height and **a** is ϕ.

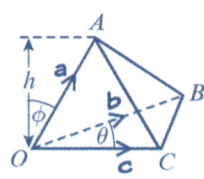

The volume of the tetrahedron is given by the formula $\frac{1}{3}$(area of base) × h
where h is the perpendicular height.
The triangular base, OBC has area $\frac{1}{2}$|**b** × **c**|
And $h = OA \cos \phi = |\mathbf{a}| \cos \phi$
So volume of tetrahedron is $\frac{1}{3} \times \frac{1}{2} |\mathbf{b} \times \mathbf{c}||\mathbf{a}| \cos \phi$
$= \frac{1}{6} \mathbf{a}.(\mathbf{b} \times \mathbf{c})$

— The volume of a pyramid is $\frac{1}{3}$(area of base) × h

— As in Example 13, **b** × **c** is in the direction of the perpendicular height, so ϕ is the angle between vector **a** and vector **b** × **c**

- If three sides of a tetrahedron are given by vectors **a**, **b** and **c** as shown in the diagram, then the volume of the tetrahedron is given by $\frac{1}{6}$|**a**.(**b** × **c**)|

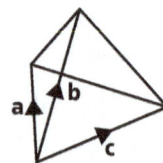

Notation **a**, **b** and **c** can be any three non-coplanar sides of the tetrahedron.

Example 15

Find the volume of a tetrahedron which has vertices at $(1, 1, -1)$, $(2, 4, -1)$, $(3, 0, -2)$ and $(0, 4, 5)$.

If the vertices are labelled A, B, C and D in the order given above and have position vectors **a**, **b**, **c** and **d** respectively, then:

$\overrightarrow{AB} = \mathbf{b} - \mathbf{a} = \mathbf{i} + 3\mathbf{j}$
$\overrightarrow{AC} = \mathbf{c} - \mathbf{a} = 2\mathbf{i} - \mathbf{j} - \mathbf{k}$
$\overrightarrow{AD} = \mathbf{d} - \mathbf{a} = -\mathbf{i} + 3\mathbf{j} + 6\mathbf{k}$

— Find expressions for the vectors describing the displacement from one of the vertices to the other three.

$$\text{Volume of tetrahedron} = \frac{1}{6}|\vec{AB}.(\vec{AC} \times \vec{AD})|$$

$$\vec{AB}.(\vec{AC} \times \vec{AD}) = \begin{vmatrix} 1 & 3 & 0 \\ 2 & -1 & -1 \\ -1 & 3 & 6 \end{vmatrix} = -36$$

So the volume is $\frac{1}{6}|-36| = 6$

Use the scalar triple product to find the volume.

Problem-solving

$\vec{AB}.(\vec{AC} \times \vec{AD})$ is negative. If you swapped any pair of vectors in this scalar triple product the answer would be 6 instead of −6.
For example, $\vec{AC}.(\vec{AB} \times \vec{AD}) = 6$

Exercise 5C   SKILLS   PROBLEM-SOLVING

1. Given that $\mathbf{a} = 5\mathbf{i} + 2\mathbf{j} - \mathbf{k}$, $\mathbf{b} = \mathbf{i} + \mathbf{j} + \mathbf{k}$ and $\mathbf{c} = 3\mathbf{i} + 4\mathbf{k}$, find
    a $\mathbf{a}.(\mathbf{b} \times \mathbf{c})$ **b** $\mathbf{b}.(\mathbf{c} \times \mathbf{a})$ **c** $\mathbf{c}.(\mathbf{a} \times \mathbf{b})$

2. Given that $\mathbf{a} = \mathbf{i} - \mathbf{j} - 2\mathbf{k}$, $\mathbf{b} = 2\mathbf{i} + \mathbf{j} - \mathbf{k}$ and $\mathbf{c} = 2\mathbf{i} - 3\mathbf{j} - 5\mathbf{k}$, find $\mathbf{a}.(\mathbf{b} \times \mathbf{c})$
   What can you deduce about the vectors **a**, **b** and **c**?

3. Find the volume of the parallelepiped $ABCDEFGH$ where the vertices A, B, D and E have coordinates $(0, 0, 0)$, $(3, 0, 1)$, $(1, 2, 0)$ and $(1, 1, 3)$ respectively.

4. Find the volume of the parallelepiped $ABCDEFGH$ where the vertices A, B, D and E have coordinates $(-1, 0, 1)$, $(3, 0, -1)$, $(2, 2, 0)$ and $(2, 1, 2)$ respectively.

5. A tetrahedron has vertices at $A(1, 2, 3)$, $B(4, 3, 4)$, $C(1, 3, 1)$ and $D(3, 1, 4)$.
   Find the volume of the tetrahedron.

6. A tetrahedron has vertices at $A(2, 2, 1)$, $B(3, -1, 2)$, $C(1, 1, 3)$ and $D(3, 1, 4)$.
    a Find the area of face BCD.
    b Find a unit vector normal to the face BCD.
    c Find the volume of the tetrahedron.

7. A tetrahedron has vertices at $A(0, 0, 0)$, $B(2, 0, 0)$, $C(1, \sqrt{3}, 0)$ and $D\left(1, \frac{\sqrt{3}}{3}, \frac{2\sqrt{6}}{3}\right)$.
    a Show that the tetrahedron is regular.
    b Find the volume of the tetrahedron.

8. A tetrahedron $OABC$ has its vertices at the points $O(0, 0, 0)$, $A(1, 2, -1)$, $B(-1, 1, 2)$ and $C(2, -1, 1)$.
    a Write down expressions for $\vec{AB}$ and $\vec{AC}$ in terms of **i**, **j** and **k** and find $\vec{AB} \times \vec{AC}$.   (3 marks)
    b Deduce the area of triangle ABC.   (2 marks)
    c Find the volume of the tetrahedron.   (3 marks)

9 The points A, B, C and D have position vectors $\mathbf{a}$, $\mathbf{b}$, $\mathbf{c}$ and $\mathbf{d}$ respectively, where

$\mathbf{a} = 2\mathbf{i} + \mathbf{j}$ $\qquad$ $\mathbf{b} = 3\mathbf{i} - \mathbf{j} + \mathbf{k}$ $\qquad$ $\mathbf{c} = -2\mathbf{j} - \mathbf{k}$ $\qquad$ $\mathbf{d} = 2\mathbf{i} - \mathbf{j} + 3\mathbf{k}$

a Find $\overrightarrow{AB} \times \overrightarrow{BC}$ and $\overrightarrow{BD} \times \overrightarrow{DC}$. **(4 marks)**

b Hence find

  i the area of triangle ABC **(2 marks)**

  ii the volume of the tetrahedron $ABCD$. **(3 marks)**

10 The edges OP, OQ and OR of a tetrahedron $OPQR$ are the vectors $\mathbf{a}$, $\mathbf{b}$ and $\mathbf{c}$ respectively, where

$\mathbf{a} = 2\mathbf{i} + 4\mathbf{j}$ $\qquad$ $\mathbf{b} = 2\mathbf{i} - \mathbf{j} + 3\mathbf{k}$ $\qquad$ $\mathbf{c} = 4\mathbf{i} - 2\mathbf{j} + 5\mathbf{k}$

a Evaluate $\mathbf{b} \times \mathbf{c}$ and deduce that OP is perpendicular to the plane OQR. **(4 marks)**

b Write down the length of OP and the area of triangle OQR and hence the volume of the tetrahedron. **(3 marks)**

c Verify your result by evaluating $\mathbf{a}.(\mathbf{b} \times \mathbf{c})$ **(2 marks)**

11 The diagram shows a parallelepiped $ABCEFDHG$.

M is the midpoint of EF. The point N lies on AB such that $AN:NB = 2:1$.

a Find the ratio of the volume of the parallelepiped to the volume of the tetrahedron $NCME$. **(6 marks)**

b State, with reasons, how this ratio varies as N moves along the line segment AB. **(2 marks)**

12 The diagram shows a pyramid with base vertices $A(-1, 0, 0)$, $B(0, 2, 1)$, $C(1, 2, 3)$ and $D(0, 0, 2)$. The vertex of the pyramid is at $E(3, 0, 1)$.

Find the exact volume of the pyramid. **(8 marks)**

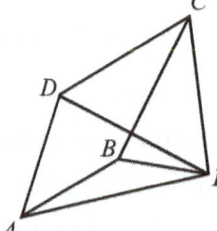

Problem-solving

Split the pyramid into two tetrahedrons.

Challenge

a Explain why $\mathbf{a}.(\mathbf{b} \times \mathbf{c}) = (\mathbf{a} \times \mathbf{b}).\mathbf{c}$

b Use the result from part **a** to show that $\mathbf{d}.(\mathbf{a} \times \mathbf{b} + \mathbf{a} \times \mathbf{c}) = \mathbf{d}.(\mathbf{a} \times (\mathbf{b} + \mathbf{c}))$

c Hence deduce that $\mathbf{a} \times \mathbf{b} + \mathbf{a} \times \mathbf{c} = \mathbf{a} \times (\mathbf{b} + \mathbf{c})$

5.4 Straight lines

You can use the vector product to write a vector equation of a line in a form that doesn't require a parameter. Suppose that **a** is the position vector of a point on a line, and that the line is parallel to the vector **b**.

Let **r** be the position vector of a general point on the line.

$\overrightarrow{AR} = \overrightarrow{OR} - \overrightarrow{OA}$
$= \mathbf{r} - \mathbf{a}$

Since $\overrightarrow{AR}$ is parallel to **b**, $\overrightarrow{AR} \times \mathbf{b} = \mathbf{0}$

So $(\mathbf{r} - \mathbf{a}) \times \mathbf{b} = \mathbf{0}$

- $(\mathbf{r} - \mathbf{a}) \times \mathbf{b} = \mathbf{0}$ is an alternative form of the vector equation of a line passing through the point A with position vector **a**, and parallel to the vector **b**.

This may also be written as $\mathbf{r} \times \mathbf{b} = \mathbf{a} \times \mathbf{b}$

Links A vector equation of a straight line passing through a point A with position vector **a**, and parallel to the vector **b**, is $\mathbf{r} = \mathbf{a} + \lambda \mathbf{b}$, where λ is a scalar parameter.
← Pure 4 Section 7.9

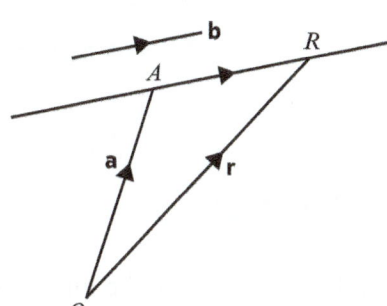

Online Explore the vector equation of a line, written using a cross product, with GeoGebra.

Example 16

Find the vector equation of the line through the points $(1, 2, -1)$ and $(3, -2, 2)$ in the form $(\mathbf{r} - \mathbf{a}) \times \mathbf{b} = \mathbf{0}$

The line is in the direction $\begin{pmatrix} 3 \\ -2 \\ 2 \end{pmatrix} - \begin{pmatrix} 1 \\ 2 \\ -1 \end{pmatrix} = \begin{pmatrix} 2 \\ -4 \\ 3 \end{pmatrix}$ — Any multiple of this vector is also parallel to the direction of the line.

So the equation is $\left(\mathbf{r} - \begin{pmatrix} 3 \\ -2 \\ 2 \end{pmatrix} \right) \times \begin{pmatrix} 2 \\ -4 \\ 3 \end{pmatrix} = \mathbf{0}$ — You could use the position vector $\begin{pmatrix} 1 \\ 2 \\ -1 \end{pmatrix}$ instead of $\begin{pmatrix} 3 \\ -2 \\ 2 \end{pmatrix}$ in this equation.

Example 17

A line has vector equation $\left(\mathbf{r} - \begin{pmatrix} 1 \\ 2 \\ -1 \end{pmatrix} \right) \times \begin{pmatrix} 4 \\ -3 \\ 2 \end{pmatrix} = \mathbf{0}$

Show that the Cartesian equation of the line can be written as $\dfrac{x-1}{l} = \dfrac{y-2}{m} = \dfrac{z+1}{n}$ where l, m and n are integers to be found.

$$\left(\mathbf{r} - \begin{pmatrix} 1 \\ 2 \\ -1 \end{pmatrix}\right) \times \begin{pmatrix} 4 \\ -3 \\ 2 \end{pmatrix} = 0$$

let $\mathbf{r} = \begin{pmatrix} x \\ y \\ z \end{pmatrix}$

So $[\mathbf{i}(x - 1) + \mathbf{j}(y - 2) + \mathbf{k}(z + 1)] \times [4\mathbf{i} - 3\mathbf{j} + 2\mathbf{k}] = 0$

$\mathbf{i}(x - 1) + \mathbf{j}(y - 2) + \mathbf{k}(z + 1) = \lambda[4\mathbf{i} - 3\mathbf{j} + 2\mathbf{k}]$

$\Rightarrow x - 1 = 4\lambda, \quad y - 2 = -3\lambda, \quad z + 1 = 2\lambda$

$\Rightarrow \dfrac{x - 1}{4} = \dfrac{y - 2}{-3} = \dfrac{z + 1}{2} = \lambda$

If the cross product of two non-zero vectors is zero, then one is a multiple of the other.

Example 18

Find the equation of the line in the form $\mathbf{r} \times \mathbf{b} = \mathbf{a} \times \mathbf{b}$ which passes through the points P and Q where $P(3, 1, -2)$ and $Q(-2, 3, 5)$

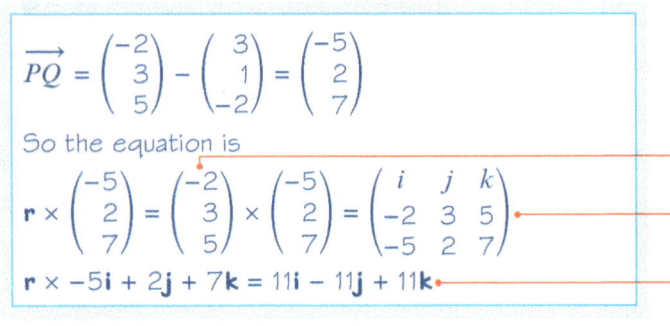

So the equation is

$\mathbf{r} \times \begin{pmatrix} -5 \\ 2 \\ 7 \end{pmatrix} = \begin{pmatrix} -2 \\ 3 \\ 5 \end{pmatrix} \times \begin{pmatrix} -5 \\ 2 \\ 7 \end{pmatrix} = \begin{pmatrix} \mathbf{i} & \mathbf{j} & \mathbf{k} \\ -2 & 3 & 5 \\ -5 & 2 & 7 \end{pmatrix}$

$\mathbf{r} \times -5\mathbf{i} + 2\mathbf{j} + 7\mathbf{k} = 11\mathbf{i} - 11\mathbf{j} + 11\mathbf{k}$

We could also use the position vector $\begin{pmatrix} 3 \\ 1 \\ -2 \end{pmatrix}$ instead of $\begin{pmatrix} -2 \\ 3 \\ 5 \end{pmatrix}$ in this equation

Writing the equation in the form $\mathbf{r} \times \mathbf{b} = \mathbf{a} \times \mathbf{b}$

Evaluating the vector product and writing the line in the required form.

Exercise 5D SKILLS PROBLEM-SOLVING

1. Find an equation of the straight line passing through the point with position vector $\mathbf{a}$ which is parallel to the vector $\mathbf{b}$, giving your answer in the form $\mathbf{r} \times \mathbf{b} = \mathbf{c}$, where $\mathbf{c}$ is a vector to be found for the following pairs $\mathbf{a}$ and $\mathbf{b}$:

   a $\mathbf{a} = 2\mathbf{i} + \mathbf{j} + 2\mathbf{k}$ $\mathbf{b} = 3\mathbf{i} + \mathbf{j} - 2\mathbf{k}$

   b $\mathbf{a} = 2\mathbf{i} - 3\mathbf{k}$ $\mathbf{b} = \mathbf{i} + \mathbf{j} + 5\mathbf{k}$

   c $\mathbf{a} = 4\mathbf{i} - 2\mathbf{j} + \mathbf{k}$ $\mathbf{b} = -\mathbf{i} - 2\mathbf{j} + 3\mathbf{k}$

2. Find a Cartesian equation for each of the lines given in question 1.

3. Find, in the form $(\mathbf{r} - \mathbf{a}) \times \mathbf{b} = 0$, an equation of the straight line passing through the points with coordinates:

   a (1, 3, 5), (6, 4, 2)   b (3, 4, 12), (4, 3, 5)

   c (−2, 2, 6), (3, 7, 11)   d (4, 2, −4), (1, 1, 1)

4. Find a Cartesian equation for each of the lines given in question 3.

5 Find, in the form $(\mathbf{r} - \mathbf{a}) \times \mathbf{b} = \mathbf{0}$, an equation of the straight line given by the following equations, where λ is a scalar parameter.

   a $\mathbf{r} = \mathbf{i} + \mathbf{j} - 2\mathbf{k} + \lambda(2\mathbf{i} - \mathbf{k})$ **b** $\mathbf{r} = \mathbf{i} + 4\mathbf{j} + \lambda(3\mathbf{i} + \mathbf{j} - 5\mathbf{k})$ **c** $\mathbf{r} = 3\mathbf{i} + 4\mathbf{j} - 4\mathbf{k} + \lambda(2\mathbf{i} - 2\mathbf{j} - 3\mathbf{k})$

6 Find the equation of the straight line with Cartesian equation

$$\frac{x-3}{2} = \frac{y+1}{5} = \frac{2z-3}{3}$$

in the form:

   a $\mathbf{r} \times \mathbf{b} = \mathbf{c}$ **b** $\mathbf{r} = \mathbf{a} + t\mathbf{b}$, where t is a scalar parameter.

7 Given that the point with coordinates $(p, q, 1)$ lies on the line with equation

$$\mathbf{r} \times \begin{pmatrix} 2 \\ 1 \\ 3 \end{pmatrix} = \begin{pmatrix} 8 \\ -7 \\ -3 \end{pmatrix}$$

find the values of p and q. **(4 marks)**

8 Given that the equation of a straight line is

$$\mathbf{r} \times \begin{pmatrix} 1 \\ 1 \\ -1 \end{pmatrix} = \begin{pmatrix} -1 \\ 2 \\ 1 \end{pmatrix}$$

Hint Let $\mathbf{a} = a_1\mathbf{i} + a_2\mathbf{j} + a_3\mathbf{k}$ and set up simultaneous equations.

find an equation for the line in the form $\mathbf{r} = \mathbf{a} + t\mathbf{b}$, where t is a scalar parameter. **(4 marks)**

5.5 Vector planes

You can write the equation of a plane in scalar form $\mathbf{r}.\mathbf{n} = p$ or in the vector form $\mathbf{r} = \mathbf{a} + s\mathbf{b} + t\mathbf{c}$, or in the Cartesian form $ax + by + cz + d = 0$

Example 18   SKILLS   PROBLEM-SOLVING

Given that the vector $\mathbf{n}$ is perpendicular to the plane Π and that Π passes through the point A with position vector $\mathbf{a}$, find an equation of the plane Π.

Let point R be a point in the plane with position vector $\mathbf{r}$,

then $\overrightarrow{AR} = \mathbf{r} - \mathbf{a}$

As $\overrightarrow{AR}$ is a vector which lies in the plane, $\overrightarrow{AR}.\mathbf{n} = 0$

So $(\mathbf{r} - \mathbf{a}).\mathbf{n} = 0$

i.e. $\mathbf{r}.\mathbf{n} = \mathbf{a}.\mathbf{n}$

So if $\mathbf{a}.\mathbf{n} = p$, where p is a scalar, then the equation of the plane Π is $\mathbf{r}.\mathbf{n} = p$

The normal to the plane is perpendicular to all lines which lie in the plane, and so $\mathbf{n}$ is perpendicular to $\overrightarrow{AR}$.

When two vectors are perpendicular their scalar product is zero.

■ **The scalar product form of the equation of a plane is $\mathbf{r}.\mathbf{n} = \mathbf{a}.\mathbf{n} = p$ where $\mathbf{n}$ is normal to the plane, $\mathbf{a}$ is the position vector of a point in the plane and $\mathbf{r}$ is the position vector of a general point on the plane. p is a scalar constant.**

Example 19

The point A with position vector $2\mathbf{i} + 3\mathbf{j} - 5\mathbf{k}$ lies in a plane. The vector $3\mathbf{i} + \mathbf{j} - \mathbf{k}$ is perpendicular to the plane. Find an equation of the plane

a in scalar product form

b in Cartesian form.

a $(\mathbf{r} - (2\mathbf{i} + 3\mathbf{j} - 5\mathbf{k})).(3\mathbf{i} + \mathbf{j} - \mathbf{k}) = 0$ — Use $(\mathbf{r} - \mathbf{a}).\mathbf{n} = 0$

So $\mathbf{r}.(3\mathbf{i} + \mathbf{j} - \mathbf{k}) = (2\mathbf{i} + 3\mathbf{j} - 5\mathbf{k}).(3\mathbf{i} + \mathbf{j} - \mathbf{k})$

$\mathbf{r}.(3\mathbf{i} + \mathbf{j} - \mathbf{k}) = 6 + 3 + 5$

So the equation of plane is $\mathbf{r}.(3\mathbf{i} + \mathbf{j} - \mathbf{k}) = 14$ — Give your answer in the form $\mathbf{r}.\mathbf{n} = p$

b This may be written

$(x\mathbf{i} + y\mathbf{j} + z\mathbf{k}).(3\mathbf{i} + \mathbf{j} - \mathbf{k}) = 14$

i.e. $3x + y - z = 14$

which is a Cartesian equation of the plane. — Replace $\mathbf{r}$ by $x\mathbf{i} + y\mathbf{j} + z\mathbf{k}$ to obtain the Cartesian equation.

Example 20

Convert the equation of a plane, $\mathbf{r}.\mathbf{n} = p$, into Cartesian form by replacing $\mathbf{r}$ by $x\mathbf{i} + y\mathbf{j} + z\mathbf{k}$ and $\mathbf{n}$ by $n_1\mathbf{i} + n_2\mathbf{j} + n_3\mathbf{k}$

$(x\mathbf{i} + y\mathbf{j} + z\mathbf{k}).(n_1\mathbf{i} + n_2\mathbf{j} + n_3\mathbf{k}) = p$

So $xn_1 + yn_2 + zn_3 = p$ or $n_1x + n_2y + n_3z - p = 0$

which is of the form $ax + by + cz + d = 0$, where a, b, c and d are constants.

- **The general Cartesian equation of a plane is $ax + by + cz + d = 0$**

Example 21 SKILLS PROBLEM-SOLVING

a Find, in the form $\mathbf{r}.\mathbf{n} = p$, an equation of the plane which contains the line l and the point with position vector $\mathbf{a}$ where l has equation

$\mathbf{r} = 3\mathbf{i} + 5\mathbf{j} - 2\mathbf{k} + \lambda(-\mathbf{i} + 2\mathbf{j} - \mathbf{k})$ and $\mathbf{a} = 4\mathbf{i} + 3\mathbf{j} + \mathbf{k}$

b Give the equation of the plane in Cartesian form.

a The vector $(-\mathbf{i} + 2\mathbf{j} - \mathbf{k})$ is perpendicular to $\mathbf{n}$ — Line l lies in the plane. The direction of l is $-\mathbf{i} + 2\mathbf{j} - \mathbf{k}$, and so this vector is perpendicular to $\mathbf{n}$.

The vector $4\mathbf{i} + 3\mathbf{j} + \mathbf{k} - (3\mathbf{i} + 5\mathbf{j} - 2\mathbf{k})$ also lies in the plane and is also perpendicular to $\mathbf{n}$

i.e. $\mathbf{i} - 2\mathbf{j} + 3\mathbf{k}$ is perpendicular to $\mathbf{n}$ — The point $(4, 3, 1)$ lies on the plane, and the point $(3, 5, -2)$ lies on the line and so also on the plane, so the vector joining these two points also lies on the plane.

This vector $\mathbf{i} - 2\mathbf{j} + 3\mathbf{k}$ is also perpendicular to $\mathbf{n}$.

So $\mathbf{n} = \begin{vmatrix} \mathbf{i} & \mathbf{j} & \mathbf{k} \\ -1 & 2 & -1 \\ 1 & -2 & 3 \end{vmatrix}$

$= 4\mathbf{i} + 2\mathbf{j}$

So $\mathbf{n}$ is in the direction of the vector product of $-\mathbf{i} + 2\mathbf{j} - \mathbf{k}$ and $\mathbf{i} - 2\mathbf{j} + 3\mathbf{k}$

So the equation of the required plane is
r.(4**i** + 2**j**) = (4**i** + 3**j** + **k**).(4**i** + 2**j**)
i.e. **r**.(4**i** + 2**j**) = 16 + 6
An equation of the plane is **r**.(4**i** + 2**j**) = 22

b In Cartesian form this may be written
$4x + 2y = 22$
i.e. $2x + y = 11$

> Replace **r** with $x\mathbf{i} + y\mathbf{j} + z\mathbf{k}$ and perform the scalar product.

Example 22

Show that the vector equation of a plane passing through the point A, with position vector **a**, may be written $\mathbf{r} = \mathbf{a} + \lambda\mathbf{b} + \mu\mathbf{c}$, where **b** and **c** are non parallel vectors in the plane and where λ and μ are scalars.

Let R be a general point on the plane.
Then $\overrightarrow{OR} = \overrightarrow{OA} + \overrightarrow{AR}$

However, $\overrightarrow{AR}$ lies in the plane and so can be written as $\lambda\mathbf{b} + \mu\mathbf{c}$, where λ and μ are scalar parameters which depend on the position of R.

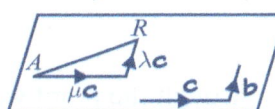

So $\mathbf{r} = \mathbf{a} + \lambda\mathbf{b} + \mu\mathbf{c}$ describes the position of R and is the vector equation of the plane.

> This follows from the triangle law.
>
> If a point lies on a plane its position is usually described as $x\mathbf{i} + y\mathbf{j}$, where **i** and **j** are unit vectors in the plane and x and y give the displacements in the **i** and **j** directions from an origin O.
>
> The two independent vectors used to describe the position of a point on a plane do not need to be unit vectors and do not need to be perpendicular. **b** and **c** are non-parallel vectors, lying in the plane, and so multiples of them may be used to define $\overrightarrow{AR}$.

- The vector equation of a plane is $\mathbf{r} = \mathbf{a} + \lambda\mathbf{b} + \mu\mathbf{c}$
  where **a** is the position vector of a point in the plane, **b** and **c** are non-parallel vectors in the plane and **r** is the position vector of a general point on the plane. λ and μ are scalars and **b** and **c** are non-zero.

Example 23

Find, in the form $\mathbf{r} = \mathbf{a} + \lambda\mathbf{b} + \mu\mathbf{c}$ an equation of the plane that passes through the points $A(2, 2, -1)$, $B(3, 2, -1)$ and $C(4, 3, 5)$.

$\overrightarrow{AB}$ and $\overrightarrow{AC}$ are vectors which lie in the plane
$\overrightarrow{AB} = \overrightarrow{OB} - \overrightarrow{OA} = \mathbf{i}$
$\overrightarrow{AC} = \overrightarrow{OC} - \overrightarrow{OA} = 2\mathbf{i} + \mathbf{j} + 6\mathbf{k}$

So an equation of the plane is
$\mathbf{r} = 2\mathbf{i} + 2\mathbf{j} - \mathbf{k} + \lambda\mathbf{i} + \mu(2\mathbf{i} + \mathbf{j} + 6\mathbf{k})$

> There are many other forms of this answer which are also correct. You could use $3\mathbf{i} + 2\mathbf{j} - \mathbf{k}$ or $4\mathbf{i} + 3\mathbf{j} + 5\mathbf{k}$ instead of $2\mathbf{i} + 2\mathbf{j} - \mathbf{k}$ in the equation.
>
> You could also use $\overrightarrow{BC}$ as a direction in the plane, instead of $\overrightarrow{AB}$ or $\overrightarrow{AC}$.

Example 24

Find a Cartesian equation of the plane that passes through the points $A(1, 0, -1)$, $B(2, 1, 0)$ and $C(2, 16, 6)$.

$\overrightarrow{AB} = \overrightarrow{OB} - \overrightarrow{OA} = \mathbf{i} + \mathbf{j} + \mathbf{k}$

$\overrightarrow{AC} = \overrightarrow{OC} - \overrightarrow{OA} = \mathbf{i} + 16\mathbf{j} + 7\mathbf{k}$

$\overrightarrow{AB} \times \overrightarrow{AC} = \begin{vmatrix} \mathbf{i} & \mathbf{j} & \mathbf{k} \\ 1 & 1 & 1 \\ 1 & 16 & 7 \end{vmatrix}$ — This is the direction of the normal to the plane.

$= -9\mathbf{i} - 6\mathbf{j} + 15\mathbf{k}$

So $\mathbf{r}.(-9\mathbf{i} - 6\mathbf{j} + 15\mathbf{k}) = (\mathbf{i} - \mathbf{k}).(29\mathbf{i} - 6\mathbf{j} + 15\mathbf{k})$ — Use $\mathbf{r}.\mathbf{n} = \mathbf{a}.\mathbf{n}$

i.e. $\mathbf{r}.(29\mathbf{i} - 6\mathbf{j} + 15\mathbf{k}) = 29 + 15 = -24$

So the equation of the plane may be written

$\mathbf{r}.(3\mathbf{i} + 2\mathbf{j} - 5\mathbf{k}) = 8$ or

$(x\mathbf{i} + y\mathbf{j} + z\mathbf{k}).(3\mathbf{i} + 2\mathbf{j} - 5\mathbf{k}) = 8$ — Replace $\mathbf{r}$ by $x\mathbf{i} + y\mathbf{j} + z\mathbf{k}$ to obtain the Cartesian equation.

i.e. $3x + 2y - 5z = 8$, which is a Cartesian equation of the plane. — You may wish to check that each point lies on this plane.

Exercise 5E   SKILLS   PROBLEM-SOLVING

1 Find, in the form $\mathbf{r}.\mathbf{n} = p$, an equation of the plane that passes through the point with position vector $\mathbf{a}$ and is perpendicular to the vector $\mathbf{n}$ where

  a $\mathbf{a} = \mathbf{i} - \mathbf{j} - \mathbf{k}$ and $\mathbf{n} = 2\mathbf{i} + \mathbf{j} + \mathbf{k}$
  b $\mathbf{a} = \mathbf{i} + 2\mathbf{j} + \mathbf{k}$ and $\mathbf{n} = 5\mathbf{i} - \mathbf{j} - 3\mathbf{k}$
  c $\mathbf{a} = 2\mathbf{i} - 3\mathbf{k}$ and $\mathbf{n} = \mathbf{i} + 3\mathbf{j} + 4\mathbf{k}$
  d $\mathbf{a} = 4\mathbf{i} - 2\mathbf{j} + \mathbf{k}$ and $\mathbf{n} = 4\mathbf{i} + \mathbf{j} - 5\mathbf{k}$

2 Find a Cartesian equation for each of the planes in question **1**.

3 Find, in the form $\mathbf{r} = \mathbf{a} + \lambda\mathbf{b} + \mu\mathbf{c}$ an equation of the plane that passes through the points
  a $(1, 2, 0), (3, 1, -1)$ and $(4, 3, 2)$ **b** $(3, 4, 1), (-1, -2, 0)$ and $(2, 1, 4)$
  c $(2, -1, -1), (3, 1, 2)$ and $(4, 0, 1)$ **d** $(-1, 1, 3), (-1, 2, 5)$ and $(0, 4, 4)$

4 Find a Cartesian equation for each of the planes in question **3**.

5 Find a Cartesian equation of the plane that passes through the points
  a $(0, 4, 2), (1, 1, 2)$ and $(-1, 5, 0)$
  b $(1, 1, 0), (2, 3, -3)$ and $(3, 7, -2)$
  c $(3, 0, 0), (2, 0, -1)$ and $(4, 1, 3)$
  d $(1, -1, 6), (3, 1, -2)$ and $(4, 1, 0)$

6 Find, in the form $\mathbf{r}.\mathbf{n} = p$, an equation of the plane which contains the line l and the point with position vector $\mathbf{a}$ where

   a l has equation $\mathbf{r} = \mathbf{i} + \mathbf{j} - 2\mathbf{k} + \lambda(2\mathbf{i} - \mathbf{k})$ and $\mathbf{a} = 4\mathbf{i} + 3\mathbf{j} + \mathbf{k}$
   b l has equation $\mathbf{r} = \mathbf{i} + 2\mathbf{j} - 2\mathbf{k} + \lambda(2\mathbf{i} + \mathbf{j} - 3\mathbf{k})$ and $\mathbf{a} = 3\mathbf{i} + 5\mathbf{j} + \mathbf{k}$
   c l has equation $\mathbf{r} = 2\mathbf{i} - \mathbf{j} + \mathbf{k} + \lambda(\mathbf{i} + 2\mathbf{j} + 2\mathbf{k})$ and $\mathbf{a} = 7\mathbf{i} + 8\mathbf{j} + 6\mathbf{k}$

7 Find a Cartesian equation of the plane which passes through the point $(1, 1, 1)$ and contains the line with equation $\dfrac{x-2}{3} = \dfrac{y+4}{1} = \dfrac{z-1}{2}$

5.6 Solving geometrical problems

You can use vectors in a variety of contexts including

- the points of intersection of lines and planes which meet
- the angle between a line and a plane or between two planes
- the shortest distance between lines and planes which do not meet
- the shortest distance from a point to a line or to a plane.

Example 25   SKILLS   PROBLEM-SOLVING

Find the coordinates of the point of intersection of the lines l_1 and l_2 where

   l_1 has equation $\mathbf{r} = 3\mathbf{i} + \mathbf{j} + \mathbf{k} + \lambda(\mathbf{i} - 2\mathbf{j} - \mathbf{k})$ and
   l_2 has equation $\mathbf{r} = -2\mathbf{j} + 3\mathbf{k} + \mu(-5\mathbf{i} + \mathbf{j} + 4\mathbf{k})$

When the lines meet

$$\begin{pmatrix} 3 + \lambda \\ 1 - 2\lambda \\ 1 - \lambda \end{pmatrix} = \begin{pmatrix} -5\mu \\ -2 + \mu \\ 3 + 4\mu \end{pmatrix}$$

         Use column matrix notation for clarity, and to help to avoid errors.

Solve the simultaneous equations

       $3 + \lambda = -5\mu$ (1)
and $1 - \lambda = 3 + 4\mu$ (2)

         Choose two of the three equations obtained by equating x, y and z components and solve the resulting simultaneous equations.

Adding gives $4 = 3 - \mu$
and so $\mu = -1$
Substituting back into equation (1) gives $\lambda = 2$
The point where the lines meet is $(5, -3, -1)$

         Check that your values of λ and μ satisfy the third equation and that the point which you obtain after substitution lies on both straight lines.

Example 26   SKILLS   ANALYSIS

Find the coordinates of the point of intersection of the line l and the plane Π where l has equation $\mathbf{r} = -\mathbf{i} + \mathbf{j} - 5\mathbf{k} + \lambda(\mathbf{i} + \mathbf{j} + 2\mathbf{k})$ and Π has equation $\mathbf{r}.(\mathbf{i} + 2\mathbf{j} + 3\mathbf{k}) = 4$

The line meets the plane when

$$\begin{pmatrix} -1 + \lambda \\ 1 + \lambda \\ -5 + 2\lambda \end{pmatrix} \cdot \begin{pmatrix} 1 \\ 2 \\ 3 \end{pmatrix} = 4$$

i.e. $-1 + \lambda + 2(1 + \lambda) + 3(-5 + 2\lambda) = 4$

i.e. $9\lambda - 14 = 4$

So $9\lambda = 18$

i.e. $\lambda = 2$

and the line meets the plane at $(1, 3, -1)$

> Write the equation of the line in column matrix form as $\begin{pmatrix} x \\ y \\ z \end{pmatrix} = \begin{pmatrix} -1 + \lambda \\ 1 + \lambda \\ -5 + 2\lambda \end{pmatrix}$
> and substitute into the equation of the plane $\begin{pmatrix} x \\ y \\ z \end{pmatrix} \cdot \begin{pmatrix} 1 \\ 2 \\ 3 \end{pmatrix} = 4$
>
> Solve to find λ and substitute its value into the equation of the line.

Example 27

Find the equation of the line of intersection of the planes Π_1 and Π_2 where Π_1 has equation $\mathbf{r}.(2\mathbf{i} - 2\mathbf{j} - \mathbf{k}) = 2$ and Π_2 has equation $\mathbf{r}.(\mathbf{i} - 3\mathbf{j} + \mathbf{k}) = 5$

In Cartesian form $\quad 2x - 2y - z = 2 \quad$ (1)

and $\quad x - 3y + z = 5 \quad$ (2)

Add equations (1) and (2)

Then $3x - 5y = 7$

So $x = \dfrac{7 + 5y}{3}$

Substituting into equation (2) $z = 5 + 3y - \dfrac{7 + 5y}{3}$

i.e. $z = \dfrac{15 + 9y - (7 + 5y)}{3} = \dfrac{8 + 4y}{3}$

So $\dfrac{x - \frac{7}{3}}{\frac{5}{3}} = y = \dfrac{z - \frac{8}{3}}{\frac{4}{3}} = \lambda$

or $\mathbf{r} = \dfrac{7}{3}\mathbf{i} + \dfrac{8}{3}\mathbf{k} + \lambda\left(\dfrac{5}{3}\mathbf{i} + \mathbf{j} + \dfrac{4}{3}\mathbf{k}\right)$

This is an equation for the line of intersection of the two planes.

> Express the equations of the planes in Cartesian form.
>
> Eliminate one of the variables from the two equations. (In this case z is the easiest to eliminate.)
>
> By eliminating z, express x in terms of y. Then substitute to give z in terms of y.
>
> Let $y = \lambda$ and make λ the subject of the formulae, to give the Cartesian equation of a straight line. This can also be written in vector form.

In Pure 4 you found that the acute angle, θ, between two straight lines was given by the formula

$$\cos \theta = \left| \dfrac{\mathbf{a}.\mathbf{b}}{||\mathbf{a}||\mathbf{b}||} \right|$$

where $\mathbf{a}$ and $\mathbf{b}$ are direction vectors of the lines. You will now extend this idea to find the angle between lines and planes, and between two planes.

Example 28

Find the acute angle between the line l with equation $\mathbf{r} = 2\mathbf{i} + \mathbf{j} - 5\mathbf{k} + \lambda(3\mathbf{i} + 4\mathbf{j} - 12\mathbf{k})$ and the plane with equation $\mathbf{r}.(2\mathbf{i} - 2\mathbf{j} - \mathbf{k}) = 2$

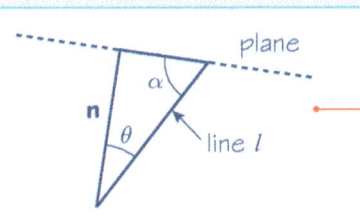

The normal to the plane is in the direction
n = 2**i** − 2**j** − **k**
The angle between this normal and the line l is θ,
where $\cos\theta = \dfrac{(3\mathbf{i} + 4\mathbf{j} - 12\mathbf{k}).(2\mathbf{i} - 2\mathbf{j} - \mathbf{k})}{\sqrt{3^2 + 4^2 + (-12)^2}\,\sqrt{2^2 + (-2)^2 + (-1)^2}}$

i.e. $\cos\theta = \dfrac{10}{13 \times 3} = \dfrac{10}{39}$

So the angle between the plane and the line l is α
where $\alpha + \theta = 90°$

So $\sin\alpha = \dfrac{10}{39}$ and $\alpha = 14.9°$

Draw a diagram showing the line, the plane and the normal to the plane. Let the required angle be α and show α and θ in your diagram.

First find the angle between the given line and the normal to the plane.

Subtract the angle θ from 90°, to give angle α, or use the trigonometric connection that $\cos\theta = \sin\alpha$

You can use the method in Example 28 to show that, in general, the angle θ between the line with equation **r** = **a** + λ**b** and the plane with equation **r.n** = p is given by the formula

■ $\sin\theta = \left|\dfrac{\mathbf{b.n}}{|\mathbf{b}||\mathbf{n}|}\right|$

Example 29

Find the acute angle between the planes with equations **r**.(4**i** + 4**j** − 7**k**) = 13 and **r**.(7**i** − 4**j** + 4**k**) = 6 respectively.

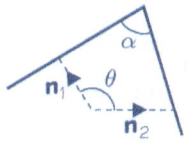

The normals to the planes are in the directions
n₁ = 4**i** + 4**j** − 7**k**, and **n**₂ = 7**i** − 4**j** + 4**k**
The angle between these normals is θ, where

$\cos\theta = \dfrac{(4\mathbf{i} + 4\mathbf{j} - 7\mathbf{k}).(7\mathbf{i} - 4\mathbf{j} + 4\mathbf{k})}{\sqrt{4^2 + 4^2 + (-7)^2}\,\sqrt{7^2 + (-4)^2 + (4)^2}}$

$= \dfrac{28 - 16 - 28}{\sqrt{16 + 16 + 49}\,\sqrt{49 + 16 + 16}}$

$= -\dfrac{16}{81}$

So $\theta = 101.4°$
So the angle between the planes is
$180 - 101.4 = 78.6°$

Draw a diagram showing the planes and the normals to the planes. Let the required angle be α and show α and θ in your diagram.

Draw a diagram showing the planes and the normals to the planes. Let the required angle be α and show α and θ in your diagram.

Subtract the angle θ from 180°, to give angle α.

You can use the method in Example 29 to show that, in general, the angle θ between the plane with equation $\mathbf{r}.\mathbf{n}_1 = p_1$ and the plane with equation $\mathbf{r}.\mathbf{n}_2 = p_2$ is given by the formula

- $\cos\theta = \left|\dfrac{\mathbf{n}_1.\mathbf{n}_2}{\|\mathbf{n}_1\|\|\mathbf{n}_2\|}\right|$

Example 30

Given that d is the length of the perpendicular from the origin to a plane Π, show that the equation of plane Π is $\mathbf{r}.\hat{\mathbf{n}} = d$, where $\hat{\mathbf{n}}$ is a unit vector perpendicular to Π.

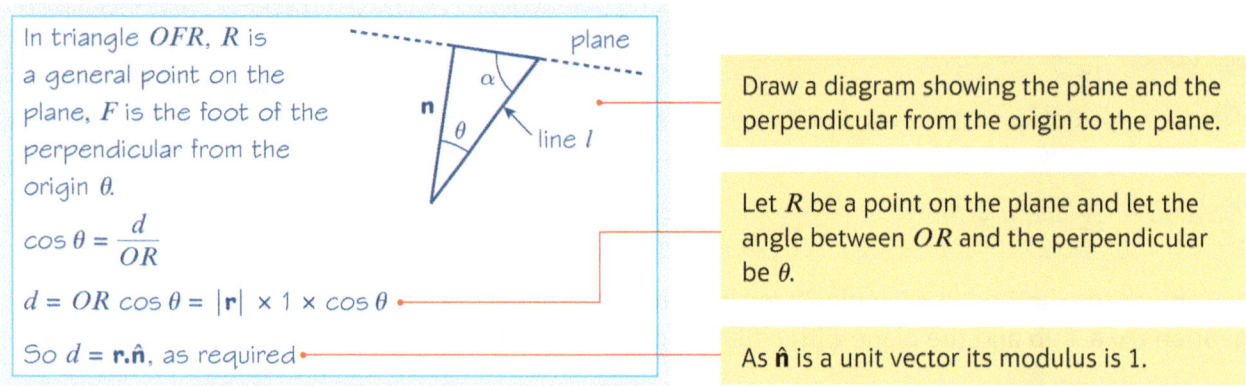

In triangle OFR, R is a general point on the plane, F is the foot of the perpendicular from the origin O.

$\cos\theta = \dfrac{d}{OR}$

$d = OR\cos\theta = |\mathbf{r}| \times 1 \times \cos\theta$

So $d = \mathbf{r}.\hat{\mathbf{n}}$, as required

Draw a diagram showing the plane and the perpendicular from the origin to the plane.

Let R be a point on the plane and let the angle between OR and the perpendicular be θ.

As $\hat{\mathbf{n}}$ is a unit vector its modulus is 1.

- d is the length of the perpendicular from the origin to a plane Π, where the equation of plane Π is written in the form $\mathbf{r}.\hat{\mathbf{n}} = d$, where $\hat{\mathbf{n}}$ is a unit vector perpendicular to Π.

Example 31

The plane Π has equation $\mathbf{r}.(\mathbf{i} + 2\mathbf{j} + 2\mathbf{k}) = 5$
a Find the perpendicular distance from the origin to plane Π.
b Find the perpendicular distance from the point $(1, 3, -2)$ to the plane Π.
c Find the perpendicular distance from the point $(3, 1, -3)$ to the plane Π.

a The unit vector parallel to $\mathbf{i} + 2\mathbf{j} + 2\mathbf{k}$ is $\frac{1}{3}(\mathbf{i} + 2\mathbf{j} + 2\mathbf{k})$

So the equation of Π may be written as
$\mathbf{r}.\hat{\mathbf{n}} = \dfrac{5}{3}$, where $\hat{\mathbf{n}} = \dfrac{1}{3}(\mathbf{i} + 2\mathbf{j} + 2\mathbf{k})$

This means the perpendicular distance from the origin to plane Π is $\dfrac{5}{3}$

The modulus of $\mathbf{i} + 2\mathbf{j} + 2\mathbf{k}$ is $\sqrt{1^2 + 2^2 + 2^2} = 3$

Divide both sides of the equation by 3 so that the equation is of the form $\mathbf{r}.\hat{\mathbf{n}} = d$, with the right hand side of the equation being the required distance.

b

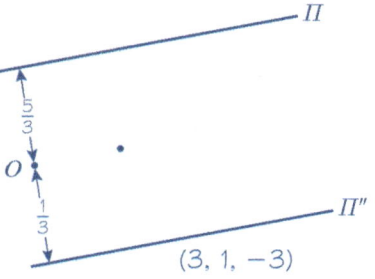

The plane Π' parallel to Π, passing through the point $(1, 3, -2)$ has equation

$$\mathbf{r}.\tfrac{1}{3}(\mathbf{i} + 2\mathbf{j} + 2\mathbf{k}) = (\mathbf{i} + 3\mathbf{j} - 2\mathbf{k}).\tfrac{1}{3}(\mathbf{i} + 2\mathbf{j} + 2\mathbf{k})$$

i.e. $\mathbf{r}.\hat{\mathbf{n}} = 1$, where $\hat{\mathbf{n}} = \tfrac{1}{3}(\mathbf{i} + 2\mathbf{j} + 2\mathbf{k})$

So the perpendicular distance from the origin to plane Π' is 1.

The distance between the two planes Π and Π' is $\tfrac{5}{3} - 1 = \tfrac{2}{3}$

So the distance from the point $(1, 3, -2)$ to the plane Π is $\tfrac{2}{3}$

- Construct a parallel plane Π' through the point $(1, 3, -2)$.
- Find its equation and hence its distance from the origin.
- Then subtract the distance 1 from Π' to O, from the distance $\tfrac{5}{3}$ from Π to O which was found in part **a**.

c

The plane Π'' parallel to Π, passing through the point $(3, 1, -3)$ has equation

$$\mathbf{r}.\tfrac{1}{3}(\mathbf{i} + 2\mathbf{j} + 2\mathbf{k}) = (3\mathbf{i} + \mathbf{j} - 3\mathbf{k}).\tfrac{1}{3}(\mathbf{i} + 2\mathbf{j} + 2\mathbf{k})$$

i.e. $\mathbf{r}.\hat{\mathbf{n}} = -\tfrac{1}{3}$, where $\hat{\mathbf{n}} = \tfrac{1}{3}(\mathbf{i} + 2\mathbf{j} + 2\mathbf{k})$

So the perpendicular distance from the origin to plane Π'' is $\tfrac{1}{3}$

The distance between the two planes Π and Π'' is $\tfrac{5}{3} - \left(-\tfrac{1}{3}\right) = 2$ and the distance from the point $(3, 1, 3)$ to the plane Π is 2.

- Construct a parallel plane Π'' through the point $(3, 1, -3)$.
- The minus sign indicates that Π'' is on the other side of the origin from the previous two planes.
- So as the planes are on opposite sides of O, the distance required is $\tfrac{5}{3} + \tfrac{1}{3}$

Example 32 — SKILLS: PROBLEM-SOLVING

Show that the shortest distance between the parallel lines with equations
$\mathbf{r} = \mathbf{i} + 2\mathbf{j} - \mathbf{k} + \lambda(5\mathbf{i} + 4\mathbf{j} + 3\mathbf{k})$ and $\mathbf{r} = 2\mathbf{i} + \mathbf{k} + \mu(5\mathbf{i} + 4\mathbf{j} + 3\mathbf{k})$
where λ and μ are scalars, is $\dfrac{21\sqrt{2}}{10}$

Method I

Let A be a general point on the first line and B be a general point on the second line, then

$$\overrightarrow{AB} = \begin{pmatrix} 1 \\ -2 \\ 2 \end{pmatrix} + t \begin{pmatrix} 5 \\ 4 \\ 3 \end{pmatrix} \text{ where } t = \mu - \lambda$$

As
$$\begin{pmatrix} 1 \\ -2 \\ 2 \end{pmatrix} = \begin{pmatrix} 2 \\ 0 \\ 1 \end{pmatrix} - \begin{pmatrix} 1 \\ 2 \\ -1 \end{pmatrix}$$

You can set $t = \mu - \lambda$ so that there is only one independent variable.

Let the distance $AB = x$ then
$x^2 = (1 + 5t)^2 + (-2 + 4t)^2 + (2 + 3t)^2 = 50t^2 + 6t + 9$

Use Pythagoras' Theorem to find an expression for the distance AB.

The minimum value of x^2 occurs when $t = -\dfrac{6}{100}$

and so $x^2 = \left(\dfrac{70}{100}\right)^2 + \left(-\dfrac{224}{100}\right)^2 + \left(\dfrac{182}{100}\right)^2$

Find the minimum value of the quadratic expression by using calculus or completion of the square.

So minimum value of $x = \left(\dfrac{14}{100}\right)\sqrt{5^2 + 16^2 + 13^2}$

i.e. $x = \left(\dfrac{7}{50}\right)\sqrt{450} = \left(\dfrac{7}{50}\right) \times 15\sqrt{2} = \dfrac{21\sqrt{2}}{10}$

Substitute and simplify to obtain the printed answer.

Method II

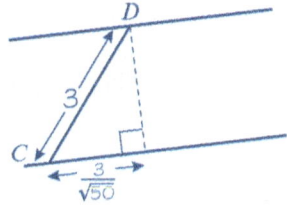

Draw a diagram and calculate the length CD.

Let C be point $(1, 2, -1)$ on the first line and D be point $(2, 0, 1)$ on the second line, then
$\overrightarrow{CD} = \mathbf{i} - 2\mathbf{j} + 2\mathbf{k}$, and $|\overrightarrow{CD}| = \sqrt{1^2 + (-2)^2 + 2^2} = 3$

The projection of CD onto one of the lines has

length $\dfrac{(\mathbf{i} - 2\mathbf{j} + 2\mathbf{k}) \cdot (5\mathbf{i} + 4\mathbf{j} + 3\mathbf{k})}{\sqrt{5^2 + 4^2 + 3^2}} = \dfrac{3}{\sqrt{50}}$

In this alternative method you project the length CD onto one of the lines (using scalar product).

Calculating the third side of the right angle triangle,
$x^2 = 3^2 - \left(\dfrac{3}{\sqrt{50}}\right)^2 = \dfrac{441}{50}$

and so $x = \dfrac{21\sqrt{2}}{10}$

Then you use Pythagoras' Theorem to calculate the required shortest distance.

i.e. the shortest distance is $\dfrac{21\sqrt{2}}{10}$

Example 33

Show that the shortest distance between the two skew lines with equations $\mathbf{r} = \mathbf{a} + \lambda\mathbf{b}$ and $\mathbf{r} = \mathbf{c} + \mu\mathbf{d}$, where λ and μ are scalars, is given by the formula $\left|\dfrac{(\mathbf{a} - \mathbf{c}).(\mathbf{b} \times \mathbf{d})}{|\mathbf{b} \times \mathbf{d}|}\right|$

The shortest distance between the lines is XY where XY is perpendicular to both lines.

The common perpendicular to the two skew lines is in the direction $\mathbf{b} \times \mathbf{d}$ and a unit vector in that direction is $\dfrac{\mathbf{b} \times \mathbf{d}}{|\mathbf{b} \times \mathbf{d}|}$

If P is a point on the line with equation $\mathbf{r} = \mathbf{a} + \lambda\mathbf{b}$ and Q is a point on the line with equation $\mathbf{r} = \mathbf{c} + \mu\mathbf{d}$ then
$$\overrightarrow{QP} = \mathbf{a} - \mathbf{c} + \lambda\mathbf{b} - \mu\mathbf{d}$$

The projection of PQ in the direction of the common perpendicular is

$(\mathbf{a} - \mathbf{c} + \lambda\mathbf{b} - \mu\mathbf{d}).\dfrac{\mathbf{b} \times \mathbf{d}}{|\mathbf{b} \times \mathbf{d}|}$ — This gives $PQ \cos\theta$, where θ is the angle between PQ and the common perpendicular.

i.e. $(\mathbf{a} - \mathbf{c}).\dfrac{\mathbf{b} \times \mathbf{d}}{|\mathbf{b} \times \mathbf{d}|} + \lambda\mathbf{b}.\dfrac{\mathbf{b} \times \mathbf{d}}{|\mathbf{b} \times \mathbf{d}|} - \mu\mathbf{d}.\dfrac{\mathbf{b} \times \mathbf{d}}{|\mathbf{b} \times \mathbf{d}|}$ — Using the distributive property.

But $\mathbf{b}.(\mathbf{b} \times \mathbf{d}) = \mathbf{d}.(\mathbf{b} \times \mathbf{d}) = 0$ and the shortest distance must be a positive quantity.

So the shortest distance is given by $\left|\dfrac{(\mathbf{a} - \mathbf{c}).(\mathbf{b} \times \mathbf{d})}{|\mathbf{b} \times \mathbf{d}|}\right|$ — Use the modulus to ensure that the result is positive.

- The shortest distance between the two skew lines with equations $\mathbf{r} = \mathbf{a} + \lambda\mathbf{b}$ and $\mathbf{r} = \mathbf{c} + \mu\mathbf{d}$, where λ and μ are scalars, is given by the formula
$$\dfrac{(\mathbf{a} - \mathbf{c}).(\mathbf{b} \times \mathbf{d})}{|\mathbf{b} \times \mathbf{d}|}$$

Example 34

Find the shortest distance between the two skew lines with equations $\mathbf{r} = \mathbf{i} + \lambda(\mathbf{j} + \mathbf{k})$ and $\mathbf{r} = -\mathbf{i} + 3\mathbf{j} - \mathbf{k} + \mu(2\mathbf{i} - \mathbf{j} - \mathbf{k})$, where λ and μ are scalars.

$\mathbf{a} - \mathbf{c} = 2\mathbf{i} - 3\mathbf{j} + \mathbf{k}$ — Using $\mathbf{a} = \mathbf{i}$ and $\mathbf{c} = -\mathbf{i} + 3\mathbf{j} - \mathbf{k}$

$\mathbf{b} \times \mathbf{d} = \begin{vmatrix} \mathbf{i} & \mathbf{j} & \mathbf{k} \\ 0 & 1 & 1 \\ 2 & -1 & -1 \end{vmatrix} = 2\mathbf{j} - 2\mathbf{k}$ — Take the vector product of the two direction vectors.

So the shortest distance is
$\left|\dfrac{(2\mathbf{i} - 3\mathbf{j} + \mathbf{k}).(2\mathbf{j} - 2\mathbf{k})}{\sqrt{2^2 + (-2)^2}}\right| = \left|\dfrac{-8}{\sqrt{8}}\right| = \sqrt{8} = 2\sqrt{2}$ — Use the formula for shortest distance obtained in Example 33.

Example 35

Find the shortest distance between the point A with coordinates $(1, 2, -1)$ and the line with equation $\mathbf{r} = \mathbf{i} + \mathbf{j} - 3\mathbf{k} + \mu(2\mathbf{i} - 2\mathbf{j} - \mathbf{k})$, where μ is a scalar.

Let B be a general point on the given line, then $\overrightarrow{AB} = \begin{pmatrix} 0 \\ -1 \\ -2 \end{pmatrix} + \mu \begin{pmatrix} 2 \\ -2 \\ -1 \end{pmatrix}$	As in Example 32, at least two methods are possible. The method shown here is the most efficient.
Let the distance $AB = x$ then $x^2 = (2\mu)^2 + (-1 - 2\mu)^2 + (-2 - \mu)^2 = 9\mu^2 + 8\mu + 5$ The minimum value of x^2 occurs when $\mu = -\frac{8}{18} = -\frac{4}{9}$	Find the minimum value of the quadratic expression by using calculus or completion of the square.
So $x^2 = \left(\frac{-8}{9}\right)^2 + \left(-\frac{1}{9}\right)^2 + \left(-\frac{14}{9}\right)^2$ So minimum value of $x = \left(\frac{1}{9}\right)\sqrt{8^2 + 1^2 + 14^2} = \frac{1}{9}\sqrt{261}$	The minimum value of x corresponds to the minimum value of x^2.
i.e. $x = \left(\frac{1}{9}\right) \times 3\sqrt{29} = \frac{\sqrt{29}}{3}$ i.e. the shortest distance is 1.80 (3 s.f.)	Give an exact answer and give your answer to 3 s.f.

Exercise 5F SKILLS PROBLEM-SOLVING

1 In each case establish whether lines l_1 and l_2 meet and if they meet find the coordinates of their point of intersection.

  a l_1 has equation $\mathbf{r} = \mathbf{i} + 3\mathbf{j} + \lambda(\mathbf{i} - \mathbf{j} + 5\mathbf{k})$ and
   l_2 has equation $\mathbf{r} = -\mathbf{i} - 3\mathbf{j} + 2\mathbf{k} + \mu(\mathbf{i} + \mathbf{j} + 2\mathbf{k})$

  b l_1 has equation $\mathbf{r} = 3\mathbf{i} + 2\mathbf{j} + \mathbf{k} + \lambda(\mathbf{i} + \mathbf{j} + 2\mathbf{k})$ and
   l_2 has equation $\mathbf{r} = 4\mathbf{i} + 3\mathbf{j} + \mu(-\mathbf{i} + \mathbf{j} - \mathbf{k})$

  c l_1 has equation $\mathbf{r} = \mathbf{i} + 3\mathbf{j} + 5\mathbf{k} + \lambda(2\mathbf{i} + 3\mathbf{j} + \mathbf{k})$ and
   l_2 has equation $\mathbf{r} = \mathbf{i} + 2\tfrac{1}{2}\mathbf{j} + 2\tfrac{1}{2}\mathbf{k} + \mu(\mathbf{i} + \mathbf{j} - 2\mathbf{k})$

  (In each of the above cases λ and μ are scalars.)

2 In each case establish whether the line l meets the plane Π and, if they meet, find the coordinates of their point of intersection.

  a $l: \mathbf{r} = \mathbf{i} + \mathbf{j} + \mathbf{k} + \lambda(-2\mathbf{i} + \mathbf{j} - 4\mathbf{k})$
   $\Pi: \mathbf{r}.(3\mathbf{i} - 4\mathbf{j} + 2\mathbf{k}) = 16$

  b $l: \mathbf{r} = 2\mathbf{i} + 3\mathbf{j} - 2\mathbf{k} + \lambda(\mathbf{i} + \mathbf{j} + \mathbf{k})$
   $\Pi: \mathbf{r}.(\mathbf{i} + \mathbf{j} - 2\mathbf{k}) = 1$

  c $l: \mathbf{r} = \mathbf{i} + \mathbf{j} + \mathbf{k} + \lambda(2\mathbf{j} - 2\mathbf{k})$
   $\Pi: \mathbf{r}.(3\mathbf{i} - \mathbf{j} - 6\mathbf{k}) = 1$

  (In each of the above cases λ is a scalar.)

3 Find the equation of the line of intersection of the planes Π_1 and Π_2 where
  a Π_1 has equation $\mathbf{r}.(3\mathbf{i} - 2\mathbf{j} - \mathbf{k}) = 5$ and Π_2 has equation $\mathbf{r}.(4\mathbf{i} - \mathbf{j} - 2\mathbf{k}) = 5$
  b Π_1 has equation $\mathbf{r}.(5\mathbf{i} - \mathbf{j} - 2\mathbf{k}) = 16$ and Π_2 has equation $\mathbf{r}.(16\mathbf{i} - 5\mathbf{j} - 4\mathbf{k}) = 53$
  c Π_1 has equation $\mathbf{r}.(\mathbf{i} - 3\mathbf{j} + \mathbf{k}) = 10$ and Π_2 has equation $\mathbf{r}.(4\mathbf{i} - 3\mathbf{j} - 2\mathbf{k}) = 1$

4 Find the acute angle between the planes with equations $\mathbf{r}.(\mathbf{i} + 2\mathbf{j} - 2\mathbf{k}) = 1$ and $\mathbf{r}.(-4\mathbf{i} + 4\mathbf{j} + 7\mathbf{k}) = 7$ respectively.

5 Find the acute angle between the planes with equations $\mathbf{r}.(3\mathbf{i} - 4\mathbf{j} + 12\mathbf{k}) = 9$ and $\mathbf{r}.(5\mathbf{i} - 12\mathbf{k}) = 7$ respectively.

6 Find the acute angle between the line with equation $\mathbf{r} = 2\mathbf{i} + \mathbf{j} - 5\mathbf{k} + \lambda(4\mathbf{i} + 4\mathbf{j} + 7\mathbf{k})$ and the plane with equation $\mathbf{r}.(2\mathbf{i} + \mathbf{j} - 2\mathbf{k}) = 13$

7 Find the acute angle between the line with equation $\mathbf{r} = -\mathbf{i} - 7\mathbf{j} + 13\mathbf{k} + \lambda(3\mathbf{i} + 4\mathbf{j} - 12\mathbf{k})$ and the plane with equation $\mathbf{r}.(4\mathbf{i} - 4\mathbf{j} - 7\mathbf{k}) = 9$

8 Find the acute angle between the line with equation $(\mathbf{r} - 3\mathbf{j}) \times (-4\mathbf{i} - 7\mathbf{j} + 4\mathbf{k}) = 0$ and the plane with equation $\mathbf{r} = \lambda(4\mathbf{i} - \mathbf{j} - \mathbf{k}) + \mu(4\mathbf{i} - 5\mathbf{j} + 3\mathbf{k})$

9 The plane Π has equation $\mathbf{r}.(10\mathbf{i} + 10\mathbf{j} + 23\mathbf{k}) = 81$
  a Find the perpendicular distance from the origin to plane Π.
  b Find the perpendicular distance from the point $(-1, -1, 4)$ to the plane Π.
  c Find the perpendicular distance from the point $(2, 1, 3)$ to the plane Π.
  d Find the perpendicular distance from the point $(6, 12, -9)$ to the plane Π.

10 Find the shortest distance between the parallel planes.
  a $\mathbf{r}.(6\mathbf{i} + 6\mathbf{j} - 7\mathbf{k}) = 55$ and $\mathbf{r}.(6\mathbf{i} + 6\mathbf{j} - 7\mathbf{k}) = 22$
  b $\mathbf{r} = 3\mathbf{i} + 4\mathbf{j} + \mathbf{k} + \lambda(4\mathbf{i} + \mathbf{k}) + \mu(8\mathbf{i} + 3\mathbf{j} + 3\mathbf{k})$ and
  $\mathbf{r} = 14\mathbf{i} + 2\mathbf{j} + 2\mathbf{k} + \lambda(3\mathbf{j} + \mathbf{k}) + \mu(8\mathbf{i} - 9\mathbf{j} - \mathbf{k})$

11 Find the shortest distance between the two skew lines with equations
$\mathbf{r} = \mathbf{i} + \lambda(-3\mathbf{i} - 12\mathbf{j} + 11\mathbf{k})$ and $\mathbf{r} = 3\mathbf{i} - \mathbf{j} + \mathbf{k} + \mu(2\mathbf{i} + 6\mathbf{j} - 5\mathbf{k})$, where λ and μ are scalars.

12 Find the shortest distance between the parallel lines with equations
$\mathbf{r} = 2\mathbf{i} - \mathbf{j} + \mathbf{k} + \lambda(-3\mathbf{i} - 4\mathbf{j} + 5\mathbf{k})$ and $\mathbf{r} = \mathbf{j} + \mathbf{k} + \mu(-3\mathbf{i} - 4\mathbf{j} + 5\mathbf{k})$, where λ and μ are scalars.

13 Determine whether the lines l_1 and l_2 meet. If they do, find their point of intersection. If they do not, find the shortest distance between them. (In each of the following cases λ and μ are scalars.)
  a l_1 has equation $\mathbf{r} = \mathbf{i} + \mathbf{j} + \lambda(2\mathbf{i} - \mathbf{j} + 5\mathbf{k})$ and
  l_2 has equation $\mathbf{r} = -\mathbf{i} + \mathbf{j} + 2\mathbf{k} + \mu(2\mathbf{i} - 5\mathbf{j} + \mathbf{k})$
  b l_1 has equation $\mathbf{r} = 2\mathbf{i} + \mathbf{j} - 2\mathbf{k} + \lambda(2\mathbf{i} - 2\mathbf{j} + 2\mathbf{k})$ and
  l_2 has equation $\mathbf{r} = \mathbf{i} - \mathbf{j} + 3\mathbf{k} + \mu(\mathbf{i} - \mathbf{j} + \mathbf{k})$
  c l_1 has equation $\mathbf{r} = \mathbf{i} + \mathbf{j} + 5\mathbf{k} + \lambda(2\mathbf{i} + \mathbf{j} - 2\mathbf{k})$ and
  l_2 has equation $\mathbf{r} = -\mathbf{i} - \mathbf{j} + 2\mathbf{k} + \mu(\mathbf{i} + \mathbf{j} + \mathbf{k})$

14 Find the shortest distance between the point with coordinates (4, 1, −1) and the line with equation
$\mathbf{r} = 3\mathbf{i} - \mathbf{j} + 2\mathbf{k} + \mu(2\mathbf{i} - \mathbf{j} - \mathbf{k})$, where μ is a scalar.

15 The plane Π has equation $\mathbf{r}.(\mathbf{i} + \mathbf{j} - \mathbf{k}) = 4$
   a Show that the line with equation $\mathbf{r} = 2\mathbf{i} + 3\mathbf{j} + \mathbf{k} + \lambda(-\mathbf{i} + 2\mathbf{j} + \mathbf{k})$ lies in the plane Π.
   b Show that the line with equation $\mathbf{r} = -\mathbf{i} + 2\mathbf{j} + 4\mathbf{k} + \lambda(-\mathbf{i} + 2\mathbf{j} + \mathbf{k})$ is parallel to the plane Π and find the shortest distance from the line to the plane.

Chapter review 5

(E) 1 Find the shortest distance between the lines with vector equations
$\mathbf{r} = 3\mathbf{i} + s\mathbf{j} - \mathbf{k}$
and $\mathbf{r} = 9\mathbf{i} - 2\mathbf{j} - \mathbf{k} + t(\mathbf{i} - 2\mathbf{j} + \mathbf{k})$
where s, t are scalars. **(5 marks)**

(E) 2 Obtain the shortest distance between the lines with equations
$\mathbf{r} = (3s - 3)\mathbf{i} - s\mathbf{j} + (s + 1)\mathbf{k}$
and $\mathbf{r} = (3 + t)\mathbf{i} + (2t - 2)\mathbf{j} + \mathbf{k}$
where s, t are parameters. **(5 marks)**

(E) 3 The position vectors of the points A, B, C and D relative to a fixed origin O, are $(-\mathbf{j} + 2\mathbf{k})$, $(\mathbf{i} - 3\mathbf{j} + 5\mathbf{k})$, $(2\mathbf{i} - 2\mathbf{j} + 7\mathbf{k})$ and $(\mathbf{j} + 2\mathbf{k})$ respectively.
   a Find $\mathbf{p} = \overrightarrow{AB} \times \overrightarrow{CD}$ **(3 marks)**
   b Calculate $\overrightarrow{AC}.\mathbf{p}$ **(2 marks)**
   c Hence determine the shortest distance between the line containing AB and the line containing CD. **(5 marks)**

(E) 4 Relative to a fixed origin O, the point M has position vector $-4\mathbf{i} + \mathbf{j} - 2\mathbf{k}$
The straight line l has equation $\mathbf{r} \times \overrightarrow{OM} = 5\mathbf{i} - 10\mathbf{k}$
   a Express the equation of the line l in the form $\mathbf{r} = \mathbf{a} + t\mathbf{b}$, where $\mathbf{a}$ and $\mathbf{b}$ are constant vectors and t is a parameter. **(4 marks)**
   b Verify that the point N with coordinates (2, −3, 1) lies on l and find the area of triangle OMN. **(5 marks)**

(E) 5 The line l_1 has equation $\mathbf{r} = \mathbf{i} - \mathbf{j} + \lambda(\mathbf{i} + 2\mathbf{j} + 3\mathbf{k})$ and the line l_2 has equation
$\mathbf{r} = 2\mathbf{i} + \mathbf{j} + \mathbf{k} + \mu(2\mathbf{i} - \mathbf{j} + \mathbf{k})$
   a Find a vector which is perpendicular to both l_1 and l_2. **(4 marks)**
   The point A lies on l_1 and the point B lies on l_2.
   Given that AB is also perpendicular to l_1 and l_2,
   b find the coordinates of A and B. **(6 marks)**

6 A plane passes through the three points A, B, C, whose position vectors, referred to an origin O, are $(\mathbf{i} + 3\mathbf{j} + 3\mathbf{k})$, $(3\mathbf{i} + \mathbf{j} + 4\mathbf{k})$, $(2\mathbf{i} + 4\mathbf{j} + \mathbf{k})$ respectively.
  a Find, in the form $(l\mathbf{i} + m\mathbf{j} + n\mathbf{k})$, a unit vector normal to this plane. **(4 marks)**
  b Find also a Cartesian equation of the plane. **(3 marks)**
  c Find the perpendicular distance from the origin to this plane. **(3 marks)**

7 a Show that the vector $\mathbf{i} + \mathbf{k}$ is perpendicular to the plane with vector equation
  $\mathbf{r} = \mathbf{i} + s\mathbf{j} + t(\mathbf{i} - \mathbf{k})$ **(3 marks)**
  b Find the perpendicular distance from the origin to this plane. **(4 marks)**
  c Hence or otherwise obtain a Cartesian equation of the plane. **(4 marks)**

8 The points A, B and C have position vectors $\mathbf{i} + \mathbf{j} + \mathbf{k}$, $5\mathbf{i} - 2\mathbf{j} + \mathbf{k}$ and $3\mathbf{i} + 2\mathbf{j} + 6\mathbf{k}$ respectively, referred to an origin O.
  a Find a vector perpendicular to the plane containing the points A, B and C. **(3 marks)**
  b Hence, or otherwise, find an equation for the plane which contains the points A, B and C, in the form $ax + by + cz + d = 0$ **(3 marks)**
  The point D has coordinates $(1, 5, 6)$.
  c Find the volume of the tetrahedron $ABCD$. **(4 marks)**

9 The plane Π passes through $A(3, -5, -1)$, $B(-1, 5, 7)$ and $C(2, -3, 0)$.
  a Find $\overrightarrow{AC} \times \overrightarrow{BC}$ **(4 marks)**
  b Hence, or otherwise, find the equation, in the form $\mathbf{r}.\mathbf{n} = p$, of the plane Π. **(3 marks)**
  c The perpendicular from the point $(2, 3, -2)$ to Π meets the plane at P. Find the coordinates of P. **(5 marks)**

10 Given that P and Q are the points with position vectors $\mathbf{p}$ and $\mathbf{q}$ respectively, relative to an origin O, and that
  $\mathbf{p} = 3\mathbf{i} - \mathbf{j} + 2\mathbf{k}$
  $\mathbf{q} = 2\mathbf{i} + \mathbf{j} - \mathbf{k}$
  a find $\mathbf{p} \times \mathbf{q}$ **(3 marks)**
  b Hence, or otherwise, find an equation of the plane containing O, P and Q in the form $ax + by + cz = d$ **(3 marks)**
  The line with equation $(\mathbf{r} - \mathbf{p}) \times \mathbf{q} = 0$ meets the plane with equation $\mathbf{r}.(\mathbf{i} + \mathbf{j} + \mathbf{k}) = 2$ at the point T.
  c Find the coordinates of the point T. **(4 marks)**

11 The planes Π_1 and Π_2 are defined by the equations $2x + 2y - z = 9$ and $x - 2y = 7$ respectively.
  a Find the acute angle between Π_1 and Π_2, giving your answer to the nearest degree. **(3 marks)**
  b Find in the form $\mathbf{r} \times \mathbf{u} = \mathbf{v}$ an equation of the line of intersection of Π_1 and Π_2. **(4 marks)**

12 A pyramid has a square base $OPQR$ and vertex S. Referred to O, the points P, Q, R and S have position vectors $\overrightarrow{OP} = 2\mathbf{i}$, $\overrightarrow{OQ} = 2\mathbf{i} + 2\mathbf{j}$, $\overrightarrow{OR} = 2\mathbf{j}$, $\overrightarrow{OS} = \mathbf{i} + \mathbf{j} + 4\mathbf{k}$
  a Express PS in terms of $\mathbf{i}$, $\mathbf{j}$ and $\mathbf{k}$. **(3 marks)**
  b Show that the vector $-4\mathbf{j} + \mathbf{k}$ is perpendicular to OS and PS. **(3 marks)**
  c Find to the nearest degree the acute angle between the line SQ and the plane OSP. **(4 marks)**

13 The plane Π has vector equation

$$\mathbf{r} = \begin{pmatrix} 1 \\ 3 \\ 4 \end{pmatrix} + u\begin{pmatrix} 4 \\ 1 \\ 2 \end{pmatrix} + v\begin{pmatrix} 3 \\ 2 \\ -1 \end{pmatrix}$$

where u and v are parameters.
The line L has vector equation

$$\mathbf{r} = \begin{pmatrix} 2 \\ 1 \\ -3 \end{pmatrix} + t\begin{pmatrix} 2 \\ 3 \\ -4 \end{pmatrix}$$

where t is a parameter.
   a Show that L is parallel to Π. (4 marks)
   b Find the shortest distance between L and Π. (3 marks)

14 Planes Π_1 and Π_2 have equations given by
   Π_1: $\mathbf{r}.(2\mathbf{i} - \mathbf{j} + \mathbf{k}) = 0$
   Π_2: $\mathbf{r}.(\mathbf{i} + 5\mathbf{j} + 3\mathbf{k}) = 1$
   a Show that the point $A(2, -2, 3)$ lies in Π_2. (2 marks)
   b Show that Π_1 is perpendicular to Π_2. (2 marks)
   c Find, in vector form, an equation of the straight line through A which is perpendicular to Π_1. (4 marks)
   d Determine the coordinates of the point where this line meets Π_1. (3 marks)
   e Find the perpendicular distance of A from Π_1. (4 marks)
   f Find a vector equation of the plane through A parallel to Π_1. (3 marks)

15 The plane Π has equation $2x + y + 3z = 21$ and the origin is O.
The line l passes through the point $P(1, 2, 1)$ and is perpendicular to Π.
   a Find a vector equation of l. (3 marks)
The line l meets the plane Π at the point M.
   b Find the coordinates of M. (3 marks)
   c Find $\overrightarrow{OP} \times \overrightarrow{OM}$. (3 marks)
   d Hence, or otherwise, find the distance from P to the line OM, giving your answer in surd form. (3 marks)
The point Q is the reflection of P in Π.
   e Find the coordinates of Q. (3 marks)

16 With respect to a fixed origin O, the straight lines l_1 and l_2 are given by
   l_1: $\mathbf{r} = \mathbf{i} - \mathbf{j} + \lambda(2\mathbf{i} + \mathbf{j} - 2\mathbf{k})$
   l_2: $\mathbf{r} = \mathbf{i} + 2\mathbf{j} + 2\mathbf{k} + \mu(-3\mathbf{i} + 4\mathbf{k})$
where λ and μ are scalar parameters.
   a Show that the lines intersect. (4 marks)
   b Find the position vector of their point of intersection. (3 marks)
   c Find the cosine of the acute angle contained between the lines. (4 marks)
   d Find a vector equation of the plane containing the lines. (3 marks)

E 17 Relative to an origin O, the points A and B have position vectors **a** metres and **b** metres respectively, where

$$\mathbf{a} = 5\mathbf{i} + 2\mathbf{j}, \mathbf{b} = 2\mathbf{i} - \mathbf{j} - 3\mathbf{k}$$

The point C moves such that the volume of the tetrahedron $OABC$ is always 5 m³.
Determine Cartesian equations of the locus of the point C. **(6 marks)**

E 18 The lines L_1 and L_2 have equations $\mathbf{r} = \mathbf{a}_1 + s\mathbf{b}_1$ and $\mathbf{r} = \mathbf{a}_2 + t\mathbf{b}_2$ respectively, where

$\mathbf{a}_1 = 3\mathbf{i} - 3\mathbf{j} - 2\mathbf{k}$ $\mathbf{b}_1 = \mathbf{j} + 2\mathbf{k}$
$\mathbf{a}_2 = 8\mathbf{i} + 3\mathbf{j}$ $\mathbf{b}_2 = 5\mathbf{i} + 4\mathbf{j} - 2\mathbf{k}$

a Verify that the point P with position vector $3\mathbf{i} - \mathbf{j} + 2\mathbf{k}$ lies on both L_1 and L_2. **(2 marks)**
b Find $\mathbf{b}_1 \times \mathbf{b}_2$ **(3 marks)**
c Find a Cartesian equation of the plane containing L_1 and L_2. **(3 marks)**

The points with position vectors $\mathbf{a}_1$ and $\mathbf{a}_2$ are A_1 and A_2 respectively.

d By expressing $\overrightarrow{A_1P}$ and $\overrightarrow{A_2P}$ as multiples of $\mathbf{b}_1$ and $\mathbf{b}_2$ respectively, or otherwise, find the area of the triangle PA_1A_2. **(4 marks)**

E 19 With respect to the origin O the points A, B, C have position vectors

$$a(5\mathbf{i} - \mathbf{j} - 3\mathbf{k}), \quad a(-4\mathbf{i} + 4\mathbf{j} - \mathbf{k}), \quad a(5\mathbf{i} - 2\mathbf{j} + 11\mathbf{k})$$

respectively, where a is a non-zero constant.
Find
  a a vector equation for the line BC **(3 marks)**
  b a vector equation for the plane OAB, **(2 marks)**
  c the cosine of the acute angle between the lines OA and OB. **(2 marks)**

Obtain, in the form $\mathbf{r}.\mathbf{n} = p$, a vector equation for Π, the plane which passes through A and is perpendicular to BC.

Find Cartesian equations for
  d the plane Π **(3 marks)**
  e the line BC. **(3 marks)**

P 20 In a tetrahedron $ABCD$ the coordinates of the vertices B, C, D are respectively $(1, 2, 3)$, $(2, 3, 3)$, $(3, 2, 4)$. Find
  a the equation of the plane BCD,
  b the sine of the angle between BC and the plane $x + 2y + 3z = 4$
  c If AC and AD are perpendicular to BD and BC respectively and if $AB = \sqrt{26}$, find the coordinates of the two possible positions of A.

Challenge

The plane Π cuts the x-, y- and z-axes at the points $(p, 0, 0)$, $(0, q, 0)$ and $(0, 0, r)$ respectively. Given that the shortest distance between the plane and the origin is d, prove that

$$\frac{1}{p^2} + \frac{1}{q^2} + \frac{1}{r^2} = \frac{1}{d^2}$$

Summary of key points

1. The **scalar** (or **dot**) **product** of two vectors **a** and **b** is written as **a.b**, and defined as

   $$\mathbf{a.b} = |\mathbf{a}||\mathbf{b}|\cos\theta$$

   where θ is the angle between **a** and **b**.

2. The **vector** (or **cross**) **product** of the vectors **a** and **b** is defined as

   $$\mathbf{a} \times \mathbf{b} = |\mathbf{a}||\mathbf{b}|\sin\theta\,\hat{\mathbf{n}}$$

   where θ is the angle between **a** and **b**.

3. $\mathbf{b} \times \mathbf{a} = -\mathbf{a} \times \mathbf{b}$

4. If **i**, **j** and **k** are unit vectors along the x-, y- and z-axes respectively, then:
   - $\mathbf{i} \times \mathbf{i} = \mathbf{0}$
   - $\mathbf{j} \times \mathbf{j} = \mathbf{0}$
   - $\mathbf{k} \times \mathbf{k} = \mathbf{0}$
   - $\mathbf{i} \times \mathbf{j} = \mathbf{k}$ and $\mathbf{j} \times \mathbf{i} = -\mathbf{k}$
   - $\mathbf{j} \times \mathbf{k} = \mathbf{i}$ and $\mathbf{k} \times \mathbf{j} = -\mathbf{i}$
   - $\mathbf{k} \times \mathbf{i} = \mathbf{j}$ and $\mathbf{i} \times \mathbf{k} = -\mathbf{j}$

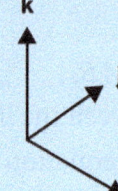

5. If $\mathbf{a} \times \mathbf{b} = \mathbf{0}$ then either $\mathbf{a} = \mathbf{0}$, $\mathbf{b} = \mathbf{0}$ or **a** and **b** are parallel.

6. $\mathbf{a} \times \mathbf{b} = (a_2 b_3 - a_3 b_2)\mathbf{i} + (a_3 b_1 - a_1 b_3)\mathbf{j} + (a_1 b_2 - a_2 b_1)\mathbf{k}$

   $$= \begin{vmatrix} \mathbf{i} & \mathbf{j} & \mathbf{k} \\ a_1 & a_2 & a_3 \\ b_1 & b_2 & b_3 \end{vmatrix} = \mathbf{i}\begin{vmatrix} a_2 & a_3 \\ b_2 & b_3 \end{vmatrix} - \mathbf{j}\begin{vmatrix} a_1 & a_3 \\ b_1 & b_3 \end{vmatrix} + \mathbf{k}\begin{vmatrix} a_1 & a_2 \\ b_1 & b_2 \end{vmatrix}$$

7. If A and B have position vectors **a** and **b** respectively, then
   Area of triangle $OAB = \frac{1}{2}|\mathbf{a} \times \mathbf{b}|$

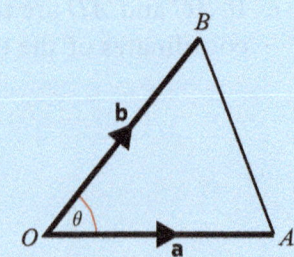

8 If A, B and C have position vectors $\mathbf{a}$, $\mathbf{b}$ and $\mathbf{c}$ respectively, then

Area of triangle $ABC = \dfrac{1}{2}\left|\overrightarrow{AB} \times \overrightarrow{AC}\right|$

$\qquad\qquad\qquad\qquad = \dfrac{1}{2}|(\mathbf{b} - \mathbf{a}) \times (\mathbf{c} - \mathbf{a})|$

$\qquad\qquad\qquad\qquad = \dfrac{1}{2}|(\mathbf{a} \times \mathbf{b}) + (\mathbf{b} \times \mathbf{c}) + (\mathbf{c} \times \mathbf{a})|$

9 If A and B have position vectors $\mathbf{a}$ and $\mathbf{b}$ respectively, then

Area of parallelogram $OABC = |\mathbf{a} \times \mathbf{b}|$

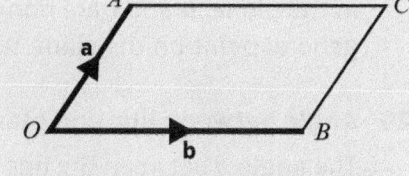

10 If A, B, C and D have position vectors $\mathbf{a}$, $\mathbf{b}$, $\mathbf{c}$ and $\mathbf{d}$ respectively, then

Area of parallelogram $ABCD = \left|\overrightarrow{AB} \times \overrightarrow{AD}\right|$

$\qquad\qquad\qquad\qquad\quad = |(\mathbf{b} - \mathbf{a}) \times (\mathbf{d} - \mathbf{a})|$

$\qquad\qquad\qquad\qquad\quad = |(\mathbf{a} \times \mathbf{b}) + (\mathbf{b} \times \mathbf{d}) + (\mathbf{d} \times \mathbf{a})|$

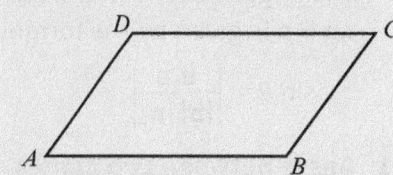

11 When $\mathbf{a} = (a_1\mathbf{i} + a_2\mathbf{j} + a_3\mathbf{k})$, $\mathbf{b} = (b_1\mathbf{i} + b_2\mathbf{j} + b_3\mathbf{k})$ and $\mathbf{c} = (c_1\mathbf{i} + c_2\mathbf{j} + c_3\mathbf{k})$,

$\mathbf{a}.(\mathbf{b} \times \mathbf{c}) = a_1(b_2c_3 - b_3c_2) + a_2(b_3c_1 - b_1c_3) + a_3(b_1c_2 - b_2c_1)$

This can also be written as

$$\mathbf{a}.(\mathbf{b} \times \mathbf{c}) = \begin{vmatrix} a_1 & a_2 & a_3 \\ b_1 & b_2 & b_3 \\ c_1 & c_2 & c_3 \end{vmatrix}$$

$\mathbf{a}.(\mathbf{b} \times \mathbf{c})$ is known as the **scalar triple product**.

12 $\mathbf{a}.(\mathbf{b} \times \mathbf{c}) = \mathbf{b}.(\mathbf{c} \times \mathbf{a}) = \mathbf{c}.(\mathbf{a} \times \mathbf{b})$

$\mathbf{a}.(\mathbf{a} \times \mathbf{p}) = \mathbf{a}.(\mathbf{p} \times \mathbf{a}) = 0$ for any vector $\mathbf{p}$.

13 If three sides of a parallelepiped are given by vectors $\mathbf{a}$, $\mathbf{b}$ and $\mathbf{c}$ as shown in the diagram, then the volume of the parallelepiped is given by $|\mathbf{a}.(\mathbf{b} \times \mathbf{c})|$.

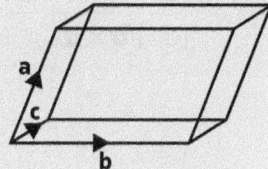

14 If three sides of a tetrahedron are given by vectors $\mathbf{a}$, $\mathbf{b}$ and $\mathbf{c}$ as shown in the diagram, then the volume of the tetrahedron is given by $\dfrac{1}{6}|\mathbf{a}.(\mathbf{b} \times \mathbf{c})|$

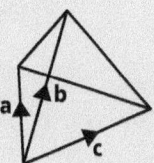

15 The vector equations of a line: $(\mathbf{r} - \mathbf{a}) \times \mathbf{b} = 0$ and $\mathbf{r} \times \mathbf{b} = \mathbf{a} \times \mathbf{b}$

16 The general Cartesian equation of a straight line is

$\dfrac{(x - x_1)}{l} = \dfrac{(y - y_1)}{m} = \dfrac{(z - z_1)}{n} = \lambda$

where the line passes through the point (x_1, y_1, z_1), has direction ratios $l : m : n$, and where λ is a parameter.

17 The scalar product form of the equation of a plane is **r.n** = **a.n** = p where **n** is normal to the plane, **a** is the position vector of a point in the plane and **r** is the position vector of a general point on the plane. p is a scalar constant.

18 The general Cartesian equation of a plane is $ax + by + cz + d = 0$

19 The vector equation of a plane is **r** = **a** + λ**b** + μ**c** where **a** is the position vector of a point in the plane, **b** and **c** are non-parallel vectors in the plane and **r** is the position vector of a general point on the plane. λ and μ are scalars and **b** and **c** are non-zero.

20 Angle between line and plane

The angle θ between the line with equation **r** = **a** + λ**b** and the plane with equation **r.n** = p is given by the formula

$$\sin\theta = \left|\frac{\mathbf{b.n}}{|\mathbf{b}||\mathbf{n}|}\right|$$

21 Angle between two planes

The angle θ between the plane with equation $\mathbf{r.n}_1 = p_1$ and the plane with equation $\mathbf{r.n}_2 = p_2$ is given by the formula

$$\cos\theta = \left|\frac{\mathbf{n}_1.\mathbf{n}_2}{|\mathbf{n}_1||\mathbf{n}_2|}\right|$$

22 When the equation of plane Π is written in the form **r.n̂** = d, where **n̂** is a unit vector perpendicular to Π, then d is the length of the perpendicular from the origin to the plane.

23 The shortest distance between the two skew lines with equations

r = **a** + λ**b** and **r** = **c** + μ**d**, where λ and μ are scalars, is given by the formula

$$\left|\frac{(\mathbf{a}-\mathbf{c}).(\mathbf{b}\times\mathbf{d})}{|(\mathbf{b}\times\mathbf{d})|}\right|$$

6 FURTHER MATRIX ALGEBRA

6.1
6.2
6.3
6.4
6.5
6.6
6.7
6.8

Learning objectives

After completing this chapter you should be able to:
- find transposes, determinants and inverses of 3 × 3 matrices → pages 138–152
- represent linear transformations by 2 × 2 and 3 × 3 matrices → pages 152–160
- use inverse matrices to reverse the effect of a linear transformation → pages 160–165
- find eigenvalues and eigenvectors of 2 × 2 and 3 × 3 matrices → pages 165–174
- reduce symmetric matrices to diagonal form. → pages 175–185

Prior knowledge check

1 Find the determinant of $\begin{pmatrix} 3 & 4 \\ 1 & 2 \end{pmatrix}$
 ← Further Pure 1 Section 5.3

2 $\mathbf{A} = \begin{pmatrix} -3 & k \\ 2 & -4 \end{pmatrix}$

 Find the value of k such that the matrix $\mathbf{A}$ is singular. ← Further Pure 1 Section 5.4

3 The matrix $\begin{pmatrix} -1 & 3 \\ 2 & -2 \end{pmatrix}$ represents a transformation.

 Show that the line with equation $3y - 2x = 0$ is invariant under this transformation.
 ← Further Pure 1 Section 6.5

This chapter extends the work on matrices which you learnt in Further Pure 1. Matrix algebra is used in many branches of mathematics, especially when large amounts of data are involved, as the rules for manipulating matrices can be efficiently implemented on computers. Eigenvalues and eigenvectors have many scientific applications. For example, they can be used in analysing population trends, atomic structures, stress and vibrations in mechanics and glacial sediments in geology.

6.1 Transposing a matrix

- Given a matrix **A**, you form the **transpose** of the matrix, $\mathbf{A}^T$, by interchanging the rows and the columns of **A**. You take the first row of **A** and write it as the first column of $\mathbf{A}^T$, you take the second row of **A** and write it as the second column of $\mathbf{A}^T$, and so on.

  If $\mathbf{A} = \begin{pmatrix} 2 & 4 \\ 0 & -3 \\ 1 & 5 \end{pmatrix}$, then $\mathbf{A}^T = \begin{pmatrix} 2 & 0 & 1 \\ 4 & -3 & 5 \end{pmatrix}$

- The transpose of a matrix of dimension $n \times m$ is a matrix of dimension $m \times n$.

- The transpose of a square matrix is another square matrix with the same dimensions. For example, the transpose of a 2×2 matrix is another 2×2 matrix.

- If $\mathbf{A} = \mathbf{A}^T$, the matrix **A** is **symmetric**.

- A 2×2 matrix is symmetric if it has the form $\begin{pmatrix} a & b \\ b & c \end{pmatrix}$. Interchanging the rows and columns leaves the matrix unchanged.

- A 3×3 matrix is symmetric if it has the form $\begin{pmatrix} a & b & c \\ b & d & e \\ c & e & f \end{pmatrix}$. Interchanging the rows and columns leaves the matrix unchanged.

- The identity 3×3 matrix is $\mathbf{I} = \begin{pmatrix} 1 & 0 & 0 \\ 0 & 1 & 0 \\ 0 & 0 & 1 \end{pmatrix}$. When a matrix or vector is multiplied by the identity matrix, the matrix or vector is unchanged.

- The zero 3×3 matrix is $\mathbf{0} = \begin{pmatrix} 0 & 0 & 0 \\ 0 & 0 & 0 \\ 0 & 0 & 0 \end{pmatrix}$

Example 1   SKILLS   REASONING, ARGUMENTATION

Write down the transposes of the following matrices. State which of the matrices is symmetric.

a $\mathbf{A} = \begin{pmatrix} 4 & 3 \\ 2 & 0 \end{pmatrix}$
b $\mathbf{B} = \begin{pmatrix} 7 & 0 & 0 \\ 3 & -1 & 3 \end{pmatrix}$
c $\mathbf{C} = \begin{pmatrix} 1 & 0 & -1 \\ 0 & 2 & 0 \\ -1 & 0 & 3 \end{pmatrix}$
d $\mathbf{D} = \begin{pmatrix} 4 & -7 & 1 \\ 0 & 3 & -5 \\ 2 & 4 & -2 \end{pmatrix}$

a $\mathbf{A}^T = \begin{pmatrix} 4 & 2 \\ 3 & 0 \end{pmatrix}$ — The first row of **A**, (4, 3) becomes the first column of $\mathbf{A}^T$, $\begin{pmatrix} 4 \\ 3 \end{pmatrix}$

The second row of **A**, (2, 0) becomes the second column of $\mathbf{A}^T$, $\begin{pmatrix} 2 \\ 0 \end{pmatrix}$

b $\mathbf{B}^T = \begin{pmatrix} 7 & 3 \\ 0 & -1 \\ 0 & 3 \end{pmatrix}$ — The first row of **B**, (7, 0, 0) becomes the first column of $\mathbf{B}^T$, $\begin{pmatrix} 7 \\ 0 \\ 0 \end{pmatrix}$

The second row of **B**, (3, −1, 3) becomes the second column of $\mathbf{B}^T$, $\begin{pmatrix} 3 \\ -1 \\ 3 \end{pmatrix}$

c $C^T = \begin{pmatrix} 1 & 0 & -1 \\ 0 & 2 & 0 \\ -1 & 0 & 3 \end{pmatrix}$, the matrix **C** is symmetric.

When you interchange the rows and columns of **C**, you obtain the same matrix. The matrix **C** is symmetric.

d $D^T = \begin{pmatrix} 4 & 0 & 2 \\ -7 & 3 & 4 \\ 1 & -5 & -2 \end{pmatrix}$

The first row of **D** becomes the first column of D^T. The second row of **D** becomes the second column of D^T, and so on.

Example 2

The matrix **M** is given by

$$M = \begin{pmatrix} 3 & 2 \\ 1 & -2 \\ 4 & 0 \end{pmatrix}$$

Find **a** M^T **b** $M^T M$ **c** $M M^T$

In each case state the dimension of your answer.

a $M^T = \begin{pmatrix} 3 & 1 & 4 \\ 2 & -2 & 0 \end{pmatrix}$

The dimension of M^T is 2 × 3.

The first row of **M** becomes the first column of M^T. The second row of **M** becomes the second column of M^T and the third row of **M** becomes the third column of M^T.

b $M^T M = \begin{pmatrix} 3 & 1 & 4 \\ 2 & -2 & 0 \end{pmatrix} \begin{pmatrix} 3 & 2 \\ 1 & -2 \\ 4 & 0 \end{pmatrix}$

$= \begin{pmatrix} 26 & 4 \\ 4 & 8 \end{pmatrix}$

The dimension of $M^T M$ is 2 × 2.

First row times the first column
$3 \times 3 + 1 \times 1 + 4 \times 4 = 26$

First row times second column
$3 \times 2 + 1 \times (-2) + 4 \times 0 = 4$

Second row times second column
$2 \times 2 + (-2) \times (-2) + 0 \times 0 = 8$

Second row times first column
$2 \times 3 + (-2) \times 1 + 0 \times 4 = 4$

c $MM^T = \begin{pmatrix} 3 & 2 \\ 1 & -2 \\ 4 & 0 \end{pmatrix} \begin{pmatrix} 3 & 1 & 4 \\ 2 & -2 & 0 \end{pmatrix}$

$= \begin{pmatrix} 3 \times 3 + 2 \times 2 & 3 \times 1 + 2 \times (-2) & 3 \times 4 + 2 \times 0 \\ 1 \times 3 + (-2) \times 2 & 1 \times 1 + (-2) \times (-2) & 1 \times 4 + (-2) \times 0 \\ 4 \times 3 + 0 \times 2 & 4 \times 1 + 0 \times (-2) & 4 \times 4 + 0 \times 0 \end{pmatrix}$

$= \begin{pmatrix} 13 & -1 & 12 \\ -1 & 5 & 4 \\ 12 & 4 & 16 \end{pmatrix}$

For example, the element in the second row and third column is the second row of **M** multiplied by the third column of M^T.

The dimension of MM^T is 3 × 3.

Notation You learnt in Further Pure 1 how to multiply matrices and that, in general, matrix multiplication is not **commutative**; in other words, that the matrix product **AB** is not equal to **BA**. Here $M^T M$ and MM^T are different and have different dimensions. You will notice, however, that both $M^T M$ and MM^T are symmetric. This property of a matrix and its transpose is always true.

Example 3 — SKILLS: PROBLEM-SOLVING

The matrix **A** is given by $\mathbf{A} = \begin{pmatrix} 2 & 2 & -1 \\ -1 & 2 & a \\ 2 & b & 2 \end{pmatrix}$, where a and b are constants.

Given that $\mathbf{AA}^T = k\mathbf{I}$, where k is a constant, find the values of k, a and b.

$\mathbf{A}^T = \begin{pmatrix} 2 & -1 & 2 \\ 2 & 2 & b \\ -1 & a & 2 \end{pmatrix}$

- You take the first row of **A** and write it as the first column of $\mathbf{A}^T$, you take the second row of **A** and write it as the second column of $\mathbf{A}^T$, and so on.

$\mathbf{AA}^T = \begin{pmatrix} 2 & 2 & -1 \\ -1 & 2 & a \\ 2 & b & 2 \end{pmatrix}\begin{pmatrix} 2 & -1 & 2 \\ 2 & 2 & b \\ -1 & a & 2 \end{pmatrix}$

$= \begin{pmatrix} 4+4+1 & -2+4-a & 4+2b-2 \\ \cdots & \cdots & \cdots \\ \cdots & \cdots & \cdots \end{pmatrix}$

- You have to find three unknowns and, in this case, three elements of $\mathbf{AA}^T$ will give you three equations to find the unknowns. You need not work out all nine elements of the matrix.

$k\mathbf{I} = k\begin{pmatrix} 1 & 0 & 0 \\ 0 & 1 & 0 \\ 0 & 0 & 1 \end{pmatrix} = \begin{pmatrix} k & 0 & 0 \\ 0 & k & 0 \\ 0 & 0 & k \end{pmatrix}$

- You have to find three unknowns and, in this case, three elements of $\mathbf{AA}^T$ will give you three equations to find the unknowns. You need not work out all nine elements of the matrix.

As $\mathbf{AA}^T = k\mathbf{I}$

$\begin{pmatrix} 9 & 2-a & 2b+2 \\ \cdots & \cdots & \cdots \\ \cdots & \cdots & \cdots \end{pmatrix} = \begin{pmatrix} k & 0 & 0 \\ 0 & k & 0 \\ 0 & 0 & k \end{pmatrix}$

- For two 3 × 3 matrices to be equal, all nine of the elements in the matrices must be individually equal. In principle, there are nine equations available but you only need three of them to solve this problem.

Equating the first element in the first row

$k = 9$

Equating the second element in the first row

$2 - a = 0 \Rightarrow a = 2$

- The second element in the first row of the left-hand matrix is $2 - a$. The second element in the first row of the right-hand matrix is 0. So $2 - a = 0$, which is an easy equation for a.

Equating the third element in the first row

$2b + 2 = 0 \Rightarrow b = -1$

$k = 9$, $a = 2$ and $b = -1$

Example 4

The matrix $\mathbf{A} = \begin{pmatrix} 3 & 0 & 2 \\ -2 & 1 & 0 \\ 1 & -2 & 4 \end{pmatrix}$ and the matrix $\mathbf{B} = \begin{pmatrix} 1 & 0 & 4 \\ 0 & 2 & 0 \\ 2 & 0 & 1 \end{pmatrix}$

a Write down $\mathbf{A}^T$ and $\mathbf{B}^T$.

b Find $\mathbf{AB}$.

c Verify that $(\mathbf{AB})^T = \mathbf{B}^T\mathbf{A}^T$

FURTHER MATRIX ALGEBRA — CHAPTER 6

a $A^T = \begin{pmatrix} 3 & -2 & 1 \\ 0 & 1 & -2 \\ 2 & 0 & 4 \end{pmatrix}$

$B^T = \begin{pmatrix} 1 & 0 & 2 \\ 0 & 2 & 0 \\ 4 & 0 & 1 \end{pmatrix}$

> You form the transposes by interchanging the rows and columns of the matrices

b $AB = \begin{pmatrix} 3 & 0 & 2 \\ -2 & 1 & 0 \\ 1 & -2 & 4 \end{pmatrix}\begin{pmatrix} 1 & 0 & 4 \\ 0 & 2 & 0 \\ 2 & 0 & 1 \end{pmatrix}$

$= \begin{pmatrix} 7 & 0 & 14 \\ -2 & 2 & -8 \\ 9 & -4 & 8 \end{pmatrix}$

> You find the element in the first row and first column by multiplying the first row by the first column.
> $3 \times 1 + 0 \times 0 + 2 \times 2 = 7$
> You repeat this process throughout the matrix. For example, you find the element in the third row and the second column by multiplying the third row by the second column.
> $1 \times 0 + (-2) \times 2 + 4 \times 0 = -4$

c $B^T A^T = \begin{pmatrix} 1 & 0 & 2 \\ 0 & 2 & 0 \\ 4 & 0 & 1 \end{pmatrix}\begin{pmatrix} 3 & -2 & 1 \\ 0 & 1 & -2 \\ 2 & 0 & 4 \end{pmatrix}$

$= \begin{pmatrix} 7 & -2 & 9 \\ 0 & 2 & -4 \\ 14 & -8 & 8 \end{pmatrix}$

Using part **b**

$(AB)^T = \begin{pmatrix} 7 & -2 & 9 \\ 0 & 2 & -4 \\ 14 & -8 & 8 \end{pmatrix}$

Hence $(AB)^T = B^T A^T$, as required.

> To verify the relation $(AB)^T = B^T A^T$, you show that it is satisfied by the specific matrices given in the question. You work out $B^T A^T$ and $(AB)^T$ separately and show that they are equal.
>
> This relation is true for any matrices **A** and **B** that can be multiplied together.

If **A** and **B** are matrices with dimensions $n \times m$ and $m \times p$, then $(AB)^T = B^T A^T$

Exercise 6A   SKILLS   PROBLEM-SOLVING

1 Write down the transposes of the following matrices.
In each case give the dimensions of the transposed matrix.

a $\begin{pmatrix} 3 & 1 & 2 \\ -1 & 0 & 4 \end{pmatrix}$ **b** $\begin{pmatrix} 0 & 2 \\ -2 & 0 \end{pmatrix}$ **c** $\begin{pmatrix} 0 & 2 & -1 \\ -2 & 0 & 3 \\ 1 & -3 & 0 \end{pmatrix}$ **d** $\begin{pmatrix} 1 \\ 2 \\ 4 \end{pmatrix}$

2 The matrix $A = \begin{pmatrix} 2 & 4 \\ -3 & 6 \end{pmatrix}$

  a Write down A^T.   **b** Find AA^T.   **c** Find $A^T A$.

3 The matrix $A = \begin{pmatrix} 3 & 2 \\ -2 & 1 \end{pmatrix}$ and the matrix $B = \begin{pmatrix} 1 & 6 \\ 0 & -4 \end{pmatrix}$

  a Find **BA**.   **b** Verify that $A^T B^T = (BA)^T$

4 The matrix $A = \begin{pmatrix} 1 & -4 & 8 \\ 4 & -7 & -4 \\ 8 & 4 & 1 \end{pmatrix}$

  a Write down A^T.   **b** Show that $AA^T = 81I$

5 The matrix $\mathbf{A} = \begin{pmatrix} 0 & 3 & 5 \\ -3 & 0 & -1 \\ -5 & 1 & 0 \end{pmatrix}$ and the matrix $\mathbf{B} = \begin{pmatrix} -4 & 1 & -1 \\ 1 & 5 & 2 \\ -3 & 0 & 3 \end{pmatrix}$

Given that $\mathbf{C} = \mathbf{AB}$

a find $\mathbf{C}$

b verify that the matrix $\mathbf{C}$ is symmetric.

6 The matrix $\mathbf{A} = \begin{pmatrix} 0 & 3 & 5 \\ 2 & 0 & -1 \\ 1 & 1 & 0 \end{pmatrix}$ and the matrix $\mathbf{B} = \begin{pmatrix} 1 & 1 & -1 \\ 0 & 1 & 0 \\ -1 & 0 & 3 \end{pmatrix}$

a Find $\mathbf{AB}$.

b Verify that $(\mathbf{AB})^\mathsf{T} = \mathbf{B}^\mathsf{T}\mathbf{A}^\mathsf{T}$

6.2 The determinant of a 3 × 3 matrix

- In Further Pure 1, you learnt that the determinant of a 2 × 2 matrix $\mathbf{A}$, written det ($\mathbf{A}$) or $|\mathbf{A}|$, is given by the formula

$$\det(\mathbf{A}) = \begin{vmatrix} a & b \\ c & d \end{vmatrix} = ad - bc$$

- You find the determinant of a 3 × 3 matrix by reducing the 3 × 3 determinant to 2 × 2 determinants using the formula

$$\begin{vmatrix} a & b & c \\ d & e & f \\ g & h & i \end{vmatrix} = a\begin{vmatrix} e & f \\ h & i \end{vmatrix} - b\begin{vmatrix} d & f \\ g & i \end{vmatrix} + c\begin{vmatrix} d & e \\ g & h \end{vmatrix}$$

Notice that, in the middle expression in this formula, there is a negative sign before the b.

The 2 × 2 determinant associated with the element a in the first row of the determinant is found by crossing out the row and column in which a lies.

$$\begin{vmatrix} \cancel{a} & \cancel{b} & \cancel{c} \\ \cancel{d} & e & f \\ \cancel{g} & h & i \end{vmatrix} \rightarrow \begin{vmatrix} e & f \\ h & i \end{vmatrix}$$

The 2 × 2 determinant associated with the element b in the first row of the determinant is found by crossing out the row and column in which b lies.

$$\begin{vmatrix} \cancel{a} & \cancel{b} & \cancel{c} \\ d & \cancel{e} & f \\ g & \cancel{h} & i \end{vmatrix} \rightarrow \begin{vmatrix} d & f \\ g & i \end{vmatrix}$$

The 2 × 2 determinant associated with the element c in the first row of the determinant is found by crossing out the row and column in which c lies.

$$\begin{vmatrix} \cancel{a} & \cancel{b} & \cancel{c} \\ d & e & \cancel{f} \\ g & h & \cancel{i} \end{vmatrix} \rightarrow \begin{vmatrix} d & e \\ g & h \end{vmatrix}$$

- As with 2 × 2 matrices, with 3 × 3 matrices,

 if det ($\mathbf{A}$) = 0, then $\mathbf{A}$ is a **singular** matrix,

 if det ($\mathbf{A}$) ≠ 0, then $\mathbf{A}$ is a **non-singular** matrix.

Example 5

Find the value of $\begin{vmatrix} 1 & 2 & 4 \\ 3 & 2 & 1 \\ -1 & 4 & 3 \end{vmatrix}$

$$\begin{vmatrix} 1 & 2 & 4 \\ 3 & 2 & 1 \\ -1 & 4 & 3 \end{vmatrix} = 1\begin{vmatrix} 2 & 1 \\ 4 & 3 \end{vmatrix} - 2\begin{vmatrix} 3 & 1 \\ -1 & 3 \end{vmatrix} + 4\begin{vmatrix} 3 & 2 \\ -1 & 4 \end{vmatrix}$$

$$= 1(6 - 4) - 2(9 + 1) + 4(12 + 2)$$
$$= 1 \times 2 - 2 \times 10 + 4 \times 14$$
$$= 2 - 20 + 56 = 38$$

This determinant is formed by crossing out the row and column in which 1 lies.

$\begin{vmatrix} \cancel{1} & \cancel{2} & \cancel{4} \\ 3 & 2 & 1 \\ -1 & 4 & 3 \end{vmatrix}$

This determinant is formed by crossing out the row and column in which 2 lies.

$\begin{vmatrix} \cancel{1} & \cancel{2} & \cancel{4} \\ 3 & \cancel{2} & 1 \\ -1 & \cancel{4} & 3 \end{vmatrix}$

This determinant is formed by crossing out the row and column in which 4 lies.

$\begin{vmatrix} \cancel{1} & \cancel{2} & \cancel{4} \\ 3 & 2 & \cancel{1} \\ -1 & 4 & \cancel{3} \end{vmatrix}$

Each 2×2 determinant is found using the formula $\begin{vmatrix} a & b \\ c & d \end{vmatrix} = ad - bc$

So $\begin{vmatrix} 2 & 1 \\ 4 & 3 \end{vmatrix} = 2 \times 3 - 1 \times 4 = 6 - 4 = 2$

$\begin{vmatrix} 3 & 1 \\ -1 & 3 \end{vmatrix} = 3 \times 3 - 1 \times (-1) = 9 + 1 = 10$

and $\begin{vmatrix} 3 & 2 \\ -1 & 4 \end{vmatrix} = 3 \times 4 - 2 \times (-1) = 12 + 2 = 14$

Example 6 SKILLS ANALYSIS

The matrix $\mathbf{A} = \begin{pmatrix} 2 & -1 & 3 \\ 8 & -2 & 7 \\ 4 & 2 & -4 \end{pmatrix}$

Show that $\mathbf{A}$ is singular.

$$\begin{vmatrix} 2 & -1 & 3 \\ 8 & -2 & 7 \\ 4 & 2 & -4 \end{vmatrix} = 2\begin{vmatrix} -2 & 7 \\ 2 & -4 \end{vmatrix} - (-1)\begin{vmatrix} 8 & 7 \\ 4 & -4 \end{vmatrix}$$
$$+ 3\begin{vmatrix} 8 & -2 \\ 4 & 2 \end{vmatrix}$$
$$= 2(8 - 14) + 1(-32 - 28) + 3(16 + 8)$$
$$= 2 \times (-6) + 1 \times (-60) + 3 \times 24$$
$$= -12 - 60 + 72 = 0$$

Hence the matrix $\mathbf{A}$ is singular.

Each 2×2 determinant is formed by crossing out the row and column in which the element in the first row lies.

In evaluating the 2×2 determinants using the formula $\begin{vmatrix} a & b \\ c & d \end{vmatrix} = ad - bc$, you need to be careful with the signs.

As det $(\mathbf{A}) = 0$, you conclude that $\mathbf{A}$ is singular.

Example 7

The matrix $\mathbf{A} = \begin{pmatrix} 3 & k & 0 \\ -2 & 1 & 2 \\ 5 & 0 & k+3 \end{pmatrix}$, where k is a constant.

a Find det (**A**) in terms of k.

Given that **A** is singular,

b find the possible values of k.

a $\begin{vmatrix} 3 & k & 0 \\ -2 & 1 & 2 \\ 5 & 0 & k+3 \end{vmatrix} = 3\begin{vmatrix} 1 & 2 \\ 0 & k+3 \end{vmatrix} - k\begin{vmatrix} -2 & 2 \\ 5 & k+3 \end{vmatrix} + 0\begin{vmatrix} -2 & 1 \\ 5 & 0 \end{vmatrix}$

$= 3(k+2) - k(-2(k+3) - 10)$

$= 3k + 9 + 2k^2 + 16k$

$= 2k^2 + 19k + 9$

*Part **b** will require you to solve det (**A**) = 0, so multiply this expression out, collect together terms and express the result as a quadratic.*

b As **A** is singular

$2k^2 + 19k + 9 = 0$

$(2k+1)(k+9) = 0$

$k = -\frac{1}{2}, -9$

*As **A** is singular, its determinant is 0. This gives a quadratic equation, which you solve, giving two possible values of k.*

Example 8

The matrix $\mathbf{A} = \begin{pmatrix} 2 & 1 & -1 \\ 1 & 0 & 4 \\ -4 & 2 & 1 \end{pmatrix}$ and the matrix $\mathbf{B} = \begin{pmatrix} 3 & 1 & 2 \\ k & 4 & 5 \\ 0 & 2 & 3 \end{pmatrix}$, where k is a constant.

a Evaluate the determinant of **A**.

Given that the determinant of **B** is 2,

b find the value of k.

Using the value of k found in part **b**,

c find **AB**

d verify that det (**AB**) = det (**A**) det (**B**)

a $\begin{vmatrix} 2 & 1 & -1 \\ 1 & 0 & 4 \\ -4 & 2 & 1 \end{vmatrix} = 2\begin{vmatrix} 0 & 4 \\ 2 & 1 \end{vmatrix} - 1\begin{vmatrix} 1 & 4 \\ -4 & 1 \end{vmatrix} + (-1)\begin{vmatrix} 1 & 0 \\ -4 & 2 \end{vmatrix}$

$= 2(0-8) - 1(1+16) + (-1)(2-0)$

$= -16 - 17 - 2 = -35$

b $\begin{vmatrix} 3 & 1 & 2 \\ k & 4 & 5 \\ 0 & 2 & 3 \end{vmatrix} = 3\begin{vmatrix} 4 & 5 \\ 2 & 3 \end{vmatrix} - 1\begin{vmatrix} k & 5 \\ 0 & 3 \end{vmatrix} + 2\begin{vmatrix} k & 4 \\ 0 & 2 \end{vmatrix}$

$= 3(12 - 10) - 1(3k - 0) + 2(2k - 0)$

$= 6 - 3k + 4k = k + 6$

$\Rightarrow k + 6 = 2 \Rightarrow k = -4$

As you are given the determinant is 2, you find an expression for det(**B**) in terms of k and then **equate** this expression to 2. You then solve the resulting linear equation to find k.

c $\mathbf{AB} = \begin{pmatrix} 2 & 1 & -1 \\ 1 & 0 & 4 \\ -4 & 2 & 1 \end{pmatrix} \begin{pmatrix} 3 & 1 & 2 \\ -4 & 4 & 5 \\ 0 & 2 & 3 \end{pmatrix}$

$= \begin{pmatrix} 6 - 4 + 0 & 2 + 4 - 2 & 4 + 5 - 3 \\ 3 + 0 + 0 & 1 + 0 + 8 & 2 + 0 + 12 \\ -12 - 8 + 0 & -4 + 8 + 2 & -8 + 10 + 3 \end{pmatrix}$

$= \begin{pmatrix} 2 & 4 & 6 \\ 3 & 9 & 14 \\ -20 & 6 & 5 \end{pmatrix}$

With $k = -4$, $\mathbf{B} = \begin{pmatrix} 2 & 1 & 2 \\ -4 & 4 & 5 \\ 0 & 2 & 3 \end{pmatrix}$

d $\det(\mathbf{AB}) = \begin{vmatrix} 2 & 4 & 6 \\ 3 & 9 & 14 \\ -20 & 6 & 5 \end{vmatrix}$

$= 2\begin{vmatrix} 9 & 14 \\ 6 & 5 \end{vmatrix} - 4\begin{vmatrix} 3 & 14 \\ -20 & 5 \end{vmatrix} + 6\begin{vmatrix} 3 & 9 \\ -20 & 6 \end{vmatrix}$

$= 2(45 - 84) - 4(15 + 280) + 6(18 + 180)$

$= -78 - 1180 + 1188$

$= -70$

$= (-35) \times 2$

$= \det(\mathbf{A}) \det(\mathbf{B})$

To verify the formula, you calculate the determinant of **AB** and show that it is equal to the determinant of **A** multiplied by the determinant of **B**. You worked out the determinant of **A** in part **a** and you were given the determinant of **B** for part **b**.

The formula $\det(\mathbf{AB}) = \det(\mathbf{A})\det(\mathbf{B})$ is true for square matrices of any size.

Exercise 6B SKILLS ANALYSIS

1 Find the values of the determinants.

a $\begin{vmatrix} 1 & 0 & 0 \\ 0 & 2 & 0 \\ 0 & 0 & 3 \end{vmatrix}$ **b** $\begin{vmatrix} 0 & 4 & 0 \\ 5 & -2 & 3 \\ 2 & 1 & 4 \end{vmatrix}$ **c** $\begin{vmatrix} 1 & 0 & 1 \\ 2 & 4 & 1 \\ 3 & 5 & 2 \end{vmatrix}$ **d** $\begin{vmatrix} 2 & -3 & 4 \\ 2 & 2 & 2 \\ 5 & 5 & 5 \end{vmatrix}$

2 Find the values of the determinants.

a $\begin{vmatrix} 4 & 3 & -1 \\ 2 & -2 & 0 \\ 0 & 4 & -2 \end{vmatrix}$ **b** $\begin{vmatrix} 3 & -2 & 1 \\ 4 & 1 & -3 \\ 7 & 2 & -4 \end{vmatrix}$ **c** $\begin{vmatrix} 5 & -2 & -3 \\ 6 & 4 & 2 \\ -2 & -4 & -3 \end{vmatrix}$

3 The matrix $\mathbf{A} = \begin{pmatrix} 2 & 1 & -4 \\ 2k + 1 & 3 & k \\ 1 & 0 & 1 \end{pmatrix}$

Given that **A** is singular, find the value of the constant k.

4 The matrix $\mathbf{A} = \begin{pmatrix} 2 & -1 & 3 \\ k & 2 & 4 \\ -2 & 1 & k+3 \end{pmatrix}$, where k is a constant.

Given that the determinant of $\mathbf{A}$ is 8, find the possible values of k.

(E) 5 The matrix $\mathbf{A} = \begin{pmatrix} 2 & 5 & 3 \\ -2 & 0 & 4 \\ 3 & 10 & 8 \end{pmatrix}$ and the matrix $\mathbf{B} = \begin{pmatrix} 1 & 1 & 0 \\ 1 & 2 & 2 \\ 0 & -2 & -1 \end{pmatrix}$

    a Show that $\mathbf{A}$ is singular.   (3 marks)

    b Find $\mathbf{AB}$.   (3 marks)

    c Show that $\mathbf{AB}$ is also singular.   (3 marks)

(E) 6 The matrix $\mathbf{A} = \begin{pmatrix} 4 & 5 & -2 \\ 2 & -3 & 2 \\ 2 & -4 & 3 \end{pmatrix}$

    a Find det ($\mathbf{A}$).   (3 marks)

    b Write down $\mathbf{A}^T$ (1 mark)

    c Verify that det ($\mathbf{A}^T$) = det ($\mathbf{A}$).   (3 marks)

7 **a** Show that, for all values of a, b and c, the matrix $\begin{pmatrix} 0 & a & -b \\ -a & 0 & c \\ b & -c & 0 \end{pmatrix}$ is singular.

  b Show that, for all real values of x, the matrix $\begin{pmatrix} 2 & -2 & 4 \\ 3 & x & -2 \\ -1 & 3 & x \end{pmatrix}$ is non-singular.

8 Find all the values of x for which the matrix $\begin{pmatrix} x-3 & -2 & 0 \\ 1 & x & -2 \\ -2 & -1 & x+1 \end{pmatrix}$ is singular.

6.3 The inverse of a 3 × 3 matrix where it exists

- In Further Pure 1 you learnt how to find the inverse of a 2 × 2 matrix. The rule for inverting a 2 × 2 matrix is that if $\mathbf{A} = \begin{pmatrix} a & b \\ c & d \end{pmatrix}$, then $\mathbf{A}^{-1} = \dfrac{1}{\det(\mathbf{A})} \begin{pmatrix} d & -b \\ -c & a \end{pmatrix}$

- If det ($\mathbf{A}$) = 0, the matrix $\mathbf{A}$ is singular and, as you cannot divide by zero, you cannot find an inverse. Similarly, with 3 × 3 matrices, you cannot find the inverse of a matrix if its determinant is zero.

- If the inverse of a matrix $\mathbf{A}$ is $\mathbf{A}^{-1}$, then $\mathbf{AA}^{-1} = \mathbf{A}^{-1}\mathbf{A} = 1$

- The **minor** of an element of a 3 × 3 matrix is the determinant of the elements which remain when the row and the column containing the element are crossed out.

Example 9

Find the minors of the elements 5 and 7 in the matrix

$$\begin{pmatrix} 5 & 0 & 2 \\ -1 & 8 & 1 \\ 6 & 7 & 3 \end{pmatrix}$$

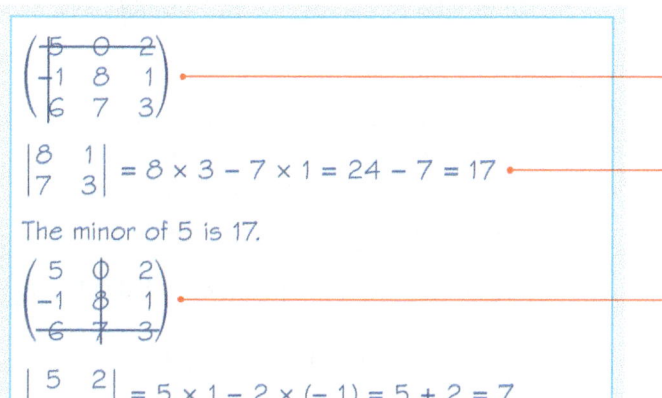

$\begin{vmatrix} 8 & 1 \\ 7 & 3 \end{vmatrix} = 8 \times 3 - 7 \times 1 = 24 - 7 = 17$

The minor of 5 is 17.

$\begin{vmatrix} 5 & 2 \\ -1 & 1 \end{vmatrix} = 5 \times 1 - 2 \times (-1) = 5 + 2 = 7$

The minor of 7 is 7.

> To find the minor of 5, you begin by crossing out the row and the column containing 5.
>
> To find the minor of 7, you begin by crossing out the row and the column containing 7.
>
> When you have crossed out the row and the column containing 5, you are left with the elements $\begin{pmatrix} 8 & 1 \\ 7 & 3 \end{pmatrix}$ and you evaluate the determinant of the matrix formed by these elements.
>
> When you have crossed out the row and the column containing 7, you are left with the elements $\begin{pmatrix} 5 & 2 \\ -1 & 1 \end{pmatrix}$ and you evaluate the determinant of the matrix formed by these elements.

- Finding the inverse of a 3×3 matrix **A** usually consists of the following 5 steps.

    Step 1   Find the determinant of **A**, det(**A**).

    Step 2   Form the matrix of the minors of **A**. In this chapter, the symbol **M** is used for the matrix of the minors unless this causes confusion with another matrix in the question.

    In forming the matrix of minors **M**, each of the nine elements of the matrix **A** is replaced by its minor.

    Step 3   From the matrix of minors, form the matrix of **cofactors** by changing the signs of some elements of the matrix of minors according to the **rule of alternating signs** illustrated by this pattern:

    $$\begin{pmatrix} + & - & + \\ - & + & - \\ + & - & + \end{pmatrix}$$

    You leave the elements of the matrix of minors corresponding to the + signs in this pattern unchanged. You change the signs of the elements corresponding to the − signs. A cofactor is a minor with its appropriate sign.

    In this chapter, the symbol **C** is used for the matrix of the cofactors unless this causes confusion with another matrix in the question.

    Step 4   Write down the transpose, $\mathbf{C}^T$, of the matrix of cofactors.

    Step 5   The inverse of the matrix **A** is given by the formula

    $$\mathbf{A}^{-1} = \frac{1}{\det(\mathbf{A})} \mathbf{C}^T$$

    Each element of the matrix $\mathbf{C}^T$ is divided by the determinant of **A**.

Example 10 — SKILLS: ANALYSIS

The matrix $\mathbf{A} = \begin{pmatrix} 1 & 3 & 1 \\ 0 & 4 & 1 \\ 2 & -1 & 0 \end{pmatrix}$. Find $\mathbf{A}^{-1}$.

Step 1

$$\det(\mathbf{A}) = 1\begin{vmatrix} 4 & 1 \\ -1 & 0 \end{vmatrix} - 3\begin{vmatrix} 0 & 1 \\ 2 & 0 \end{vmatrix} + 1\begin{vmatrix} 0 & 4 \\ 2 & -1 \end{vmatrix}$$

$$= 1(0 + 1) - 3(0 - 2) + 1(0 - 8)$$

$$= 1 + 6 - 8 = -1$$

The first step of finding the inverse of a matrix is to evaluate its determinant. If the determinant is 0, the matrix does not have an inverse.

Step 2

$$\mathbf{M} = \begin{bmatrix} \begin{vmatrix} 4 & 1 \\ -1 & 0 \end{vmatrix} & \begin{vmatrix} 0 & 1 \\ 2 & 0 \end{vmatrix} & \begin{vmatrix} 0 & 4 \\ 2 & -1 \end{vmatrix} \\ \begin{vmatrix} 3 & 1 \\ -1 & 0 \end{vmatrix} & \begin{vmatrix} 1 & 1 \\ 2 & 0 \end{vmatrix} & \begin{vmatrix} 1 & 3 \\ 2 & -1 \end{vmatrix} \\ \begin{vmatrix} 3 & 1 \\ 4 & 1 \end{vmatrix} & \begin{vmatrix} 1 & 1 \\ 0 & 1 \end{vmatrix} & \begin{vmatrix} 1 & 3 \\ 0 & 4 \end{vmatrix} \end{bmatrix}$$

$$= \begin{pmatrix} 1 & -2 & -8 \\ 1 & -2 & -7 \\ -1 & 1 & 4 \end{pmatrix}$$

The second step is to form the matrix of minors. The minor of an element is found by deleting the row and the column in which the element lies. For example, to find the minor of 4 in $\begin{pmatrix} 1 & 3 & 1 \\ 0 & 4 & 1 \\ 2 & -1 & 0 \end{pmatrix}$, delete the row and column containing 4, $\begin{pmatrix} 1 & 3 & 1 \\ 0 & 4 & 1 \\ 2 & -1 & 0 \end{pmatrix}$. The minor is the determinant of the elements left, $\begin{vmatrix} 1 & 1 \\ 2 & 0 \end{vmatrix} = -2$

Step 3

$$\mathbf{C} = \begin{pmatrix} 1 & 2 & -8 \\ -1 & -2 & 7 \\ -1 & -1 & 4 \end{pmatrix}$$

You find the matrix of cofactors by adjusting the signs of the minors using the pattern $\begin{pmatrix} + & - & + \\ - & + & - \\ + & - & + \end{pmatrix}$. Here you leave the elements $\begin{pmatrix} 1 & & -8 \\ & -2 & \\ -1 & & 4 \end{pmatrix}$ unchanged but change the signs of $\begin{pmatrix} & -2 & \\ 1 & & -7 \\ & 1 & \end{pmatrix}$

Step 4

$$\mathbf{C}^T = \begin{pmatrix} 1 & -1 & -1 \\ 2 & -2 & -1 \\ -8 & 7 & 4 \end{pmatrix}$$

Step 5

$$\mathbf{A}^{-1} = \frac{1}{\det(\mathbf{A})}\mathbf{C}^T = \frac{1}{-1}\begin{pmatrix} 1 & -1 & -1 \\ 2 & -2 & -1 \\ -8 & 7 & 4 \end{pmatrix}$$

$$= \begin{pmatrix} -1 & 1 & 1 \\ -2 & 2 & 1 \\ 8 & -7 & -4 \end{pmatrix}$$

You divide each term of the matrix of cofactors, $\mathbf{C}^T$, by the determinant of $\mathbf{A}$, which is -1.

Hint Some calculators will find inverse matrices and you may wish to use your calculator to check your answer. You may be asked to find the inverse of a matrix in algebraic form.

FURTHER MATRIX ALGEBRA — CHAPTER 6

Example 11

The matrix $\mathbf{A} = \begin{pmatrix} 3 & 2 & -2 \\ -2 & k & 0 \\ -1 & -3 & 3 \end{pmatrix}$, $k \neq 0$. Find $\mathbf{A}^{-1}$

Step 1

$$\det(\mathbf{A}) = 3\begin{vmatrix} k & 0 \\ -3 & 3 \end{vmatrix} - 2\begin{vmatrix} -2 & 0 \\ -1 & 3 \end{vmatrix} + (-2)\begin{vmatrix} -2 & k \\ -1 & -3 \end{vmatrix}$$

$$= 3(3k - 0) - 2(-6 - 0) - 2(6 + k)$$

$$= 9k + 12 - 12 - 2k = 7k$$

As you are given that $k \neq 0$, the matrix is non-singular and the inverse can be found.

Step 2

$$\mathbf{M} = \begin{pmatrix} \begin{vmatrix} k & 0 \\ -3 & 3 \end{vmatrix} & \begin{vmatrix} -2 & 0 \\ -1 & 3 \end{vmatrix} & \begin{vmatrix} -2 & k \\ -1 & -3 \end{vmatrix} \\ \begin{vmatrix} 2 & -2 \\ -3 & 3 \end{vmatrix} & \begin{vmatrix} 3 & -2 \\ -1 & 3 \end{vmatrix} & \begin{vmatrix} 3 & 2 \\ -1 & -3 \end{vmatrix} \\ \begin{vmatrix} 2 & -2 \\ k & 0 \end{vmatrix} & \begin{vmatrix} 3 & -2 \\ -2 & 0 \end{vmatrix} & \begin{vmatrix} 3 & 2 \\ -2 & k \end{vmatrix} \end{pmatrix}$$

$$= \begin{pmatrix} 3k & -6 & k+6 \\ 0 & 7 & -7 \\ 2k & -4 & 3k+4 \end{pmatrix}$$

The second step is to find the matrix of the minors in terms of k.

Step 3

$$\mathbf{C} = \begin{pmatrix} 3k & 6 & k+6 \\ 0 & 7 & 7 \\ 2k & 4 & 3k+4 \end{pmatrix}$$

You obtain the matrix of the cofactors from the matrix of the minors by changing the signs of the elements corresponding to the $-$ signs in the pattern $\begin{pmatrix} + & - & + \\ - & + & - \\ + & - & + \end{pmatrix}$

Step 4

$$\mathbf{C}^{\mathsf{T}} = \begin{pmatrix} 3k & 0 & 2k \\ 6 & 7 & 4 \\ k+6 & 7 & 3k+4 \end{pmatrix}$$

Step 5

$$\mathbf{A}^{-1} = \frac{1}{\det(\mathbf{A})}\mathbf{C}^{\mathsf{T}} = \frac{1}{7k}\begin{pmatrix} 3k & 0 & 2k \\ 6 & 7 & 4 \\ k+6 & 7 & 3k+4 \end{pmatrix}$$

You can leave the answer in this form or write the inverse matrix as

$$\begin{pmatrix} \frac{3}{7} & 0 & \frac{2}{7} \\ \frac{6}{7k} & \frac{1}{k} & \frac{4}{7k} \\ \frac{k+6}{7k} & \frac{1}{k} & \frac{3k+4}{7k} \end{pmatrix}$$

Example 12

A and **B** are non-singular matrices. Prove that $(AB)^{-1} = B^{-1}A^{-1}$

Let $P = (AB)^{-1}$,
then
$$(AB)P = I$$
$$A^{-1}ABP = A^{-1}I$$
$$(A^{-1}A)BP = A^{-1}$$
$$IBP = A^{-1}$$
$$BP = A^{-1}$$
$$B^{-1}BP = B^{-1}A^{-1}$$
$$IP = B^{-1}A^{-1}$$
$$P = B^{-1}A^{-1}$$
Hence
$$(AB)^{-1} = B^{-1}A^{-1}$$

- Use the definition of the inverse $XX^{-1} = I$ with $X = AB$
- Multiply both sides of the previous equation on the left by A^{-1}
- As matrix multiplication is associative you can multiply any two of the matrices together first provided the order is kept.
- Using $A^{-1}A = I$ and $IB = B$
- Multiply both sides of the previous equation on the left by B^{-1}
- Using $B^{-1}B = I$ and $IP = P$

■ If **A** and **B** are non-singular matrices then $(AB)^{-1} = B^{-1}A^{-1}$

In Section 6.1, you learnt a very similar formula for transposes, $(AB)^T = B^T A^T$. The formula for transposes is, however, true if **A** and **B** are singular.

Example 13

The matrix $A = \begin{pmatrix} -2 & 3 & -3 \\ 0 & 1 & 0 \\ 1 & -1 & 2 \end{pmatrix}$ and the matrix **B** is such that $(AB)^{-1} = \begin{pmatrix} 8 & -17 & 9 \\ -5 & 10 & -6 \\ -3 & 5 & -4 \end{pmatrix}$.

a Show that $A^{-1} = A$

b Find B^{-1}

a $A^2 = \begin{pmatrix} -2 & 3 & -3 \\ 0 & 1 & 0 \\ 1 & -1 & 2 \end{pmatrix}\begin{pmatrix} -2 & 3 & -3 \\ 0 & 1 & 0 \\ 1 & -1 & 2 \end{pmatrix}$

$= \begin{pmatrix} 4+0-3 & -6+3+3 & 6+0-6 \\ 0+0+0 & 0+1+0 & 0+0+0 \\ -2+0+2 & 3-1-2 & -3+0+4 \end{pmatrix}$

$= \begin{pmatrix} 1 & 0 & 0 \\ 0 & 1 & 0 \\ 0 & 0 & 1 \end{pmatrix} = I$

Comparing $A^2 = AA = I$ with the definition of an inverse $AA^{-1} = I$, then $A^{-1} = A$, as required.

Given that $A^{-1} = A$, then multiplying both sides of the equation by **A**,
$$AA^{-1} = A^2$$
$$I = A^2$$
$A^2 = I$ and $A^{-1} = A$ are equivalent statements. If you prove one of them, the other follows.

Multiply both sides of this formula on the right by **A** and use $A^{-1}A = I$ to obtain an expression for B^{-1} in terms of $(AB)^{-1}$ and **A**, both of which you already know.

b $(AB)^{-1} = B^{-1}A^{-1}$

$(AB)^{-1}A = B^{-1}A^{-1}A = B^{-1}I = B^{-1}$

$B^{-1} = (AB)^{-1}A$

$$= \begin{pmatrix} 8 & -17 & 9 \\ -5 & 10 & -6 \\ -3 & 5 & -4 \end{pmatrix} \begin{pmatrix} -2 & 3 & -3 \\ 0 & 1 & 0 \\ 1 & -1 & 2 \end{pmatrix}$$

$$= \begin{pmatrix} -16+0+9 & 24-17-9 & -24+0+18 \\ 10+0-6 & -15+10+6 & 15+0-12 \\ 6+0-4 & -9+5+4 & 9+0-8 \end{pmatrix}$$

$$= \begin{pmatrix} -7 & -2 & -6 \\ 4 & 1 & 3 \\ 2 & 0 & 1 \end{pmatrix}$$

Finding the inverse of a matrix using the five steps given earlier can be a lengthy process. For particular matrices, other methods may be available and this example illustrates two possibilities.

Exercise 6C   SKILLS   ANALYSIS

1 Find the inverses of these matrices.

a $\begin{pmatrix} 1 & 0 & 0 \\ 0 & 2 & 1 \\ 0 & 1 & 2 \end{pmatrix}$ **b** $\begin{pmatrix} 1 & 0 & 0 \\ 0 & 2 & 0 \\ 0 & 0 & 3 \end{pmatrix}$ **c** $\begin{pmatrix} 1 & 0 & 0 \\ 0 & \frac{3}{5} & -\frac{4}{5} \\ 0 & \frac{4}{5} & \frac{3}{5} \end{pmatrix}$

2 Find the inverses of these matrices.

a $\begin{pmatrix} 1 & -3 & 2 \\ 0 & -2 & 1 \\ 3 & 0 & 2 \end{pmatrix}$ **b** $\begin{pmatrix} 2 & 3 & 2 \\ 3 & -2 & 1 \\ 2 & 1 & 1 \end{pmatrix}$ **c** $\begin{pmatrix} 3 & 2 & -7 \\ 1 & -3 & 1 \\ 0 & 2 & -2 \end{pmatrix}$

3 The matrix $A = \begin{pmatrix} 1 & 0 & 1 \\ 0 & 1 & 0 \\ 2 & 0 & 1 \end{pmatrix}$ and the matrix $B = \begin{pmatrix} 2 & 1 & -1 \\ 1 & 0 & 1 \\ 1 & 2 & 1 \end{pmatrix}$.

a Find A^{-1}

b Find B^{-1}

Given that $(AB)^{-1} = \begin{pmatrix} -\frac{2}{3} & \frac{1}{2} & \frac{1}{2} \\ 1 & -\frac{1}{2} & -\frac{1}{2} \\ \frac{2}{3} & \frac{1}{2} & -\frac{1}{2} \end{pmatrix}$

c verify that $B^{-1}A^{-1} = (AB)^{-1}$

4 The matrix $A = \begin{pmatrix} 2 & 0 & 3 \\ k & 1 & 1 \\ 1 & 1 & 4 \end{pmatrix}$.

a Show that det $(A) = 3(k+1)$ **b** Given that $k \neq -1$, find A^{-1}

5 The matrix $A = \begin{pmatrix} 5 & a & 4 \\ b & -7 & 8 \\ 2 & -2 & c \end{pmatrix}$

Given that $A = A^{-1}$, find the values of the constants a, b and c.

6 The matrix $\mathbf{A} = \begin{pmatrix} 2 & -1 & 1 \\ 4 & -3 & 0 \\ -3 & 3 & 1 \end{pmatrix}$

    a Show that $\mathbf{A}^3 = \mathbf{I}$ **b** Hence find $\mathbf{A}^{-1}$

7 The matrix $\mathbf{A} = \begin{pmatrix} 1 & 1 & 0 \\ 3 & -3 & 1 \\ 0 & 3 & 2 \end{pmatrix}$

    a Show that $\mathbf{A}^3 = 13\mathbf{A} - 15\mathbf{I}$ **(4 marks)**

    b Deduce that $15\mathbf{A}^{-1} = 13\mathbf{I} - \mathbf{A}^2$ **(3 marks)**

    c Hence find $\mathbf{A}^{-1}$ **(3 marks)**

8 The matrix $\mathbf{A} = \begin{pmatrix} 2 & 0 & 1 \\ 4 & 3 & -2 \\ 0 & 3 & -4 \end{pmatrix}$

    a Show that $\mathbf{A}$ is singular.   **(3 marks)**

    The matrix $\mathbf{C}$ is the matrix of the cofactors of $\mathbf{A}$.

    b Find $\mathbf{C}$.   **(3 marks)**

    c Show that $\mathbf{AC}^T = 0$ **(3 marks)**

6.4 Using matrices to represent linear transformations in 3 dimensions

- A linear transformation T in three dimensions is a mapping which transforms a point with position vector $\begin{pmatrix} x \\ y \\ z \end{pmatrix}$ to another point according to a rule with the following properties.

$$\mathbf{T}\begin{pmatrix} kx \\ ky \\ kz \end{pmatrix} = k\mathbf{T}\begin{pmatrix} x \\ y \\ z \end{pmatrix}$$

$$\mathbf{T}\begin{pmatrix} x_1+x_2 \\ y_1+y_2 \\ z_1+z_2 \end{pmatrix} = \mathbf{T}\begin{pmatrix} x_1 \\ y_1 \\ z_1 \end{pmatrix} + \mathbf{T}\begin{pmatrix} x_2 \\ y_2 \\ z_2 \end{pmatrix}$$

- In Further Pure 1, you learnt that linear transformations in two dimensions can be represented by 2×2 matrices. In this chapter, you will extend this work by representing linear transformations in three dimensions by 3×3 matrices.

- The new point to which the point $\begin{pmatrix} x \\ y \\ z \end{pmatrix}$ is moved is called the **image** of $\begin{pmatrix} x \\ y \\ z \end{pmatrix}$.

- To identify the matrix representing a particular **transformation** you consider the effect of the matrix or the transformation on three simple vectors; $\begin{pmatrix} 1 \\ 0 \\ 0 \end{pmatrix}$ (sometimes denoted as **i**), $\begin{pmatrix} 0 \\ 1 \\ 0 \end{pmatrix}$ (sometimes denoted as **j**) and $\begin{pmatrix} 0 \\ 0 \\ 1 \end{pmatrix}$ (sometimes denoted as **k**).

- Given any matrix $\mathbf{M} = \begin{pmatrix} a & b & c \\ d & e & f \\ g & h & i \end{pmatrix}$, then

$$\mathbf{M}\begin{pmatrix} 1 \\ 0 \\ 0 \end{pmatrix} = \begin{pmatrix} a \\ d \\ g \end{pmatrix}, \mathbf{M}\begin{pmatrix} 0 \\ 1 \\ 0 \end{pmatrix} = \begin{pmatrix} b \\ e \\ h \end{pmatrix} \text{ and } \mathbf{M}\begin{pmatrix} 0 \\ 0 \\ 1 \end{pmatrix} = \begin{pmatrix} c \\ f \\ i \end{pmatrix}$$

So the image of $\begin{pmatrix} 1 \\ 0 \\ 0 \end{pmatrix}$ is the first column of $\mathbf{M}$, the image of $\begin{pmatrix} 0 \\ 1 \\ 0 \end{pmatrix}$ is the second column of $\mathbf{M}$ and the image of $\begin{pmatrix} 0 \\ 0 \\ 1 \end{pmatrix}$ is the third column of $\mathbf{M}$.

- If the transformation T is represented by the matrix $\mathbf{T}$ and the transformation U is represented by the matrix $\mathbf{U}$, then the matrix $\mathbf{UT}$ represents the combined transformation of the transformation of T followed by the transformation U.

- The symbol $\mathbb{R}^3$ is used to represent three-dimensional space. You read $T: \mathbb{R}^3 \rightarrow \mathbb{R}^3$ as 'the transformation T transforms points in three-dimensional space to points in three-dimensional space'.

Example 14

Given that $T: \begin{pmatrix} x \\ y \\ z \end{pmatrix} \mapsto \begin{pmatrix} x + z \\ y \\ x + 2y + z \end{pmatrix}$ and $U: \begin{pmatrix} x \\ y \\ z \end{pmatrix} \mapsto \begin{pmatrix} x - y + z \\ 3x - 2z \\ 3x - 4y - z \end{pmatrix}$, find matrices representing

a T **b** U **c** UT **d** TU

a $T: \begin{pmatrix} 1 \\ 0 \\ 0 \end{pmatrix} \mapsto \begin{pmatrix} 1 + 0 \\ 0 \\ 1 + 0 + 0 \end{pmatrix} = \begin{pmatrix} 1 \\ 0 \\ 1 \end{pmatrix}$ — You find the first column of the matrix by finding the image of $\begin{pmatrix} 1 \\ 0 \\ 0 \end{pmatrix}$ under T.

You find this image by substituting $x = 1$, $y = 0$ and $z = 0$ into the rule for T.

$T: \begin{pmatrix} 0 \\ 1 \\ 0 \end{pmatrix} \mapsto \begin{pmatrix} 0 + 0 \\ 1 \\ 0 + 2 + 0 \end{pmatrix} = \begin{pmatrix} 0 \\ 1 \\ 2 \end{pmatrix}$ — This is the second column of the matrix.

$T: \begin{pmatrix} 0 \\ 0 \\ 1 \end{pmatrix} \mapsto \begin{pmatrix} 0 + 1 \\ 0 \\ 0 + 0 + 1 \end{pmatrix} = \begin{pmatrix} 1 \\ 0 \\ 1 \end{pmatrix}$ — This is the third column of the matrix.

The matrix representing T is $\begin{pmatrix} 1 & 0 & 1 \\ 0 & 1 & 0 \\ 1 & 2 & 1 \end{pmatrix}$

b $U: \begin{pmatrix} 1 \\ 0 \\ 0 \end{pmatrix} \mapsto \begin{pmatrix} 1 - 0 + 0 \\ 3 - 0 \\ 3 - 0 - 0 \end{pmatrix} = \begin{pmatrix} 1 \\ 3 \\ 3 \end{pmatrix}$

$U: \begin{pmatrix} 0 \\ 1 \\ 0 \end{pmatrix} \mapsto \begin{pmatrix} 0 - 1 + 0 \\ 0 - 0 \\ 0 - 4 - 0 \end{pmatrix} = \begin{pmatrix} -1 \\ 0 \\ -4 \end{pmatrix}$

$U: \begin{pmatrix} 0 \\ 0 \\ 1 \end{pmatrix} \mapsto \begin{pmatrix} 0 - 0 + 1 \\ 0 - 2 \\ 0 - 0 - 1 \end{pmatrix} = \begin{pmatrix} 1 \\ -2 \\ -1 \end{pmatrix}$

The matrix representing U is $\begin{pmatrix} 1 & -1 & 1 \\ 3 & 0 & -2 \\ 3 & -4 & -1 \end{pmatrix}$

> With practice, you should be able to write down such a matrix without showing intermediate working.
> $\begin{pmatrix} x - y + z \\ 3x - 2z \\ 3x - 4y - z \end{pmatrix}$ can just be recognised
> as $\begin{pmatrix} 1 & -1 & 1 \\ 3 & 0 & -2 \\ 3 & -4 & -1 \end{pmatrix} \begin{pmatrix} x \\ y \\ z \end{pmatrix}$

c The matrix representing UT is

$\begin{pmatrix} 1 & -1 & 1 \\ 3 & 0 & -2 \\ 3 & -4 & -1 \end{pmatrix} \begin{pmatrix} 1 & 0 & 1 \\ 0 & 1 & 0 \\ 1 & 2 & 1 \end{pmatrix}$

$= \begin{pmatrix} 1 + 0 + 1 & 0 - 1 + 2 & 1 + 0 + 1 \\ 3 + 0 - 2 & 0 + 0 - 4 & 3 + 0 - 2 \\ 3 + 0 - 1 & 0 - 4 - 2 & 3 + 0 - 1 \end{pmatrix}$

$= \begin{pmatrix} 2 & 1 & 2 \\ 1 & -4 & 1 \\ 2 & -6 & 2 \end{pmatrix}$

> The matrix representing UT is found by multiplying the matrix representing U by the matrix representing T in that order. **Remember**, the transformation UT applies T first and U second.

d The matrix representing TU is

$\begin{pmatrix} 1 & 0 & 1 \\ 0 & 1 & 0 \\ 1 & 2 & 1 \end{pmatrix} \begin{pmatrix} 1 & -1 & 1 \\ 3 & 0 & -2 \\ 3 & -4 & -1 \end{pmatrix}$

$= \begin{pmatrix} 1 + 0 + 3 & -1 + 0 - 4 & 1 + 0 - 1 \\ 0 + 3 + 0 & 0 + 0 + 0 & 0 - 2 + 0 \\ 1 + 6 + 3 & -1 + 0 - 4 & 1 - 4 - 1 \end{pmatrix}$

$= \begin{pmatrix} 4 & -5 & 0 \\ 3 & 0 & -2 \\ 10 & -5 & -4 \end{pmatrix}$

> The transformation TU applies U first and T second.

Example 15 SKILLS ANALYSIS

The transformation $T: \mathbb{R}^3 \to \mathbb{R}^3$ is represented by the matrix **T**.

The point with position vector $\begin{pmatrix} 1 \\ 0 \\ 0 \end{pmatrix}$ is transformed by T to the point with position vector $\begin{pmatrix} 3 \\ 4 \\ 2 \end{pmatrix}$

The point with position vector $\begin{pmatrix} 1 \\ 1 \\ 0 \end{pmatrix}$ is transformed by T to the point with position vector $\begin{pmatrix} 6 \\ 1 \\ 5 \end{pmatrix}$

The point with position vector $\begin{pmatrix} 2 \\ 1 \\ -4 \end{pmatrix}$ is transformed by T to the point with position vector $\begin{pmatrix} 1 \\ 1 \\ -1 \end{pmatrix}$

a Find T.

b Find the image of the point with position vector $\begin{pmatrix} -1 \\ 3 \\ 0 \end{pmatrix}$

a Let $T = \begin{pmatrix} a & b & c \\ d & e & f \\ g & h & i \end{pmatrix}$

$\begin{pmatrix} a & b & c \\ d & e & f \\ g & h & i \end{pmatrix} \begin{pmatrix} 1 \\ 0 \\ 0 \end{pmatrix} = \begin{pmatrix} 3 \\ 4 \\ 2 \end{pmatrix}$

$\begin{pmatrix} a \\ d \\ g \end{pmatrix} = \begin{pmatrix} 3 \\ 4 \\ 2 \end{pmatrix}$

If T transforms the column vector $\mathbf{x}_1$ to the column vector $\mathbf{x}_2$, then $\mathbf{T}\mathbf{x}_1 = \mathbf{x}_2$

Equating the elements $a = 3$, $d = 4$, $g = 2$

$\begin{pmatrix} 3 & b & c \\ 4 & e & f \\ 2 & h & i \end{pmatrix} \begin{pmatrix} 1 \\ 1 \\ 0 \end{pmatrix} = \begin{pmatrix} 6 \\ 1 \\ 5 \end{pmatrix}$

If two vectors or matrices are equal, then the corresponding elements of the vectors or matrices must be individually equal.

$\begin{pmatrix} 3 + b \\ 4 + e \\ 2 + h \end{pmatrix} = \begin{pmatrix} 6 \\ 1 \\ 5 \end{pmatrix}$

As you now know the values of a, d and g, you can substitute these values into $\mathbf{T}$.

Equating the top elements
$3 + b = 6 \Rightarrow b = 3$
Equating the middle elements
$4 + e = 1 \Rightarrow e = -3$

Equating the three elements enables you to find the values in the second column of $\mathbf{T}$.

Equating the lowest elements
$2 + h = 5 \Rightarrow h = 3$

$\begin{pmatrix} 3 & 3 & c \\ 4 & -3 & f \\ 2 & 3 & i \end{pmatrix} \begin{pmatrix} 2 \\ 1 \\ -4 \end{pmatrix} = \begin{pmatrix} 1 \\ 1 \\ -1 \end{pmatrix}$

$\begin{pmatrix} 9 - 4c \\ 5 - 4f \\ 7 - 4i \end{pmatrix} = \begin{pmatrix} 1 \\ 1 \\ -1 \end{pmatrix}$

Equating the top elements
$9 - 4c = 1 \Rightarrow c = 2$
Equating the middle elements
$5 - 4f = 1 \Rightarrow f = 1$
Equating the lowest elements
$7 - 4i = -1 \Rightarrow i = 2$

$T = \begin{pmatrix} 3 & 3 & 2 \\ 4 & -3 & 1 \\ 2 & 3 & 2 \end{pmatrix}$

b $T\begin{pmatrix}-1\\3\\0\end{pmatrix} = \begin{pmatrix}3 & 3 & 2\\4 & -3 & 1\\2 & 3 & 2\end{pmatrix}\begin{pmatrix}-1\\3\\0\end{pmatrix}$

$= \begin{pmatrix}-3+9+0\\-4-9+0\\-2+9+0\end{pmatrix} = \begin{pmatrix}6\\-13\\7\end{pmatrix}$

The image of $\begin{pmatrix}-1\\3\\0\end{pmatrix}$ is $\begin{pmatrix}6\\-13\\7\end{pmatrix}$

> To find the image of the column vector **x** under T, you evaluate **Tx**.

Example 16

The transformation $T: \mathbb{R}^3 \to \mathbb{R}^3$ is represented by the matrix **T** where $\mathbf{T} = \begin{pmatrix}3 & -2 & 1\\1 & 3 & 4\\2 & -1 & 1\end{pmatrix}$

The line l_1 is transformed by T to the line l_2

The line l_1 has vector equation $\mathbf{r} = \begin{pmatrix}2\\0\\-3\end{pmatrix} + t\begin{pmatrix}3\\-2\\1\end{pmatrix}$, where t is a real parameter.

Find a vector equation of l_2

$\mathbf{r} = \begin{pmatrix}2\\0\\-3\end{pmatrix} + t\begin{pmatrix}3\\-2\\1\end{pmatrix} = \begin{pmatrix}2+3t\\-2t\\-3+t\end{pmatrix}$

The image of l_1 is given by

$\mathbf{r} = \begin{pmatrix}3 & -2 & 1\\1 & 3 & 4\\2 & -1 & 1\end{pmatrix}\begin{pmatrix}2+3t\\-2t\\-3+t\end{pmatrix} = \begin{pmatrix}3(2+3t)+4t-3+t\\2+3t-6t+4(-3+t)\\2(2+3t)+2t-3+t\end{pmatrix}$

$= \begin{pmatrix}14t+3\\t-10\\9t+1\end{pmatrix} = \begin{pmatrix}3\\-10\\1\end{pmatrix} + t\begin{pmatrix}14\\1\\9\end{pmatrix}$

A vector equation of l_2 is

$\mathbf{r} = \begin{pmatrix}3\\-10\\1\end{pmatrix} + t\begin{pmatrix}14\\1\\9\end{pmatrix}$

> $\begin{pmatrix}2+3t\\-2t\\-3+t\end{pmatrix}$ is a general form of a point on l_1

> **Hint** When investigating transformations, it is often a good start to express the coordinates of a general point on a line or plane in parametric form.

> There are an infinite number of possible answers to this question but, in any equation, the vector giving the direction of l_2 must be parallel to $\begin{pmatrix}14\\1\\9\end{pmatrix}$

FURTHER MATRIX ALGEBRA — CHAPTER 6

Example 17

The transformation $T: \mathbb{R}^3 \to \mathbb{R}^3$ is represented by the matrix **T** where $\mathbf{T} = \begin{pmatrix} 2 & 5 & -1 \\ -6 & 0 & 2 \\ 0 & 4 & 3 \end{pmatrix}$

The plane Π_1 is transformed by T to the plane Π_2. The plane Π_1 has vector equation

$\mathbf{r} = \begin{pmatrix} 2 \\ 1 \\ -3 \end{pmatrix} + s\begin{pmatrix} 1 \\ -2 \\ 3 \end{pmatrix} + t\begin{pmatrix} -3 \\ 1 \\ -1 \end{pmatrix}$, where s and t are real parameters.

Find a vector equation of Π_2.

$\mathbf{r} = \begin{pmatrix} 2 \\ 1 \\ -3 \end{pmatrix} + s\begin{pmatrix} 1 \\ -2 \\ 3 \end{pmatrix} + t\begin{pmatrix} -3 \\ 1 \\ -1 \end{pmatrix} = \begin{pmatrix} 2 + s - 3t \\ 1 - 2s + t \\ -3 + 3s - t \end{pmatrix}$

- $\begin{pmatrix} 2 + s - 3t \\ 1 - 2s + t \\ -3 + 3s - t \end{pmatrix}$ is general form of a point on Π_1.

$\begin{pmatrix} 2 & 5 & -1 \\ -6 & 0 & 2 \\ 0 & 4 & 3 \end{pmatrix}\begin{pmatrix} 2 + s - 3t \\ 1 - 2s + t \\ -3 + 3s - t \end{pmatrix}$

- To find the image of the column vector **x** under T, you work out **Tx**.

$= \begin{pmatrix} 2(2 + s - 3t) + 5(1 - 2s + t) - 1(-3 + 3s - t) \\ -6(2 + s - 3t) + 0(1 - 2s + t) + 2(-3 + 3s - t) \\ 0(2 + s - 3t) + 4(1 - 2s + t) + 3(-3 + 3s - t) \end{pmatrix}$

$= \begin{pmatrix} 12 - 11s \\ -18 + 16t \\ -5 + s + t \end{pmatrix} = \begin{pmatrix} 12 \\ -18 \\ -5 \end{pmatrix} + s\begin{pmatrix} -11 \\ 0 \\ 1 \end{pmatrix} + t\begin{pmatrix} 0 \\ 16 \\ 1 \end{pmatrix}$

- Here, the simplest equation of a plane to find is one of the form $\mathbf{r} = \mathbf{a} + s\mathbf{b} + t\mathbf{c}$, so you separate out the terms in s and the terms in t.

A vector equation of Π_2 is

$\mathbf{r} = \begin{pmatrix} 12 \\ -18 \\ -5 \end{pmatrix} + s\begin{pmatrix} -11 \\ 0 \\ 1 \end{pmatrix} + t\begin{pmatrix} 0 \\ 16 \\ 1 \end{pmatrix}$

Example 18

The transformation $T: \mathbb{R}^3 \to \mathbb{R}^3$ is represented by the matrix **T** where $\mathbf{T} = \begin{pmatrix} 2 & -1 & 3 \\ -1 & 4 & -2 \\ 3 & 2 & 4 \end{pmatrix}$

The plane Π has equation $2x - y + 3z = -6$

Show that the image of Π under T is a line and find Cartesian equations of this line.

$2x - y + 3z = -6$
Let $x = s$ and $z = t$, then $y = 6 + 2s + 3t$

The general point on Π is $\begin{pmatrix} s \\ 6 + 2s + 3t \\ t \end{pmatrix}$

$\begin{pmatrix} 2 & -1 & 3 \\ -1 & 4 & -2 \\ 3 & 2 & 4 \end{pmatrix}\begin{pmatrix} s \\ 6 + 2s + 3t \\ t \end{pmatrix}$

- Change the equation of Π into parametric form by putting two of the variables equal to s and t and then finding the third variable in terms of s and t.

$$= \begin{pmatrix} 2s - 6 - 2s - 3t + 3t \\ -s + 24 + 8s + 12t - 2t \\ 3s + 12 + 4s + 6t + 4t \end{pmatrix}$$

$$= \begin{pmatrix} -6 \\ 24 + 7s + 10t \\ 12 + 7s + 10t \end{pmatrix}$$

Let $\begin{pmatrix} x \\ y \\ z \end{pmatrix} = \begin{pmatrix} -6 \\ 24 + 7s + 10t \\ 12 + 7s + 10t \end{pmatrix}$

$x = -6$

$y = 24 + 7s + 10t$ (1)

$z = 12 + 7s + 10t$ (2)

Subtract (1) from (2)

$y - z = 12$

Cartesian equations of the image of Π are

$x = -6, y = z + 12$

Let $y = z + 12 = \lambda$

Then, as $x = -6$ can be written as $x + 6 = 0\lambda$,

the Cartesian equations can be written as

$$\frac{x + 6}{0} = \frac{y}{1} = \frac{z + 12}{1} = \lambda$$

The image of Π under T is a line.

> $x = -6$ and $y - z = 12$ are both planes and the image of Π is the intersection of these planes, which is a line. In general, under linear transformations, points are transformed to points, lines to lines and planes to planes. However, this is not necessarily true when, as here, the matrix representing the transformation is singular.

> In Chapter 5, you learnt that the general form of the Cartesian equation of a straight line is
> $$\frac{x - x_1}{l} = \frac{y - y_1}{m} = \frac{z - z_1}{n}$$
> Anything in this form is a straight line.

Exercise 6D   SKILLS   ANALYSIS

1 Given that $T : \begin{pmatrix} x \\ y \\ z \end{pmatrix} \mapsto \begin{pmatrix} x - y \\ y + z \\ 2x - 3z \end{pmatrix}$ and $U : \begin{pmatrix} x \\ y \\ z \end{pmatrix} \mapsto \begin{pmatrix} 2x - 3y - z \\ 2y + 3z \\ 5z \end{pmatrix}$, find matrices representing

    a T **b** U **c** TU

2 The point with position vector $\begin{pmatrix} 1 \\ 3 \\ a \end{pmatrix}$ is transformed by the linear transformation represented by

the matrix $\begin{pmatrix} 4 & -1 & 0 \\ -2 & 2 & 3 \\ 5 & -2 & 1 \end{pmatrix}$ to the point with position vector $\begin{pmatrix} b \\ -5 \\ c \end{pmatrix}$

Find the values of the constants a, b and c.

3 The transformation $T : \mathbb{R}^3 \to \mathbb{R}^3$ is represented by the matrix **T**.

The vector $\begin{pmatrix} 2 \\ 0 \\ 0 \end{pmatrix}$ is transformed by T to the vector $\begin{pmatrix} 6 \\ 2 \\ 4 \end{pmatrix}$

The vector $\begin{pmatrix} 3 \\ 0 \\ -1 \end{pmatrix}$ is transformed by T to the vector $\begin{pmatrix} -2 \\ 3 \\ 5 \end{pmatrix}$

The vector $\begin{pmatrix} 0 \\ 1 \\ -1 \end{pmatrix}$ is transformed by T to the vector $\begin{pmatrix} 2 \\ -1 \\ -2 \end{pmatrix}$

Find **T**.

P 4 The transformation $T: \mathbb{R}^3 \to \mathbb{R}^3$ is represented by the matrix **T** where $\mathbf{T} = \begin{pmatrix} 0 & -1 & 2 \\ 2 & 5 & -4 \\ 3 & 2 & 1 \end{pmatrix}$

The line l_1 is transformed by T to the line l_2. The line l_1 has vector equation

$\mathbf{r} = \begin{pmatrix} 2 \\ 4 \\ 1 \end{pmatrix} + t \begin{pmatrix} -1 \\ -2 \\ 3 \end{pmatrix}$, where t is a real parameter.

Find a vector equation of l_2.

P 5 The points A and B have position vectors $\begin{pmatrix} 2 \\ 1 \\ 0 \end{pmatrix}$ and $\begin{pmatrix} -2 \\ 3 \\ 4 \end{pmatrix}$ respectively. The points A and B are transformed by the linear transformation T to the points A' and B' respectively.

The transformation T is represented by the matrix **T**, where $\mathbf{T} = \begin{pmatrix} 1 & -3 & 4 \\ 2 & 3 & -2 \\ 0 & 2 & 5 \end{pmatrix}$

a Find the position vectors of A' and B'.
b Hence find a vector equation of the line $A'B'$.

P 6 The transformation $T: \mathbb{R}^3 \to \mathbb{R}^3$ is represented by the matrix **T** where $\mathbf{T} = \begin{pmatrix} 3 & -2 & -2 \\ -2 & -8 & 4 \\ -2 & 4 & 0 \end{pmatrix}$

The plane Π_1 is transformed by T to the plane Π_2.
The plane Π_1 has Cartesian equation $x - 2y + z = 0$
Find a Cartesian equation of Π_2

P 7 The transformation $T: \mathbb{R}^3 \to \mathbb{R}^3$ is represented by the matrix **T** where $\mathbf{T} = \begin{pmatrix} 4 & 5 & -3 \\ -1 & 2 & 1 \\ 1 & 0 & 1 \end{pmatrix}$

The plane Π_1 is transformed by T to the plane Π_2. The plane Π_1 has vector equation

$\mathbf{r} = \begin{pmatrix} 0 \\ 1 \\ 1 \end{pmatrix} + s \begin{pmatrix} 1 \\ -1 \\ 2 \end{pmatrix} + t \begin{pmatrix} 3 \\ 0 \\ 4 \end{pmatrix}$, where s and t are real parameters.

Find an equation of Π_2 in the form $\mathbf{r}.\mathbf{n} = p$

8 The transformation $T: \mathbb{R}^3 \to \mathbb{R}^3$ is represented by the matrix $\mathbf{T}$ where $\mathbf{T} = \begin{pmatrix} 4 & 1 & -2 \\ -2 & 3 & 4 \\ -1 & 0 & 2 \end{pmatrix}$

There is a line through the origin for which every point on the line is mapped onto itself under T. Find a vector equation of this line.

6.5 Using inverse matrices to reverse the effect of a linear transformation

- If the column vector $\mathbf{x}_1$ is transformed to the column vector $\mathbf{x}_2$ by the linear transformation represented by the matrix $\mathbf{T}$ then

$$\mathbf{T}\mathbf{x}_1 = \mathbf{x}_2$$

Multiplying throughout on the left by $\mathbf{T}^{-1}$

$$\mathbf{T}^{-1}\mathbf{T}\mathbf{x}_1 = \mathbf{T}^{-1}\mathbf{x}_2$$

As $\mathbf{T}^{-1}\mathbf{T} = \mathbf{I}$ and $\mathbf{I}\mathbf{x}_1 = \mathbf{x}_1$

$$\mathbf{x}_1 = \mathbf{T}^{-1}\mathbf{x}_2$$

You can use this relation to find $\mathbf{x}_1$ when you know the image of $\mathbf{x}_1$.

Example 19   SKILLS   PROBLEM-SOLVING

The transformation $T: \mathbb{R}^3 \to \mathbb{R}^3$ is represented by the matrix $\mathbf{T}$ where $\mathbf{T} = \begin{pmatrix} 1 & 0 & 2 \\ 0 & 2 & 0 \\ 1 & 0 & 1 \end{pmatrix}$

a Find $\mathbf{T}^{-1}$.

The vector $\begin{pmatrix} a \\ b \\ c \end{pmatrix}$ is transformed by T to the vector $\begin{pmatrix} -3 \\ 4 \\ 2 \end{pmatrix}$

b Find the values of the constants a, b and c.

a Step 1

$$\det(\mathbf{T}) = 1\begin{vmatrix} 2 & 0 \\ 0 & 1 \end{vmatrix} - 0\begin{vmatrix} 0 & 0 \\ 1 & 1 \end{vmatrix} + 2\begin{vmatrix} 0 & 2 \\ 1 & 0 \end{vmatrix}$$

$$= 2 - 4 = -2$$

Step 2

The matrix of minors $\mathbf{M}$ is given by

$$\mathbf{M} = \begin{pmatrix} \begin{vmatrix} 2 & 0 \\ 0 & 1 \end{vmatrix} & \begin{vmatrix} 0 & 0 \\ 1 & 1 \end{vmatrix} & \begin{vmatrix} 0 & 2 \\ 1 & 0 \end{vmatrix} \\ \begin{vmatrix} 0 & 2 \\ 0 & 1 \end{vmatrix} & \begin{vmatrix} 1 & 2 \\ 1 & 1 \end{vmatrix} & \begin{vmatrix} 1 & 0 \\ 1 & 0 \end{vmatrix} \\ \begin{vmatrix} 0 & 2 \\ 2 & 0 \end{vmatrix} & \begin{vmatrix} 1 & 2 \\ 0 & 0 \end{vmatrix} & \begin{vmatrix} 1 & 0 \\ 0 & 2 \end{vmatrix} \end{pmatrix}$$

$$= \begin{pmatrix} 2 & 0 & -2 \\ 0 & -1 & 0 \\ -4 & 0 & 2 \end{pmatrix}$$

All of the elements of $\mathbf{M}$ corresponding to the negative signs in the pattern $\begin{pmatrix} + & - & + \\ - & + & - \\ + & - & + \end{pmatrix}$ are 0, so in this case the matrix of minors and the matrix of cofactors are identical.

Step 3
The matrix of cofactors is given by

$$\mathbf{C} = \begin{pmatrix} 2 & 0 & -2 \\ 0 & -1 & 0 \\ -4 & 0 & 2 \end{pmatrix}$$

> All of the elements of **M** corresponding to the negative signs in the pattern $\begin{pmatrix} + & - & + \\ - & + & - \\ + & - & + \end{pmatrix}$ are 0, so in this case the matrix of minors and the matrix of cofactors are identical.

Step 4

$$\mathbf{C}^T = \begin{pmatrix} 2 & 0 & -4 \\ 0 & -1 & 0 \\ -2 & 0 & 2 \end{pmatrix}$$

Step 5

$$\mathbf{T}^{-1} = \frac{1}{\det(\mathbf{T})} \mathbf{C}^T = \frac{1}{-2} \begin{pmatrix} 2 & 0 & -4 \\ 0 & -1 & 0 \\ -2 & 0 & 2 \end{pmatrix}$$

$$= \begin{pmatrix} -1 & 0 & 2 \\ 0 & \frac{1}{2} & 0 \\ 1 & 0 & -1 \end{pmatrix}$$

$$\mathbf{b} \begin{pmatrix} a \\ b \\ c \end{pmatrix} = \begin{pmatrix} -1 & 0 & 2 \\ 0 & \frac{1}{2} & 0 \\ 1 & 0 & -1 \end{pmatrix} \begin{pmatrix} -3 \\ 4 \\ 2 \end{pmatrix}$$

> As $\begin{pmatrix} -3 \\ 4 \\ 2 \end{pmatrix}$ is the image of $\begin{pmatrix} a \\ b \\ c \end{pmatrix}$ then $\mathbf{T}\begin{pmatrix} a \\ b \\ c \end{pmatrix} = \begin{pmatrix} -3 \\ 4 \\ 2 \end{pmatrix}$. You find $\begin{pmatrix} a \\ b \\ c \end{pmatrix}$ by reversing the transformation using the inverse matrix, $\begin{pmatrix} a \\ b \\ c \end{pmatrix} = \mathbf{T}^{-1} \begin{pmatrix} -3 \\ 4 \\ 2 \end{pmatrix}$

$$= \begin{pmatrix} 3 + 4 \\ 2 \\ -3 - 2 \end{pmatrix} = \begin{pmatrix} 7 \\ 2 \\ -5 \end{pmatrix}$$

$a = 7, b = 2, c = -5$

Example 20

The matrix $\begin{pmatrix} 1 & 3 & -1 \\ 0 & 2 & 1 \\ -1 & -2 & 1 \end{pmatrix}$ and the matrix $\mathbf{B} = \begin{pmatrix} -4 & 1 & -5 \\ 1 & 0 & 1 \\ -2 & 1 & -2 \end{pmatrix}$

a Find $\mathbf{AB} = \mathbf{I}$

The transformation $T: \mathbb{R}^3 \rightarrow \mathbb{R}^3$ is represented by the matrix **A**. The line l_1 is transformed by T to the line l_2. The line l_2 has vector equation $\mathbf{r} = \begin{pmatrix} 3 \\ 4 \\ -1 \end{pmatrix} + t \begin{pmatrix} -1 \\ 2 \\ 2 \end{pmatrix}$, where t is a real parameter.

b Find a vector equation of l_1

a $\mathbf{AB} = \begin{pmatrix} 1 & 3 & -1 \\ 0 & 2 & 1 \\ -1 & -2 & 1 \end{pmatrix} \begin{pmatrix} -4 & 1 & -5 \\ 1 & 0 & 1 \\ -2 & 1 & -2 \end{pmatrix}$

$= \begin{pmatrix} -4+3+2 & 1+0-1 & -5+3+2 \\ 0+2-2 & 0+0+1 & 0+2-2 \\ 4-2-2 & -1+0+1 & 5-2-2 \end{pmatrix}$

$= \begin{pmatrix} 1 & 0 & 0 \\ 0 & 1 & 0 \\ 0 & 0 & 1 \end{pmatrix} = \mathbf{I}$, as required

*As $\mathbf{AB} = \mathbf{I}$, $\mathbf{B} = \mathbf{A}^{-1}$ and you can use $\mathbf{B}$ to reverse the effect of T in part **b**.*

b The general point on l_2 is

$\begin{pmatrix} 3 \\ 4 \\ -1 \end{pmatrix} + t\begin{pmatrix} -1 \\ 2 \\ 2 \end{pmatrix} = \begin{pmatrix} 3-t \\ 4+2t \\ -1+2t \end{pmatrix}$

$\mathbf{B}\begin{pmatrix} 3-t \\ 4+2t \\ -1+2t \end{pmatrix} = \begin{pmatrix} -4 & 1 & -5 \\ 1 & 0 & 1 \\ -2 & 1 & -2 \end{pmatrix}\begin{pmatrix} 3-t \\ 4+2t \\ -1+2t \end{pmatrix}$

$= \begin{pmatrix} -12+4t+4+2t+5-10t \\ 3-t-1+2t \\ -6+2t+4+2t+2-4t \end{pmatrix}$

Simplify this column vector and rearrange it into the form $\mathbf{a} + t\mathbf{b}$

$= \begin{pmatrix} -3-4t \\ 2+t \\ 0 \end{pmatrix} = \begin{pmatrix} -3 \\ 2 \\ 0 \end{pmatrix} + t\begin{pmatrix} -4 \\ 1 \\ 0 \end{pmatrix}$

A vector equation of l_1 is

$\mathbf{r} = \begin{pmatrix} -3 \\ 2 \\ 0 \end{pmatrix} + t\begin{pmatrix} -4 \\ 1 \\ 0 \end{pmatrix}$

- Finding the inverse of a matrix can be a long process. If a particular question does not ask you to find the inverse or give a clue that enables you to find it quickly, it is possible to reverse the effect of a transformation using simultaneous equations.

Example 21 illustrates this method. If no method is specified in a question, you are free to choose any appropriate method.

Example 21

The transformation $T: \mathbb{R}^3 \to \mathbb{R}^3$ is represented by the matrix $\mathbf{T}$ where $\mathbf{T} = \begin{pmatrix} 2 & 4 & -1 \\ 3 & -2 & 1 \\ 1 & -3 & 2 \end{pmatrix}$

The vector $\begin{pmatrix} a \\ b \\ c \end{pmatrix}$ is transformed by T to the vector $\begin{pmatrix} -13 \\ 17 \\ 21 \end{pmatrix}$

Find the values of the constants a, b and c.

$$\begin{pmatrix} 2 & 4 & -1 \\ 3 & -2 & 1 \\ 1 & -3 & 2 \end{pmatrix} \begin{pmatrix} a \\ b \\ c \end{pmatrix} = \begin{pmatrix} -13 \\ 17 \\ 21 \end{pmatrix}$$

$$\begin{pmatrix} 2a + 4b - c \\ 3a - 2b + c \\ a - 3b + 2c \end{pmatrix} = \begin{pmatrix} -13 \\ 17 \\ 21 \end{pmatrix}$$

Equating the elements

$2a + 4b - c = -13$ (1)

$3a - 2b + c = 17$ (2)

$a - 3b + 2c = 21$ (3)

(1) + (2)

$5a + 2b = 4$ (4)

2 × (1) + (3)

$5a + 5b = -5$ (5)

(5) − (4)

$3b = -9 \Rightarrow b = -3$

Substituting $b = -3$ into (4)

$5a - 6 = 4 \Rightarrow a = 2$

Substituting $a = 2$ and $b = -3$ into (2)

$6 + 6 + c = 17 \Rightarrow c = 5$

$a = 2$, $b = -3$ and $c = 5$

> Equating the elements gives you 3 equations in 3 unknowns. You solve these by eliminating one variable, here c, between the equations. This leaves 2 equations in 2 unknowns, which you can solve by any appropriate method.

Exercise 6E   SKILLS   PROBLEM-SOLVING

1 A transformation $T: \mathbb{R}^3 \to \mathbb{R}^3$ is represented by the matrix **T** where $\mathbf{T}^{-1} = \begin{pmatrix} 2 & 3 & 3 \\ -1 & 4 & 5 \\ 2 & 1 & 1 \end{pmatrix}$

The point with position vector $\begin{pmatrix} a \\ b \\ c \end{pmatrix}$ is transformed by T to the point with position vector $\begin{pmatrix} -12 \\ -7 \\ 8 \end{pmatrix}$

a Find the values of the constants a, b and c.

A line l_1 which passes through the origin is transformed by T to the line l_2

A vector equation of l_1 is $\mathbf{r} = t\begin{pmatrix} 2 \\ -2 \\ 1 \end{pmatrix}$

b Find a vector equation of l_2

2 The transformation $T: \mathbb{R}^3 \to \mathbb{R}^3$ is represented by the matrix **T** where $\mathbf{T} = \begin{pmatrix} 2 & 0 & -3 \\ 0 & 1 & 2 \\ -3 & 2 & 8 \end{pmatrix}$

a Find $\mathbf{T}^{-1}$.

The vector $\begin{pmatrix} a \\ b \\ c \end{pmatrix}$ is transformed by T to the vector $\begin{pmatrix} -5 \\ 5 \\ 16 \end{pmatrix}$

b Find the values of the constants a, b and c.

3 The transformation $T: \mathbb{R}^3 \to \mathbb{R}^3$ is represented by the matrix

$\mathbf{T}$, where $\mathbf{T} = \begin{pmatrix} 1 & 1 & 2 \\ 0 & 2 & 2 \\ -3 & 0 & -4 \end{pmatrix}$

a Find $\mathbf{T}^{-1}$. (3 marks)

The line l_1 is transformed by T to the line l_2. The line l_1 has vector equation

$\mathbf{r} = \begin{pmatrix} 2 \\ 4 \\ 1 \end{pmatrix} + t \begin{pmatrix} -1 \\ 0 \\ 1 \end{pmatrix}$, where t is a real parameter.

b Find a vector equation of l_2 (4 marks)

4 The matrix $\mathbf{T} = \begin{pmatrix} a & 1 & 2 \\ 4 & 0 & 0 \\ 0 & 0 & -1 \end{pmatrix}$, where a is a constant.

a Find $\mathbf{T}^{-1}$, in terms of a. (3 marks)

The transformation $T: \mathbb{R}^3 \to \mathbb{R}^3$ is represented by the matrix $\mathbf{T}$. The point with position vector $\begin{pmatrix} p \\ q \\ r \end{pmatrix}$ is transformed by T to the point with position vector $\begin{pmatrix} 2 \\ 3 \\ -1 \end{pmatrix}$

b Find the values of the constants p, q and r. (5 marks)

5 The matrix $\mathbf{S} = \begin{pmatrix} 1 & -\sqrt{2} & 1 \\ \sqrt{2} & 0 & -\sqrt{2} \\ 1 & \sqrt{2} & 1 \end{pmatrix}$

a Show that $\mathbf{SS}^T = k\mathbf{I}$, stating the value of k.

The transformation $S: \mathbb{R}^3 \to \mathbb{R}^3$ is represented by the matrix $\mathbf{S}$.

The vector $\begin{pmatrix} a \\ b \\ c \end{pmatrix}$ is transformed by S to the vector $\begin{pmatrix} 2\sqrt{2} \\ \sqrt{2} \\ -2\sqrt{2} \end{pmatrix}$

b Find the values of the constants a, b and c.

6 The matrix $\mathbf{A} = \begin{pmatrix} 3 & 5 & 1 \\ -2 & 3 & 0 \\ 4 & 3 & 1 \end{pmatrix}$ and the matrix $\mathbf{B} = \begin{pmatrix} 3 & a & -3 \\ b & -1 & -2 \\ -18 & 11 & c \end{pmatrix}$. Given that $\mathbf{AB} = \mathbf{I}$,

a find the values of the constants a, b and c. (4 marks)

The transformation $A: \mathbb{R}^3 \to \mathbb{R}^3$ is represented by the matrix $\mathbf{A}$.

The plane Π_1 is transformed by A to the plane Π_2. The plane Π_2 has vector equation

$\mathbf{r} = \begin{pmatrix} 1 \\ 1 \\ 0 \end{pmatrix} + s \begin{pmatrix} -1 \\ 0 \\ 2 \end{pmatrix} + t \begin{pmatrix} 0 \\ 1 \\ 1 \end{pmatrix}$, where s and t are real parameters.

b Find a vector equation of the plane Π_1. (5 marks)

7 The transformation $T: \mathbb{R}^3 \to \mathbb{R}^3$ is represented by the matrix **T** where $\mathbf{T} = \begin{pmatrix} -1 & 3 & 6 \\ 1 & 4 & 2 \\ 2 & -5 & 1 \end{pmatrix}$

The vector $\begin{pmatrix} a \\ b \\ c \end{pmatrix}$ is transformed by T to the vector $\begin{pmatrix} -8 \\ 0 \\ 3 \end{pmatrix}$

Find the values of the constants a, b and c.

8 The matrix $\mathbf{S} = \begin{pmatrix} 2 & -1 & 2 \\ 0 & 2 & 1 \\ 1 & 0 & 1 \end{pmatrix}$ and the matrix $\mathbf{T} = \begin{pmatrix} 3 & 4 & 4 \\ -6 & -7 & -6 \\ 4 & 4 & 3 \end{pmatrix}$

a Find $\mathbf{S}^{-1}$. **(3 marks)**

b Show that $\mathbf{T}^2 = \mathbf{I}$ **(3 marks)**

The transformation $S: \mathbb{R}^3 \to \mathbb{R}^3$ is represented by the matrix **S** and the transformation $T: \mathbb{R}^3 \to \mathbb{R}^3$ is represented by the matrix **T**.

The transformation U is the transformation T followed by the transformation S. The point $\begin{pmatrix} a \\ b \\ c \end{pmatrix}$ is transformed by U to the point $\begin{pmatrix} 6 \\ -3 \\ 2 \end{pmatrix}$

c Find the values of the constants a, b and c. **(5 marks)**

6.6 The eigenvalues and eigenvectors of 2 × 2 and 3 × 3 matrices

- An **eigenvector** of a matrix **A** is a non-zero column vector **x** which satisfies the equation $\mathbf{Ax} = \lambda\mathbf{x}$, where λ is a scalar.

  The value of the scalar λ is the **eigenvalue** of the matrix corresponding to the eigenvector **x**.

  The word **eigen** is German and means particular or special. Eigenvectors are also known as characteristic or latent vectors.

- The magnitude of an eigenvector may be changed by the linear transformation represented by the matrix but the direction of the eigenvector is unchanged or **invariant**. The eigenvalue can be interpreted as the magnification factor of the eigenvector under the transformation.

- If **x** is an eigenvector of the matrix **A** then, by definition

  $\mathbf{Ax} = \lambda\mathbf{x} = \lambda\mathbf{Ix}$

  Watch out Under the transformation, the eigenvector **x** maps to the vector $\lambda\mathbf{x}$

  Rearranging,

  $\mathbf{Ax} - \lambda\mathbf{Ix} = (\mathbf{A} - \lambda\mathbf{I})\mathbf{x} = \mathbf{0}$

  As by definition **x** is non-zero, the matrix $(\mathbf{A} - \lambda\mathbf{I})$ is singular and has determinant zero, that is

  $\det(\mathbf{A} - \lambda\mathbf{I}) = 0$

- The equation $\det(\mathbf{A} - \lambda\mathbf{I}) = 0$ is called the **characteristic equation** of **A**.

  You solve the characteristic equation of a matrix to find its eigenvalues.

  If the matrix is a 2 × 2 matrix, then the characteristic equation is a quadratic equation.

  If the matrix is a 3 × 3 matrix, then the characteristic equation is a cubic equation. Often questions will give you a hint which enables you to factorise the cubic. However, if a hint is not given, you may have to search for one of the eigenvalues using the factor theorem you learnt in Pure 2.

By definition, an eigenvector is non-zero, but zero can be an eigenvalue of a matrix. Although complex eigenvalues are possible and have important applications in Physics, only real eigenvalues are used in this specification.

- A **normalised vector** is a vector of unit magnitude. In three dimensions, you normalise the vector $\begin{pmatrix} a \\ b \\ c \end{pmatrix}$ by dividing each of the elements by the magnitude of the vector, which is $\sqrt{(a^2 + b^2 + c^2)}$.

Any vector can be normalised, but in this chapter you will be mainly concerned with **normalised eigenvectors**.

If $\begin{pmatrix} a \\ b \end{pmatrix}$ is an eigenvector of a 2 × 2 matrix, then $\begin{pmatrix} \dfrac{a}{\sqrt{(a^2 + b^2)}} \\ \dfrac{b}{\sqrt{(a^2 + b^2)}} \end{pmatrix}$ is the corresponding normalised eigenvector.

If $\begin{pmatrix} a \\ b \\ c \end{pmatrix}$ is an eigenvector of a 3 × 3 matrix, then $\begin{pmatrix} \dfrac{a}{\sqrt{(a^2 + b^2 + c^2)}} \\ \dfrac{b}{\sqrt{(a^2 + b^2 + c^2)}} \\ \dfrac{c}{\sqrt{(a^2 + b^2 + c^2)}} \end{pmatrix}$ is the corresponding normalised eigenvector.

Example 22

Find the eigenvalues and corresponding eigenvectors of the matrix $\mathbf{A} = \begin{pmatrix} 2 & 5 \\ -1 & -4 \end{pmatrix}$

$$\mathbf{A} - \lambda \mathbf{I} = \begin{pmatrix} 2 & 5 \\ -1 & -4 \end{pmatrix} - \lambda \begin{pmatrix} 1 & 0 \\ 0 & 1 \end{pmatrix}$$

$$= \begin{pmatrix} 2 & 5 \\ -1 & -4 \end{pmatrix} - \begin{pmatrix} \lambda & 0 \\ 0 & \lambda \end{pmatrix} = \begin{pmatrix} 2 - \lambda & 5 \\ -1 & -4 - \lambda \end{pmatrix}$$

$$\det(\mathbf{A} - \lambda \mathbf{I}) = \begin{vmatrix} 2 - \lambda & 5 \\ -1 & -4 - \lambda \end{vmatrix}$$

$$= (2 - \lambda)(-4 - \lambda) - 5 \times (-1)$$

$$= -8 - 2\lambda + 4\lambda + \lambda^2 + 5$$

$$= \lambda^2 + 2\lambda - 3$$

$$\det(\mathbf{A} - \lambda \mathbf{I}) = 0 \Rightarrow \lambda^2 + 2\lambda - 3 = 0$$

$$(\lambda - 1)(\lambda + 3) = 0$$

$$\lambda = 1, -3$$

The eigenvalues of **A** are 1 and −3.

> The eigenvalues are the solutions of $\det(\mathbf{A} - \lambda \mathbf{I}) = 0$. You begin by finding $\mathbf{A} - \lambda \mathbf{I}$ and finding its determinant as a polynomial in λ.

> This equation is the **characteristic equation of A**.

FURTHER MATRIX ALGEBRA — CHAPTER 6

To find an eigenvector of A corresponding to the eigenvalue 1

$$\begin{pmatrix} 2 & 5 \\ -1 & -4 \end{pmatrix} \begin{pmatrix} x \\ y \end{pmatrix} = 1 \begin{pmatrix} x \\ y \end{pmatrix}$$

$$\begin{pmatrix} 2x + 5y \\ -x - 4y \end{pmatrix} = \begin{pmatrix} x \\ y \end{pmatrix}$$

Equating the upper elements

$2x + 5y = x$
$\quad x = -5y$

Let $y = 1$, then $x = -5 \times 1 = -5$

An eigenvector corresponding to 1 is $\begin{pmatrix} -5 \\ 1 \end{pmatrix}$.

To find an eigenvector of A corresponding to the eigenvalue −3

$$\begin{pmatrix} 2 & 5 \\ -1 & -4 \end{pmatrix} \begin{pmatrix} x \\ y \end{pmatrix} = -3 \begin{pmatrix} x \\ y \end{pmatrix}$$

$$\begin{pmatrix} 2x + 5y \\ -x - 4y \end{pmatrix} = \begin{pmatrix} -3x \\ -3y \end{pmatrix}$$

Equating the upper elements

$2x + 5y = -3x$
$5x + 5y = 0 \Rightarrow y = -x$

Let $x = 1$, then $y = -1$

An eigenvector corresponding to −3 is $\begin{pmatrix} 1 \\ -1 \end{pmatrix}$.

An eigenvector is a solution of $\mathbf{Ax} = \lambda \mathbf{x}$. In this case, you have to find a column vector $x = \begin{pmatrix} x \\ y \end{pmatrix}$ satisfying the equation when $\lambda = 1$

Equating the lower elements gives $-x - 4y = y$, which leads to $x = -5y$. This is the same equation as you obtained from the upper elements and so gives you no extra information. With 2 × 2 matrices, one equation gives sufficient information to find an eigenvector.

Here you have a free choice of one variable. You can choose any non-zero value of y and then evaluate x. It is sensible to choose a simple number that avoids fractions.

Any non-zero multiple of this vector is also a correct eigenvector of the matrix. For example, $\begin{pmatrix} 5 \\ -1 \end{pmatrix}$ is also correct.

You repeat the procedure used for $\lambda = 1$ with $\lambda = -3$

The lower elements would give $-x - 4y = -3y$ which is equivalent to $y = -x$ so, again, this gives you no additional information.

Any multiple of this vector is also correct. The normalised eigenvector $\begin{pmatrix} \frac{1}{\sqrt{2}} \\ -\frac{1}{\sqrt{2}} \end{pmatrix}$ is sometimes asked for.

Example 23 SKILLS ANALYSIS

Find the eigenvalues and corresponding eigenvectors of the matrix $\mathbf{A} = \begin{pmatrix} 2 & 1 & -3 \\ 0 & 2 & 1 \\ 0 & -4 & -3 \end{pmatrix}$

$$A - \lambda I = \begin{pmatrix} 2 & 1 & -3 \\ 0 & 2 & 1 \\ 0 & -4 & -3 \end{pmatrix} - \lambda \begin{pmatrix} 1 & 0 & 0 \\ 0 & 1 & 0 \\ 0 & 0 & 1 \end{pmatrix}$$

$$= \begin{pmatrix} 2 & 1 & -3 \\ 0 & 2 & 1 \\ 0 & -4 & -3 \end{pmatrix} - \begin{pmatrix} \lambda & 0 & 0 \\ 0 & \lambda & 0 \\ 0 & 0 & \lambda \end{pmatrix} = \begin{pmatrix} 2-\lambda & 1 & -3 \\ 0 & 2-\lambda & 1 \\ 0 & -4 & -3-\lambda \end{pmatrix}$$

$$\det(A - \lambda I) = \begin{vmatrix} 2-\lambda & 1 & -3 \\ 0 & 2-\lambda & 1 \\ 0 & -4 & -3-\lambda \end{vmatrix}$$

$$= (2-\lambda)\begin{vmatrix} 2-\lambda & 1 \\ -4 & -3-\lambda \end{vmatrix} - 1\begin{vmatrix} 0 & 1 \\ 0 & -3-\lambda \end{vmatrix} + (-3)\begin{vmatrix} 0 & 2-\lambda \\ 0 & -4 \end{vmatrix}$$

$$= (2-\lambda)((2-\lambda)(-3-\lambda) + 4) - 0 + 0$$

$$= (2-\lambda)(-6 - 2\lambda + 3\lambda + \lambda^2 + 4)$$

$$= (2-\lambda)(\lambda^2 + \lambda - 2) = (2-\lambda)(\lambda + 2)(\lambda - 1)$$

$\det(A - \lambda I) = 0 \Rightarrow (2-\lambda)(\lambda + 2)(\lambda - 1) = 0$

$\lambda = 2, -2, 1$

The eigenvalues of **A** are −2, 1 and 2.

> As with 2 × 2 matrices, the eigenvalues are the solutions of det (**A** − λ**I**) = 0. You begin by finding **A** − λ**I** and finding its determinant. With a 3 × 3 matrix the characteristic equation is a cubic which can have 3 real **roots** and, hence, 3 eigenvalues.

To find an eigenvector of A corresponding to the eigenvalue −2

$$\begin{pmatrix} 2 & 1 & -3 \\ 0 & 2 & 1 \\ 0 & -4 & -3 \end{pmatrix} \begin{pmatrix} x \\ y \\ z \end{pmatrix} = -2 \begin{pmatrix} x \\ y \\ z \end{pmatrix}$$

$$\begin{pmatrix} 2x + y - 3z \\ 2y + z \\ -4y - 3z \end{pmatrix} = \begin{pmatrix} -2x \\ -2y \\ -2z \end{pmatrix}$$

> An eigenvector is a solution of **Ax** = λ**x**. In this case, you have to find a column vector $\mathbf{x} = \begin{pmatrix} x \\ y \\ z \end{pmatrix}$ satisfying the equation when $\lambda = -2$

Equating the middle elements

$2y + z = -2y \Rightarrow z = -4y$

Let $y = 1$, then $z = -4$.

> Here you have a free choice of one variable. You can choose any non-zero value for y or x and then evaluate the other variable.

Equating the top elements and substituting $y = 1$ and $z = -4$

$2x + y - 3z = -2x$

$4x = -y + 3z = -1 - 12 = -13 \Rightarrow x = -\frac{13}{4}$

An eigenvector corresponding to −2 is $\begin{pmatrix} -\frac{13}{4} \\ 1 \\ -4 \end{pmatrix}$.

> Equating the lowest elements gives an equivalent equation to the one you obtained from the middle elements and so gives you no extra information. With 3 × 3 matrices usually two equations will give you all the information you need to find an eigenvector.

To find an eigenvector of A corresponding to the eigenvalue 1

$$\begin{pmatrix} 2 & 1 & -3 \\ 0 & 2 & 1 \\ 0 & -4 & -3 \end{pmatrix} \begin{pmatrix} x \\ y \\ z \end{pmatrix} = 1 \begin{pmatrix} x \\ y \\ z \end{pmatrix}$$

$$\begin{pmatrix} 2x + y - 3z \\ 2y + z \\ -4y - 3z \end{pmatrix} = \begin{pmatrix} x \\ y \\ z \end{pmatrix}$$

> You repeat the procedure used for $\lambda = -2$ with $\lambda = 1$

Equating the middle elements

$2y + z = y \Rightarrow y = -z$

Let $z = 1$, the $y = -1$

Equating the top elements and substituting $y = -1$ and $z = 1$

$2x + y - 3z = x$

$x = -y + 3z = 1 + 3 = 4$

An eigenvector corresponding to 1 is $\begin{pmatrix} 4 \\ -1 \\ 1 \end{pmatrix}$ ← Any non-zero multiple of this eigenvector is also a correct eigenvector.

To find an eigenvector of A corresponding to the eigenvalue 2

$\begin{pmatrix} 2 & 1 & -3 \\ 0 & 2 & 1 \\ 0 & -4 & -3 \end{pmatrix} \begin{pmatrix} x \\ y \\ z \end{pmatrix} = 2 \begin{pmatrix} x \\ y \\ z \end{pmatrix}$

$\begin{pmatrix} 2x + y - 3z \\ 2y + z \\ -4y - 3z \end{pmatrix} = \begin{pmatrix} 2x \\ 2y \\ 2z \end{pmatrix}$

Equating the middle elements

$2y + z = 2y \Rightarrow z = 0$ ← This calculation differs from the calculation for the other two eigenvalues in that these two equations give you that $y = z = 0$ and there is no choice of values.

Equating the lowest elements and using $z = 0$

$-4y - 3z = 2z \Rightarrow 4y = -5z = 0 \Rightarrow y = 0$

Equating the top elements

$2x + y - 3z = 2x \Rightarrow y = 3z$ ← It is sensible now to equate the top elements to attempt to find out something about x but the terms in x cancel out.

Let $x = 1$

An eigenvector corresponding to 2 is $\begin{pmatrix} 1 \\ 0 \\ 0 \end{pmatrix}$

The variable x appears in no equation and so can take any value. 1 is the simplest value to take.

Example 24

The matrix $\mathbf{A} = \begin{pmatrix} 2 & -1 & 1 \\ 0 & 3 & 5 \\ 2 & 1 & 0 \end{pmatrix}$

a Show that -2 is the only real eigenvalue of **A**.

b Find a normalised eigenvector of **A** corresponding to the eigenvalue -2.

a $A - \lambda I = \begin{pmatrix} 2 & -1 & 1 \\ 0 & 3 & 5 \\ 2 & 1 & 0 \end{pmatrix} - \begin{pmatrix} \lambda & 0 & 0 \\ 0 & \lambda & 0 \\ 0 & 0 & \lambda \end{pmatrix} = \begin{pmatrix} 2-\lambda & -1 & 1 \\ 0 & 3-\lambda & 5 \\ 2 & 1 & -\lambda \end{pmatrix}$

$\det(A - \lambda I) = \begin{vmatrix} 2-\lambda & -1 & 1 \\ 0 & 3-\lambda & 5 \\ 2 & 1 & -\lambda \end{vmatrix}$

$= (2-\lambda)\begin{vmatrix} 3-\lambda & 5 \\ 1 & -\lambda \end{vmatrix} - (-1)\begin{vmatrix} 0 & 5 \\ 2 & -\lambda \end{vmatrix} + 1\begin{vmatrix} 0 & 3-\lambda \\ 2 & 1 \end{vmatrix}$

$= (2-\lambda)(-3\lambda + \lambda^2 - 5) - 10 - 2(3-\lambda)$

$= -6\lambda + 2\lambda^2 - 10 + 3\lambda^2 - \lambda^3 + 5\lambda - 10 - 6 + 2\lambda$

$= -\lambda^3 + 5\lambda^2 + \lambda - 26$

$\det(A - \lambda I) = 0 \Rightarrow -\lambda^3 + 5\lambda^2 + \lambda - 26 = 0$

$\lambda^3 - 5\lambda^2 - \lambda + 26 = 0$

$(\lambda + 2)(\lambda^2 + k\lambda + 13) = 0$

Equating coefficients of λ^2

$-5 = 2 + k \Rightarrow k = -7$

$(\lambda + 2)(\lambda^2 - 7\lambda + 13) = 0$

The discriminant of $\lambda^2 - 7\lambda + 13 = 0$ is

$b^2 - 4ac = 49 - 52 = -3 < 0$

The equation $\lambda^2 - 7\lambda + 13 = 0$ has no real solutions.

Hence -2 is the only real eigenvalue of **A**.

> The question implies that $\lambda = -2$ is a root of the characteristic equation and so $(\lambda + 2)$ must be a factor of the cubic. Here, equating a coefficient has been used to complete the factorisation but you can use any appropriate method.

> To show that there is only one real root of the cubic, you show that the discriminant of the quadratic factor is negative.

b $\begin{pmatrix} 2 & -1 & 1 \\ 0 & 3 & 5 \\ 2 & 1 & 0 \end{pmatrix}\begin{pmatrix} x \\ y \\ z \end{pmatrix} = -2\begin{pmatrix} x \\ y \\ z \end{pmatrix}$

$\begin{pmatrix} 2x - y + z \\ 3y + 5z \\ 2x + y \end{pmatrix} = \begin{pmatrix} -2x \\ -2y \\ -2z \end{pmatrix}$

Equating the middle elements

$3y + 5z = -2y \Rightarrow y = -z$

Let $z = 1$, then $y = -1$

Equating the lowest elements

$2x + y = -2z \Rightarrow x = \dfrac{-y - 2z}{2}$

Substituting $y = -1$ and $z = 1$

$x = \dfrac{1 - 2}{2} = -\dfrac{1}{2}$

FURTHER MATRIX ALGEBRA — CHAPTER 6

An eigenvector of $\mathbf{A}$ is $2\begin{pmatrix} -\frac{1}{2} \\ -1 \\ 1 \end{pmatrix} = \begin{pmatrix} -1 \\ -2 \\ 2 \end{pmatrix}$

The working gives $\begin{pmatrix} -\frac{1}{2} \\ -1 \\ 1 \end{pmatrix}$ as the eigenvector but as any multiple of this is also an eigenvector, it is sensible to multiply this by 2, or −2, to avoid working in fractions.

The magnitude of this eigenvector is
$$\sqrt{((-1)^2 + (-2)^2 + 2^2)} = 3$$

A normalised eigenvector of $\mathbf{A}$ is
$$\frac{1}{3}\begin{pmatrix} -1 \\ -2 \\ 2 \end{pmatrix} = \begin{pmatrix} -\frac{1}{3} \\ -\frac{2}{3} \\ \frac{2}{3} \end{pmatrix}$$

A normalised eigenvector is found by dividing all of the terms by the magnitude of the original eigenvector.

$\begin{pmatrix} \frac{1}{3} \\ \frac{2}{3} \\ -\frac{2}{3} \end{pmatrix}$ is also correct. If a column vector $\mathbf{x}$ is a normalised eigenvector, then $-\mathbf{x}$ is also a normalised eigenvector.

Example 25

A transformation $T: \mathbb{R}^2 \to \mathbb{R}^2$ is represented by the matrix
$$\mathbf{A} = \begin{pmatrix} 4 & -5 \\ 1 & -2 \end{pmatrix}$$

a Find the eigenvalues of $\mathbf{A}$.

b Find Cartesian equations of the two lines passing through the origin which are invariant under T.

a $\mathbf{A} - \lambda\mathbf{I} = \begin{pmatrix} 4 & -5 \\ 1 & -2 \end{pmatrix} - \begin{pmatrix} \lambda & 0 \\ 0 & \lambda \end{pmatrix} = \begin{pmatrix} 4-\lambda & -5 \\ 1 & -2-\lambda \end{pmatrix}$

$\det(\mathbf{A} - \lambda\mathbf{I}) = \begin{vmatrix} 4-\lambda & -5 \\ 1 & -2-\lambda \end{vmatrix}$

With practice, you can write down this line without the previous working.

$= (4-\lambda)(-2-\lambda) + 5$

$= -8 - 4\lambda + 2\lambda + \lambda^2 + 5$

$= \lambda^2 - 2\lambda - 3 = (\lambda - 3)(\lambda + 1)$

$\det(\mathbf{A} - \lambda\mathbf{I}) = 0 \Rightarrow \lambda = 3, -1$

The eigenvalues of $\mathbf{A}$ are 3 and −1.

b $\begin{pmatrix} 4 & -5 \\ 1 & -2 \end{pmatrix}\begin{pmatrix} x \\ y \end{pmatrix} = 3\begin{pmatrix} x \\ y \end{pmatrix}$

$\begin{pmatrix} 4x - 5y \\ x - 2y \end{pmatrix} = \begin{pmatrix} 3x \\ 3y \end{pmatrix}$

Equating the lower elements

$x - 2y = 3y \Rightarrow y = \frac{1}{5}x$

$\begin{pmatrix} 4 & -5 \\ 1 & -2 \end{pmatrix} \begin{pmatrix} x \\ y \end{pmatrix} = -1 \begin{pmatrix} x \\ y \end{pmatrix}$

$\begin{pmatrix} 4x - 5y \\ x - 2y \end{pmatrix} = \begin{pmatrix} -x \\ -y \end{pmatrix}$

Equating the lower elements

$x - 2y = -y \Rightarrow y = x$

The equations of the invariant lines are

$y = x$ and $y = \frac{1}{5}x$

> An eigenvector is a vector whose direction is unchanged or invariant under a transformation. The line of action of the eigenvector is invariant under the transformation and a Cartesian equation for an eigenvector is an equation of an invariant line.

- An **invariant line** is one where any point on the line is mapped to another point on the same line. Except for the origin, the image point will usually be different from the original point as the distance of the point from the origin will be changed by a **scale factor** which is the eigenvalue.

  However if the eigenvalue is 1, then every point on the line of action of the corresponding eigenvector is mapped to itself. This gives a **line of invariant points**.

Example 26

The matrix $\mathbf{A} = \begin{pmatrix} 2 & 1 & 1 \\ 1 & 2 & 1 \\ 1 & 1 & 2 \end{pmatrix}$

a Show that 1 is an eigenvalue of **A** and that there is only one other distinct eigenvalue.

b Find an eigenvector corresponding to the eigenvalue 1.

a $\mathbf{A} - \lambda\mathbf{I} = \begin{pmatrix} 2-\lambda & 1 & 1 \\ 1 & 2-\lambda & 1 \\ 1 & 1 & 2-\lambda \end{pmatrix}$

$\det(\mathbf{A} - \lambda\mathbf{I}) = \begin{vmatrix} 2-\lambda & 1 & 1 \\ 1 & 2-\lambda & 1 \\ 1 & 1 & 2-\lambda \end{vmatrix}$

$= (2-\lambda) \begin{vmatrix} 2-\lambda & 1 \\ 1 & 2-\lambda \end{vmatrix} - 1 \begin{vmatrix} 1 & 1 \\ 1 & 2-\lambda \end{vmatrix} + 1 \begin{vmatrix} 1 & 2-\lambda \\ 1 & 1 \end{vmatrix}$

$= (2-\lambda)((2-\lambda)^2 - 1) - (1-\lambda) + (\lambda - 1)$

$= (2-\lambda)(\lambda^2 - 4\lambda + 3) + 2\lambda - 2$

$= (2-\lambda)(\lambda - 3)(\lambda - 1) + 2(\lambda - 1)$

$= (\lambda - 1)(2\lambda - 6 - \lambda^2 + 3\lambda + 2)$

$= -(\lambda - 1)(\lambda^2 - 5\lambda + 4) = -(\lambda - 1)^2(\lambda - 4)$

> There are a number of ways of factorising the cubic, but as you know that 1 is a root of the cubic, it is sensible to look for the factor $(\lambda - 1)$.

The solutions of $\det(\mathbf{A} - \lambda\mathbf{I}) = 0$ are 1 repeated and 4. 1 is an eigenvalue and the only other distinct eigenvalue is 4.

b $\begin{pmatrix} 2x + y + z \\ x + 2y + z \\ x + y + 2z \end{pmatrix} = \begin{pmatrix} x \\ y \\ z \end{pmatrix}$

Equating the top elements

$2x + y + z = x \Rightarrow x + y + z = 0$

Let $y = 1$ and $z = 1$, then $x = -2$

An eigenvector corresponding to the eigenvalue 1 is $\begin{pmatrix} -2 \\ 1 \\ 1 \end{pmatrix}$

Equating the middle and the lowest elements both give you the same equation, $x + y + z = 0$. So, the elements of the eigenvector only need to satisfy this one equation. You can choose any value you like for two of the values and then work the third one out. This extra element of choice has arisen because of the repeated root. However, not every case of a repeated root gives this extra choice.

Exercise 6F SKILLS ANALYSIS

1 Find the eigenvalues and corresponding eigenvectors of the matrices

  a $\begin{pmatrix} 2 & 4 \\ 1 & 5 \end{pmatrix}$ **b** $\begin{pmatrix} 4 & -1 \\ -1 & 4 \end{pmatrix}$ **c** $\begin{pmatrix} 3 & -2 \\ 0 & 4 \end{pmatrix}$

E/P **2** A transformation $T: \mathbb{R}^2 \rightarrow \mathbb{R}^2$ is represented by the matrix

$$\mathbf{A} = \begin{pmatrix} 3 & 4 \\ -2 & 9 \end{pmatrix}$$

  a Find the eigenvalues of $\mathbf{A}$. (4 marks)

  b Find Cartesian equations of the two lines passing through the origin which are invariant under T. (4 marks)

3 Find the eigenvalues and corresponding eigenvectors of the matrices

  a $\begin{pmatrix} 3 & 0 & 0 \\ 2 & 4 & 2 \\ -2 & 0 & 1 \end{pmatrix}$ **b** $\begin{pmatrix} 4 & -2 & -4 \\ 2 & 3 & 0 \\ 2 & -5 & -4 \end{pmatrix}$

4 The matrix $\mathbf{A} = \begin{pmatrix} 2 & 2 & -2 \\ -3 & 2 & 0 \\ 1 & 4 & -3 \end{pmatrix}$

  a Show that -1 is the only real eigenvalue of $\mathbf{A}$.

  b Find an eigenvector corresponding to the eigenvalue -1.

E/P **5** The matrix $\mathbf{A} = \begin{pmatrix} 2 & -1 & 3 \\ 0 & 2 & 4 \\ 0 & 2 & 0 \end{pmatrix}$

  a Show that 4 is an eigenvalue of $\mathbf{A}$ and find the other two eigenvalues of $\mathbf{A}$. (4 marks)

  b Find an eigenvector corresponding to the eigenvalue 4. (3 marks)

6 The matrix $A = \begin{pmatrix} 1 & 1 & 3 \\ 2 & 4 & -1 \\ 4 & 4 & 3 \end{pmatrix}$

Given that 3 is an eigenvalue of **A**,

a find the other two eigenvalues of **A** (4 marks)

b find eigenvectors corresponding to each of the eigenvalues of **A**. (4 marks)

7 The matrix $A = \begin{pmatrix} 2 & 2 & 1 \\ -2 & 4 & 0 \\ 4 & 2 & 5 \end{pmatrix}$

a Show that 2 is an eigenvalue of **A**. (2 marks)

b Find the other two eigenvalues of **A**. (2 marks)

c Find a normalised eigenvector of **A** corresponding to the eigenvalue 2. (2 marks)

8 The matrix $A = \begin{pmatrix} 4 & 2 & 1 \\ -2 & 0 & 5 \\ 0 & 3 & 4 \end{pmatrix}$

a Show that -2 is an eigenvalue of **A** and that there is only one other distinct eigenvalue. (4 marks)

b Find an eigenvector corresponding to each of the eigenvalues. (4 marks)

9 The matrix $A = \begin{pmatrix} 1 & -1 & 0 \\ -1 & 0 & 1 \\ 1 & 2 & 1 \end{pmatrix}$

Given that 2 is an eigenvalue of **A**,

a find the other two eigenvalues of **A** (4 marks)

b find eigenvectors corresponding to each of the eigenvalues of **A**. (4 marks)

10 Given that $\begin{pmatrix} 2 \\ 2 \\ -1 \end{pmatrix}$ is an eigenvector of the matrix **A** where

$$A = \begin{pmatrix} 4 & 1 & 2 \\ 1 & a & 0 \\ -1 & 1 & b \end{pmatrix}$$

a find the eigenvalue of **A** corresponding to $\begin{pmatrix} 2 \\ 2 \\ -1 \end{pmatrix}$ (2 marks)

b find the value of a and the value of b (4 marks)

c show that **A** has only one real eigenvalue. (2 marks)

Challenge

The linear transformation T is represented by the matrix $\begin{pmatrix} -1 & 0 \\ -2 & 1 \end{pmatrix}$

Explain why T has infinitely many invariant lines, and fully describe all such invariant lines.

6.7 Reducing a symmetric matrix to diagonal form

- If $\mathbf{M}$ is a square matrix such that $\mathbf{MM}^T = \mathbf{I}$, then $\mathbf{M}$ is called an **orthogonal matrix**.

- If $\mathbf{M}$ is an orthogonal matrix, then $\mathbf{M}^{-1} = \mathbf{M}^T$

- Two eigenvectors $\mathbf{x}_1$ and $\mathbf{x}_2$ are **orthogonal** if their scalar product $\mathbf{x}_1.\mathbf{x}_2 = 0$. You learnt in Pure 4 that if non-zero vectors have scalar product zero, they are perpendicular. Perpendicular and orthogonal have the same meaning in this context but it is customary to use the word orthogonal when referring to eigenvectors.

- If $\mathbf{M}$ is an orthogonal matrix consisting of the normalised column vectors $\mathbf{x}_1$, $\mathbf{x}_2$ and $\mathbf{x}_3$, then $\mathbf{x}_1.\mathbf{x}_2 = \mathbf{x}_2.\mathbf{x}_3 = \mathbf{x}_3.\mathbf{x}_1 = 0$ The three vectors are mutually orthogonal.

> **Problem-solving**
>
> To show that a matrix $\mathbf{M}$ is orthogonal, you can either show that $\mathbf{MM}^T = \mathbf{I}$ or that it consists of 3 mutually orthogonal normalised vectors.

- A **diagonal matrix** is a square matrix in which all of the elements which are not on the diagonal from the top left to the bottom right of the matrix are zero. The diagonal from the top left to the bottom right of the matrix is called the **leading diagonal**.

 leading diagonal

The general 2 × 2 diagonal matrix is $\begin{pmatrix} a & 0 \\ 0 & b \end{pmatrix}$

The general 3 × 3 diagonal matrix is $\begin{pmatrix} a & 0 & 0 \\ 0 & b & 0 \\ 0 & 0 & c \end{pmatrix}$

It is possible for some of the elements on the leading diagonal to be 0.

- You reduce a symmetric matrix $\mathbf{A}$ to a diagonal matrix $\mathbf{D}$ using the following procedure.
  1. Find normalised eigenvectors of $\mathbf{A}$.
  2. Form a matrix $\mathbf{P}$ with columns consisting of the normalised eigenvectors of $\mathbf{A}$.
  3. Write down $\mathbf{P}^T$, the transpose of the matrix $\mathbf{P}$.
  4. A diagonal matrix $\mathbf{D}$ is given by $\mathbf{P}^T\mathbf{AP} = \mathbf{D}$

- When you reduce a symmetric matrix $\mathbf{A}$ to a diagonal matrix $\mathbf{D}$, the elements on the diagonal are the eigenvalues of $\mathbf{A}$.

- For any symmetric matrix $\mathbf{A}$, the matrix $\mathbf{P}$ is an orthogonal matrix and so $\mathbf{P}^T = \mathbf{P}^{-1}$ It follows that $\mathbf{P}^T\mathbf{AP}$ and $\mathbf{P}^{-1}\mathbf{AP}$ are identical expressions.

Example 27 — SKILLS: CRITICAL THINKING

The matrix $\mathbf{M} = \begin{pmatrix} a & \frac{2}{3} & \frac{2}{3} \\ -\frac{2}{3} & b & c \\ -\frac{2}{3} & -\frac{1}{3} & \frac{2}{3} \end{pmatrix}$

Given that $\mathbf{MM}^T = \mathbf{I}$,

a find the values of the constants a, b and c.

Given that $\mathbf{x}_1 = \begin{pmatrix} a \\ -\frac{2}{3} \\ -\frac{2}{3} \end{pmatrix}$, $\mathbf{x}_2 = \begin{pmatrix} \frac{2}{3} \\ b \\ -\frac{1}{3} \end{pmatrix}$ and $\mathbf{x}_3 = \begin{pmatrix} \frac{2}{3} \\ c \\ \frac{2}{3} \end{pmatrix}$, using the values of a, b and c found in part **a**,

b show that each of the vectors $\mathbf{x}_1$, $\mathbf{x}_2$ and $\mathbf{x}_3$ is orthogonal to the other two.

a $\mathbf{MM}^T = \begin{pmatrix} a & \frac{2}{3} & \frac{2}{3} \\ -\frac{2}{3} & b & c \\ -\frac{2}{3} & -\frac{1}{3} & \frac{2}{3} \end{pmatrix} \begin{pmatrix} a & -\frac{2}{3} & -\frac{2}{3} \\ \frac{2}{3} & b & -\frac{1}{3} \\ \frac{2}{3} & c & \frac{2}{3} \end{pmatrix}$

$= \begin{pmatrix} \cdots & -\frac{2}{3}a + \frac{2}{3}b + \frac{2}{3}c & -\frac{2}{3}a - \frac{2}{9} + \frac{4}{9} \\ \cdots & \cdots & \frac{4}{9} - \frac{1}{3}b + \frac{2}{3}c \\ \cdots & \cdots & \cdots \end{pmatrix} = \begin{pmatrix} 1 & 0 & 0 \\ 0 & 1 & 0 \\ 0 & 0 & 1 \end{pmatrix}$

> As you only need three equations to find three variables, there is no need to work all of the matrix product out.
>
> It is sensible not to use the elements on the leading diagonal as two of these will contain squared terms which lead to ambiguous signs.

Equating the third elements in the first row

$-\frac{2}{3}a - \frac{2}{9} + \frac{4}{9} = 0 \Rightarrow \frac{2}{3}a = \frac{2}{9} \Rightarrow a = \frac{1}{3}$

Equating the second elements in the first row and substituting $a = \frac{1}{3}$

$-\frac{2}{9} + \frac{2}{3}b + \frac{2}{3}c = 0$

$\frac{2}{3}b + \frac{2}{3}c = \frac{2}{9}$ (1)

Equating the third elements in the second row

$\frac{4}{9} - \frac{1}{3}b + \frac{2}{3}c = 0$

$-\frac{1}{3}b + \frac{2}{3}c = -\frac{4}{9}$ (2)

(2) − (1)

$b = \frac{2}{9} + \frac{4}{9} = \frac{2}{3}$

Substituting $b = \frac{2}{3}$ into (1)

$\frac{4}{9} + \frac{2}{3}c = \frac{2}{9} \Rightarrow \frac{2}{3}c = \frac{2}{9} - \frac{4}{9} = -\frac{2}{9} \Rightarrow c = -\frac{1}{3}$

$a = \frac{1}{3}, b = \frac{2}{3}, c = -\frac{1}{3}$

> Although you are not asked to state it explicitly (i.e. clearly and directly), part **a** establishes that the matrix
>
> $\begin{pmatrix} \frac{1}{3} & \frac{2}{3} & \frac{2}{3} \\ -\frac{2}{3} & \frac{2}{3} & -\frac{1}{3} \\ -\frac{2}{3} & -\frac{1}{3} & \frac{2}{3} \end{pmatrix}$ is an orthogonal matrix.

b $\mathbf{x}_1 \cdot \mathbf{x}_2 = \begin{pmatrix} \frac{1}{3} \\ -\frac{2}{3} \\ -\frac{2}{3} \end{pmatrix} \cdot \begin{pmatrix} \frac{2}{3} \\ \frac{2}{3} \\ -\frac{1}{3} \end{pmatrix} = \frac{1}{3} \times \frac{2}{3} + \left(-\frac{2}{3}\right) \times \frac{2}{3} + \left(-\frac{2}{3}\right) \times \left(-\frac{1}{3}\right)$

$= \frac{2}{9} - \frac{4}{9} + \frac{2}{9} = 0$

Hence $\mathbf{x}_1$ and $\mathbf{x}_2$ are orthogonal.

$\mathbf{x}_2 \cdot \mathbf{x}_3 = \begin{pmatrix} \frac{2}{3} \\ \frac{2}{3} \\ -\frac{1}{3} \end{pmatrix} \cdot \begin{pmatrix} \frac{2}{3} \\ -\frac{1}{3} \\ \frac{2}{3} \end{pmatrix} = \frac{2}{3} \times \frac{2}{3} + \frac{2}{3} \times \left(-\frac{1}{3}\right) + \left(-\frac{1}{3}\right) \times \frac{2}{3}$

$= \frac{4}{9} - \frac{2}{9} - \frac{2}{9} = 0$

Hence $\mathbf{x}_2$ and $\mathbf{x}_3$ are orthogonal.

$\mathbf{x}_1 \cdot \mathbf{x}_3 = \begin{pmatrix} \frac{1}{3} \\ -\frac{2}{3} \\ -\frac{2}{3} \end{pmatrix} \cdot \begin{pmatrix} \frac{2}{3} \\ -\frac{1}{3} \\ \frac{2}{3} \end{pmatrix} = \frac{1}{3} \times \frac{2}{3} + \left(-\frac{2}{3}\right) \times \left(-\frac{1}{3}\right) + \left(-\frac{2}{3}\right) \times \frac{2}{3}$

$= \frac{2}{9} + \frac{2}{9} - \frac{4}{9} = 0$

Hence $\mathbf{x}_2$ and $\mathbf{x}_3$ are orthogonal.

Each of the vectors $\mathbf{x}_1$, $\mathbf{x}_2$ and $\mathbf{x}_3$ is orthogonal to the other two.

> Part b illustrates the general property that the thee column vectors making up a 3 × 3 orthogonal matrix are orthogonal to each other.

Example 28

The matrix $\mathbf{A} = \begin{pmatrix} 2 & -1 \\ -1 & 2 \end{pmatrix}$. Reduce $\mathbf{A}$ to a diagonal matrix.

$\mathbf{A} - \lambda \mathbf{I} = \begin{pmatrix} 2 & -1 \\ -1 & 2 \end{pmatrix} - \begin{pmatrix} \lambda & 0 \\ 0 & \lambda \end{pmatrix} = \begin{pmatrix} 2 - \lambda & -1 \\ -1 & 2 - \lambda \end{pmatrix}$

$\begin{vmatrix} 2 - \lambda & -1 \\ -1 & 2 - \lambda \end{vmatrix} = (2 - \lambda)(2 - \lambda) - 1$

$= \lambda^2 - 4\lambda + 3 = (\lambda - 1)(\lambda - 3)$

$\det(\mathbf{A} - \lambda \mathbf{I}) = 0 \Rightarrow (\lambda - 1)(\lambda - 3) = 0 \Rightarrow \lambda = 1, 3$

> To diagonalise a symmetric matrix, you find normalised eigenvectors of the matrix. You begin by finding the eigenvalues.

For $\lambda = 1$

$\begin{pmatrix} 2 & -1 \\ -1 & 2 \end{pmatrix} \begin{pmatrix} x \\ y \end{pmatrix} = 1 \begin{pmatrix} x \\ y \end{pmatrix}$

$\begin{pmatrix} 2x - y \\ -x + 2y \end{pmatrix} = \begin{pmatrix} x \\ y \end{pmatrix}$

Equating the upper elements

$2x - y = x \Rightarrow y = x$

Let $x = 1$, then $y = 1$

An eigenvector corresponding to the eigenvalue 1 is $\begin{pmatrix} 1 \\ 1 \end{pmatrix}$

The magnitude of $\begin{pmatrix} 1 \\ 1 \end{pmatrix}$ is $\sqrt{(1^2 + 1^2)} = \sqrt{2}$

A normalised eigenvector corresponding to the eigenvalue 1 is

$$\begin{pmatrix} \frac{1}{\sqrt{2}} \\ \frac{1}{\sqrt{2}} \end{pmatrix}$$

> To convert an eigenvector **x** to a normalised eigenvector, you divide each of the elements of **x** by the magnitude of **x**.

For $\lambda = 3$

$$\begin{pmatrix} 2 & -1 \\ -1 & 2 \end{pmatrix}\begin{pmatrix} x \\ y \end{pmatrix} = 3\begin{pmatrix} x \\ y \end{pmatrix}$$

$$\begin{pmatrix} 2x - y \\ -x + 2y \end{pmatrix} = \begin{pmatrix} 3x \\ 3y \end{pmatrix}$$

Equating the upper elements

$$2x - y = 3x \Rightarrow y = -x$$

Let $x = 1$, then $y = -1$

An eigenvector corresponding to the eigenvalue 3 is $\begin{pmatrix} 1 \\ -1 \end{pmatrix}$

The magnitude of $\begin{pmatrix} 1 \\ -1 \end{pmatrix}$ is $\sqrt{(1^2 + (-1)^2)} = \sqrt{2}$

A normalised eigenvector corresponding to the eigenvalue 3 is

$$\begin{pmatrix} \frac{1}{\sqrt{2}} \\ -\frac{1}{\sqrt{2}} \end{pmatrix}$$

> The negative of this vector $\begin{pmatrix} -\frac{1}{\sqrt{2}} \\ \frac{1}{\sqrt{2}} \end{pmatrix}$ is also correct and would be just as appropriate for diagonalising the matrix.

$$\mathbf{P} = \begin{pmatrix} \frac{1}{\sqrt{2}} & \frac{1}{\sqrt{2}} \\ \frac{1}{\sqrt{2}} & -\frac{1}{\sqrt{2}} \end{pmatrix}$$

> You form the orthogonal matrix **P** from the normalised eigenvectors by using the eigenvectors as the columns of the matrix.

$$\mathbf{P}^T = \begin{pmatrix} \frac{1}{\sqrt{2}} & \frac{1}{\sqrt{2}} \\ \frac{1}{\sqrt{2}} & -\frac{1}{\sqrt{2}} \end{pmatrix}$$

> In this case, as **P** is symmetric, $\mathbf{P}^T = \mathbf{P}$

$$\mathbf{P}^T\mathbf{AP} = \begin{pmatrix} \frac{1}{\sqrt{2}} & \frac{1}{\sqrt{2}} \\ \frac{1}{\sqrt{2}} & -\frac{1}{\sqrt{2}} \end{pmatrix}\begin{pmatrix} 2 & -1 \\ -1 & 2 \end{pmatrix}\begin{pmatrix} \frac{1}{\sqrt{2}} & \frac{1}{\sqrt{2}} \\ \frac{1}{\sqrt{2}} & -\frac{1}{\sqrt{2}} \end{pmatrix}$$

$$= \begin{pmatrix} \frac{1}{\sqrt{2}} & \frac{1}{\sqrt{2}} \\ \frac{1}{\sqrt{2}} & -\frac{1}{\sqrt{2}} \end{pmatrix}\begin{pmatrix} \frac{2}{\sqrt{2}} - \frac{1}{\sqrt{2}} & \frac{2}{\sqrt{2}} + \frac{1}{\sqrt{2}} \\ -\frac{1}{\sqrt{2}} + \frac{2}{\sqrt{2}} & -\frac{1}{\sqrt{2}} - \frac{2}{\sqrt{2}} \end{pmatrix}$$

$$= \begin{pmatrix} \frac{1}{\sqrt{2}} & \frac{1}{\sqrt{2}} \\ \frac{1}{\sqrt{2}} & -\frac{1}{\sqrt{2}} \end{pmatrix}\begin{pmatrix} \frac{1}{\sqrt{2}} & \frac{3}{\sqrt{2}} \\ \frac{1}{\sqrt{2}} & -\frac{3}{\sqrt{2}} \end{pmatrix}$$

$$= \begin{pmatrix} \frac{1}{2} + \frac{1}{2} & \frac{3}{2} - \frac{3}{2} \\ \frac{1}{2} - \frac{1}{2} & \frac{3}{2} + \frac{3}{2} \end{pmatrix} = \begin{pmatrix} 1 & 0 \\ 0 & 3 \end{pmatrix}$$

> The non-zero number in the first column, 1, is the eigenvalue corresponding to the eigenvector $\begin{pmatrix} \frac{1}{\sqrt{2}} \\ \frac{1}{\sqrt{2}} \end{pmatrix}$ used as the first column of **P**.
>
> The non-zero number in the second column, 3 is the eigenvalue corresponding to the eigenvector $\begin{pmatrix} \frac{1}{\sqrt{2}} \\ -\frac{1}{\sqrt{2}} \end{pmatrix}$ used as the first column of **P**.

The diagonal matrix is given by

$$\mathbf{D} = \begin{pmatrix} 1 & 0 \\ 0 & 3 \end{pmatrix}$$

> A similar result is true for 3 × 3 matrices. If you take **P** as $\begin{pmatrix} \frac{1}{\sqrt{2}} & \frac{1}{\sqrt{2}} \\ -\frac{1}{\sqrt{2}} & \frac{1}{\sqrt{2}} \end{pmatrix}$, which is a valid choice, then $\mathbf{D} = \begin{pmatrix} 3 & 0 \\ 0 & 1 \end{pmatrix}$

FURTHER MATRIX ALGEBRA — CHAPTER 6

Example 29 SKILLS CRITICAL THINKING

The matrix $\mathbf{A} = \begin{pmatrix} 2 & 0 & 1 \\ 0 & -2 & 0 \\ 1 & 0 & 2 \end{pmatrix}$. Reduce $\mathbf{A}$ to a diagonal matrix.

$$\mathbf{A} - \lambda\mathbf{I} = \begin{pmatrix} 2 & 0 & 1 \\ 0 & -2 & 0 \\ 1 & 0 & 2 \end{pmatrix} - \begin{pmatrix} \lambda & 0 & 0 \\ 0 & \lambda & 0 \\ 0 & 0 & \lambda \end{pmatrix} = \begin{pmatrix} 2-\lambda & 0 & 1 \\ 0 & -2-\lambda & 0 \\ 1 & 0 & 2-\lambda \end{pmatrix}$$

$$\begin{vmatrix} 2-\lambda & 0 & 1 \\ 0 & -2-\lambda & 0 \\ 1 & 0 & 2-\lambda \end{vmatrix} = (2-\lambda)\begin{vmatrix} -2-\lambda & 0 \\ 0 & 2-\lambda \end{vmatrix} - 0\begin{vmatrix} 0 & 0 \\ 1 & 2-\lambda \end{vmatrix} + 1\begin{vmatrix} 0 & -2-\lambda \\ 1 & 0 \end{vmatrix}$$

$$= (2-\lambda)(-2-\lambda)(2-\lambda) + 1(2-\lambda)$$

$$= (2+\lambda)(-(2-\lambda)^2 + 1) = -(2+\lambda)((2-\lambda)^2 - 1)$$

$$= -(\lambda + 2)(\lambda^2 - 4\lambda + 3) = -(\lambda + 2)(\lambda - 1)(\lambda - 3)$$

$\det(\mathbf{A} - \lambda\mathbf{I}) = 0 \Rightarrow -(\lambda + 2)(\lambda - 1)(\lambda - 3) = 0 \Rightarrow \lambda = -2, 1, 3$

For $\lambda = -2$

$$\begin{pmatrix} 2 & 0 & 1 \\ 0 & -2 & 0 \\ 1 & 0 & 2 \end{pmatrix}\begin{pmatrix} x \\ y \\ z \end{pmatrix} = -2\begin{pmatrix} x \\ y \\ z \end{pmatrix}$$

$$\begin{pmatrix} 2x + z \\ -2y \\ x + 2z \end{pmatrix} = \begin{pmatrix} -2x \\ -2y \\ -2z \end{pmatrix}$$

> The orthogonal matrix $\mathbf{P}$ is formed with normalised eigenvectors, so you begin by finding these.

$2x + z = -2x \Rightarrow z = -4x$

Equating the lowest elements and using $z = -4x$

$x + 2z = -2z \Rightarrow x = -4z = 16x \Rightarrow x = 0$

As $z = -4x$, $z = 0$

Equating the middle elements

$-2y = -2y$, an identity

> As the only equation in y is an identity then y can take any non-zero value. The simplest value to take is 1.

Equating the top elements

Let $y = 1$

An eigenvector corresponding to -2 is $\begin{pmatrix} 0 \\ 1 \\ 0 \end{pmatrix}$

> As this vector has magnitude 1, it is already normalised and can be used to form the orthogonal matrix $\mathbf{P}$ without modification.

For $\lambda = 1$

$$\begin{pmatrix} 2 & 0 & 1 \\ 0 & -2 & 0 \\ 1 & 0 & 2 \end{pmatrix}\begin{pmatrix} x \\ y \\ z \end{pmatrix} = 1\begin{pmatrix} x \\ y \\ z \end{pmatrix}$$

$$\begin{pmatrix} 2x + z \\ -2y \\ x + 2z \end{pmatrix} = \begin{pmatrix} x \\ y \\ z \end{pmatrix}$$

Equating the top elements

$2x + z = x \Rightarrow z = -x$

Let $x = 1$, then $z = -1$

Equating the middle elements

$-2y = y \Rightarrow y = 0$

An eigenvector corresponding to 1 is $\begin{pmatrix} 1 \\ 0 \\ -1 \end{pmatrix}$

The magnitude of $\begin{pmatrix} 1 \\ 0 \\ -1 \end{pmatrix}$ is $\sqrt{(1^2 + 0^2 + (-1)^2)} = \sqrt{2}$

A normalised eigenvector corresponding to 1 is $\begin{pmatrix} \frac{1}{\sqrt{2}} \\ 0 \\ -\frac{1}{\sqrt{2}} \end{pmatrix}$

> To convert an eigenvector **x** to a normalised eigenvector, you divide each of the elements of **x** by the magnitude of **x**.

For $\lambda = 3$

$\begin{pmatrix} 2 & 0 & 1 \\ 0 & -2 & 0 \\ 1 & 0 & 2 \end{pmatrix} \begin{pmatrix} x \\ y \\ z \end{pmatrix} = 3 \begin{pmatrix} x \\ y \\ z \end{pmatrix}$

$\begin{pmatrix} 2x + z \\ -2y \\ x + 2z \end{pmatrix} = \begin{pmatrix} 3x \\ 3y \\ 3z \end{pmatrix}$

Equating the top elements

$2x + z = 3x \Rightarrow z = x$

Let $x = 1$, then $z = 1$

Equating the middle elements

$-2y = 3y \Rightarrow y = 0$

An eigenvector corresponding to 3 is $\begin{pmatrix} 1 \\ 0 \\ 1 \end{pmatrix}$

The magnitude of $\begin{pmatrix} 1 \\ 0 \\ 1 \end{pmatrix}$ is $\sqrt{(1^2 + 0^2 + 1^2)} = \sqrt{2}$

A normalised eigenvector corresponding to 3 is $\begin{pmatrix} \frac{1}{\sqrt{2}} \\ 0 \\ \frac{1}{\sqrt{2}} \end{pmatrix}$

> If you have carried out the calculation correctly, the normalised eigenvectors of a symmetric matrix are orthogonal and, as a check on your work, it is worth checking that this is so. In this case
>
> $\begin{pmatrix} \frac{1}{\sqrt{2}} \\ 0 \\ -\frac{1}{\sqrt{2}} \end{pmatrix} \cdot \begin{pmatrix} \frac{1}{\sqrt{2}} \\ 0 \\ \frac{1}{\sqrt{2}} \end{pmatrix} = \frac{1}{\sqrt{2}} \times \frac{1}{\sqrt{2}} + 0 \times 0 + \left(-\frac{1}{\sqrt{2}}\right) \times \frac{1}{\sqrt{2}} = 0,$
>
> which confirms that the vectors are orthogonal and that the working is correct.

$\mathbf{P} = \begin{pmatrix} 0 & \frac{1}{\sqrt{2}} & \frac{1}{\sqrt{2}} \\ 1 & 0 & 0 \\ 0 & -\frac{1}{\sqrt{2}} & \frac{1}{\sqrt{2}} \end{pmatrix}$

$\mathbf{P}^T = \begin{pmatrix} 0 & 1 & 0 \\ \frac{1}{\sqrt{2}} & 0 & -\frac{1}{\sqrt{2}} \\ \frac{1}{\sqrt{2}} & 0 & \frac{1}{\sqrt{2}} \end{pmatrix}$

> You form the orthogonal matrix **P** from the normalised eigenvectors by using the eigenvectors as the columns of the matrix.

$$P^TAP = \begin{pmatrix} 0 & 1 & 0 \\ \frac{1}{\sqrt{2}} & 0 & -\frac{1}{\sqrt{2}} \\ \frac{1}{\sqrt{2}} & 0 & \frac{1}{\sqrt{2}} \end{pmatrix} \begin{pmatrix} 2 & 0 & 1 \\ 0 & -2 & 0 \\ 1 & 0 & 2 \end{pmatrix} \begin{pmatrix} 0 & \frac{1}{\sqrt{2}} & \frac{1}{\sqrt{2}} \\ 1 & 0 & 0 \\ 0 & -\frac{1}{\sqrt{2}} & \frac{1}{\sqrt{2}} \end{pmatrix}$$

$$= \begin{pmatrix} 0 & 1 & 0 \\ \frac{1}{\sqrt{2}} & 0 & -\frac{1}{\sqrt{2}} \\ \frac{1}{\sqrt{2}} & 0 & \frac{1}{\sqrt{2}} \end{pmatrix} \begin{pmatrix} 0 & \frac{2}{\sqrt{2}} - \frac{1}{\sqrt{2}} & \frac{2}{\sqrt{2}} + \frac{1}{\sqrt{2}} \\ -2 & 0 & 0 \\ 0 & \frac{1}{\sqrt{2}} - \frac{2}{\sqrt{2}} & \frac{1}{\sqrt{2}} + \frac{2}{\sqrt{2}} \end{pmatrix}$$

$$= \begin{pmatrix} 0 & 1 & 0 \\ \frac{1}{\sqrt{2}} & 0 & -\frac{1}{\sqrt{2}} \\ \frac{1}{\sqrt{2}} & 0 & \frac{1}{\sqrt{2}} \end{pmatrix} \begin{pmatrix} 0 & \frac{1}{\sqrt{2}} & \frac{3}{\sqrt{2}} \\ -2 & 0 & 0 \\ 0 & -\frac{1}{\sqrt{2}} & \frac{3}{\sqrt{2}} \end{pmatrix}$$

$$= \begin{pmatrix} -2 & 0 & 0 \\ 0 & \frac{1}{2} + \frac{1}{2} & 0 \\ 0 & 0 & \frac{3}{2} + \frac{3}{2} \end{pmatrix} = \begin{pmatrix} -2 & 0 & 0 \\ 0 & 1 & 0 \\ 0 & 0 & 3 \end{pmatrix}$$

$$D = \begin{pmatrix} -2 & 0 & 0 \\ 0 & 1 & 0 \\ 0 & 0 & 3 \end{pmatrix}$$

> The non-zero elements in the diagonal matrix are the eigenvalues of the matrix **A** in the order corresponding to the order in which you used the normalised eigenvectors to form the orthogonal matrix **P**.

Example 30   SKILLS   REASONING, ARGUMENTION

The matrix $A = \begin{pmatrix} 7 & 5 & 5 \\ 5 & -2 & 4 \\ 5 & 4 & -2 \end{pmatrix}$

a Verify that $\begin{pmatrix} 2 \\ 1 \\ 1 \end{pmatrix}$ is an eigenvector of **A** and find the corresponding eigenvalue.

b Show that -6 is another eigenvalue of **A** and find the corresponding eigenvector.

c Given that the third eigenvector of **A** is $\begin{pmatrix} 1 \\ -1 \\ -1 \end{pmatrix}$, find a matrix **P** and a diagonal matrix **D** such that $P^{-1}AP = D$

a $\begin{pmatrix} 7 & 5 & 5 \\ 5 & -2 & 4 \\ 5 & 4 & -2 \end{pmatrix} \begin{pmatrix} 2 \\ 1 \\ 1 \end{pmatrix} = \begin{pmatrix} 14+5+5 \\ 10-2+4 \\ 10+4-2 \end{pmatrix} = \begin{pmatrix} 24 \\ 12 \\ 12 \end{pmatrix} = 12 \begin{pmatrix} 2 \\ 1 \\ 1 \end{pmatrix}$

> To show that a column vector **x** is an eigenvector of **A**, you have to find a constant such that $Ax = \lambda x$

Hence $\begin{pmatrix} 2 \\ 1 \\ 1 \end{pmatrix}$ is an eigenvector of **A** corresponding to the eigenvalue 12.

b $A - \lambda I = \begin{pmatrix} 7 & 5 & 5 \\ 5 & -2 & 4 \\ 5 & 4 & -2 \end{pmatrix} - \begin{pmatrix} \lambda & 0 & 0 \\ 0 & \lambda & 0 \\ 0 & 0 & \lambda \end{pmatrix} = \begin{pmatrix} 7-\lambda & 5 & 5 \\ 5 & -2-\lambda & 4 \\ 5 & 4 & -2-\lambda \end{pmatrix}$

When $\lambda = -6$

$\det(A - \lambda I) = \begin{vmatrix} 7-(-6) & 5 & 5 \\ 5 & -2-(-6) & 4 \\ 5 & 4 & -2-(-6) \end{vmatrix} = \begin{vmatrix} 13 & 5 & 5 \\ 5 & 4 & 4 \\ 5 & 4 & 4 \end{vmatrix}$

$= 13 \begin{vmatrix} 4 & 4 \\ 4 & 4 \end{vmatrix} - 5 \begin{vmatrix} 5 & 4 \\ 5 & 4 \end{vmatrix} + 5 \begin{vmatrix} 5 & 4 \\ 5 & 4 \end{vmatrix}$

$= 13(16 - 16) - 5(20 - 20) + 5(20 - 20) = 0$

Hence -6 is an eigenvalue of **A**.

> To show that -6 is an eigenvalue, it is sufficient to show that substituting $\lambda = -6$ into the determinant of $A - \lambda I$ gives 0. You do not have to solve the cubic characteristic equation completely.

To find an eigenvector corresponding to -6

$$\begin{pmatrix} 7 & 5 & 5 \\ 5 & -2 & 4 \\ 5 & 4 & -2 \end{pmatrix} \begin{pmatrix} x \\ y \\ z \end{pmatrix} = -6 \begin{pmatrix} x \\ y \\ z \end{pmatrix}$$

$$\begin{pmatrix} 7x + 5y + 5z \\ 5x - 2y + 4z \\ 5x + 4y - 2z \end{pmatrix} = \begin{pmatrix} -6x \\ -6y \\ -6z \end{pmatrix}$$

Equating the top elements

$7x + 5y + 5z = -6x \Rightarrow 5y + 5z = -13x$

$y + z = -\frac{13}{5}x$ (1)

Equating the middle elements

$5x - 2y + 4z = -6y \Rightarrow 4y + 4z = -5x$

$y + z = -\frac{5}{4}x$ (2)

From (1) and (2)

$-\frac{13}{5}x = -\frac{5}{4}x \Rightarrow x = 0$

Substituting $x = 0$ into (2)

$4y + 4z = 0 \Rightarrow z = -y$

Let $y = 1$, then $z = -1$

An eigenvector corresponding to the eigenvalue -6 is $\begin{pmatrix} 0 \\ 1 \\ -1 \end{pmatrix}$

c To find the eigenvalue corresponding to $\begin{pmatrix} 0 \\ -1 \\ -1 \end{pmatrix}$

$$\begin{pmatrix} 7 & 5 & 5 \\ 5 & -2 & 4 \\ 5 & 4 & -2 \end{pmatrix} \begin{pmatrix} 1 \\ -1 \\ -1 \end{pmatrix} = \begin{pmatrix} 7 - 5 - 5 \\ 5 + 2 - 4 \\ 5 - 4 + 2 \end{pmatrix} = \begin{pmatrix} -3 \\ 3 \\ 3 \end{pmatrix} = -3 \begin{pmatrix} 1 \\ -1 \\ -1 \end{pmatrix}$$

The corresponding eigenvalue is -3.

The magnitude of $\begin{pmatrix} 2 \\ 1 \\ 1 \end{pmatrix}$ is $\sqrt{(2^2 + 1^2 + 1^2)} = \sqrt{6}$

A normalised eigenvector corresponding to 12 is $\begin{pmatrix} \frac{2}{\sqrt{6}} \\ \frac{1}{\sqrt{6}} \\ \frac{1}{\sqrt{6}} \end{pmatrix}$

The magnitude of $\begin{pmatrix} 0 \\ 1 \\ -1 \end{pmatrix}$ is $\sqrt{(0^2 + 1^2 + (-1)^2)} = \sqrt{2}$

A normalised eigenvector corresponding to -6 is $\begin{pmatrix} 0 \\ \frac{1}{\sqrt{2}} \\ -\frac{1}{\sqrt{2}} \end{pmatrix}$

> You will need the eigenvalue corresponding to this eigenvector for the third non-zero element of the diagonal matrix **D**. You already know that the other two elements are 12 and -6.

> The matrix **P** is made up of columns of normalised eigenvalues. **P** is an orthogonal matrix and so $\mathbf{P}^T = \mathbf{P}^{-1}$. Hence there is no difference between the expression $\mathbf{P}^{-1}\mathbf{AP}$, used to diagonalise **A** in this example and the expression $\mathbf{P}^T\mathbf{AP}$, used in Examples 28 and 29.

The magnitude of $\begin{pmatrix} 1 \\ 1 \\ -1 \end{pmatrix}$ is $\sqrt{(1^2 + (-1)^2 + (-1)^2)} = \sqrt{3}$

A normalised eigenvector corresponding to -3 is $\begin{pmatrix} \frac{1}{\sqrt{3}} \\ -\frac{1}{\sqrt{3}} \\ -\frac{1}{\sqrt{3}} \end{pmatrix}$

$\mathbf{P} = \begin{pmatrix} \frac{2}{\sqrt{6}} & 0 & \frac{1}{\sqrt{3}} \\ \frac{1}{\sqrt{6}} & \frac{1}{\sqrt{2}} & -\frac{1}{\sqrt{3}} \\ -\frac{1}{\sqrt{6}} & -\frac{1}{\sqrt{2}} & -\frac{1}{\sqrt{3}} \end{pmatrix}$

$\mathbf{D} = \begin{pmatrix} 12 & 0 & 0 \\ 0 & -6 & 0 \\ 0 & 0 & -3 \end{pmatrix}$

> You know that **P** is a matrix made up with normalised eigenvectors as its columns and that **D** is the diagonal matrix with corresponding eigenvalues as the elements of the leading diagonals. Multiplying the matrices out is a laborious process and you should not do this unless the question requires it.

Exercise 6G

SKILLS PROBLEM-SOLVING AND REASONING/ARGUMENTATION

1 Reduce the following matrices to diagonal matrices.

   a $\begin{pmatrix} 1 & 3 \\ 3 & 1 \end{pmatrix}$ **b** $\begin{pmatrix} 1 & -2 \\ -2 & 4 \end{pmatrix}$

2 The matrix $\mathbf{A} = \begin{pmatrix} 3 & \sqrt{2} \\ \sqrt{2} & 4 \end{pmatrix}$

   a Find the eigenvalues of **A**.   **(3 marks)**

   b Find normalised eigenvectors of **A** corresponding to each of the two eigenvalues of **A**.   **(4 marks)**

   c Write down a matrix **P** and a diagonal matrix **D** such that $\mathbf{P^TAP} = \mathbf{D}$ **(2 marks)**

3 The matrix $\mathbf{P} = \begin{pmatrix} \frac{1}{\sqrt{6}} & -\frac{1}{\sqrt{3}} & \frac{1}{\sqrt{2}} \\ \frac{1}{\sqrt{6}} & -\frac{1}{\sqrt{3}} & -\frac{1}{\sqrt{2}} \\ \frac{2}{\sqrt{6}} & \frac{1}{\sqrt{3}} & 0 \end{pmatrix}$

   a Show that **P** is an orthogonal matrix.

The matrix $\mathbf{A} = \begin{pmatrix} \frac{3}{2} & -\frac{3}{2} & 1 \\ -\frac{3}{2} & \frac{3}{2} & 1 \\ 1 & 1 & 1 \end{pmatrix}$

   b Show that $\mathbf{P^TAP}$ is a diagonal matrix.

4 The matrix $\mathbf{A} = \begin{pmatrix} 2 & 0 & 2 \\ 0 & 2 & 0 \\ 2 & 0 & 2 \end{pmatrix}$. Reduce **A** to a diagonal matrix.

E/P **5** The matrix $\mathbf{A} = \begin{pmatrix} 5 & 3 & 3 \\ 3 & 1 & 1 \\ 3 & 1 & 1 \end{pmatrix}$

The eigenvalues of $\mathbf{A}$ are 0, −1 and 8.

a Find a normalised eigenvector corresponding to the eigenvalue 0. **(2 marks)**

Given that $\begin{pmatrix} -1 \\ 1 \\ 1 \end{pmatrix}$ is an eigenvector of $\mathbf{A}$ corresponding to the eigenvalue −1 and that $\begin{pmatrix} 2 \\ 1 \\ 1 \end{pmatrix}$ is an eigenvector of $\mathbf{A}$ corresponding to the eigenvalue 8,

b find a matrix $\mathbf{P}$ and a diagonal matrix $\mathbf{D}$ such that $\mathbf{P}^{-1}\mathbf{AP} = \mathbf{D}$ **(3 marks)**

E/P **6** The matrix $\mathbf{A} = \begin{pmatrix} 7 & 0 & 2 \\ 0 & 5 & -2 \\ -2 & -2 & 6 \end{pmatrix}$

a Given that 9 is an eigenvalue of $\mathbf{A}$, find the other two eigenvalues of $\mathbf{A}$. **(4 marks)**

b Find eigenvectors of $\mathbf{A}$ corresponding to each of the three eigenvalues of $\mathbf{A}$. **(4 marks)**

c Find a matrix $\mathbf{P}$ and a diagonal matrix $\mathbf{D}$ such that $\mathbf{P}^T\mathbf{AP} = \mathbf{D}$ **(2 marks)**

E/P **7** The matrix $\mathbf{A} = \begin{pmatrix} 1 & 2 & 0 \\ 2 & 1 & \sqrt{5} \\ 0 & \sqrt{5} & 1 \end{pmatrix}$

a Show that 4 is an eigenvalue of $\mathbf{A}$ and find the other two eigenvalues of $\mathbf{A}$. **(4 marks)**

b Find a normalised eigenvector of $\mathbf{A}$ corresponding to the eigenvalue 4. **(3 marks)**

Given that $\begin{pmatrix} -2 \\ 3 \\ -\sqrt{5} \end{pmatrix}$ and $\begin{pmatrix} \sqrt{5} \\ 0 \\ -2 \end{pmatrix}$ are eigenvectors of $\mathbf{A}$,

c find a matrix $\mathbf{P}$ and a diagonal matrix $\mathbf{D}$ such that $\mathbf{P}^{-1}\mathbf{AP} = \mathbf{D}$ **(2 marks)**

E/P **8** The eigenvalues of the matrix

$$\mathbf{A} = \begin{pmatrix} 2 & 2 & -3 \\ 2 & 2 & 3 \\ -3 & 3 & 3 \end{pmatrix}$$

are $\lambda_1, \lambda_2, \lambda_3$, where $\lambda_1 > \lambda_2 > \lambda_3$

a Show that $\lambda_1 = 6$ and find the other two eigenvalues λ_2 and λ_3. **(4 marks)**

b Verify that $\det(\mathbf{A}) = \lambda_1\lambda_2\lambda_3$ **(2 marks)**

c Find an eigenvector corresponding to the value $\lambda_1 = 6$ **(2 marks)**

Given that $\begin{pmatrix} 1 \\ 1 \\ 0 \end{pmatrix}$ and $\begin{pmatrix} 1 \\ -1 \\ 1 \end{pmatrix}$ are eigenvectors corresponding to λ_2 and λ_3 respectively,

d write down a matrix $\mathbf{P}$ such that $\mathbf{P}^T\mathbf{AP}$ is a diagonal matrix. **(3 marks)**

> **Challenge**
>
> SKILLS CREATIVITY
>
> The matrix $\mathbf{M} = \begin{pmatrix} \frac{1}{9} & \frac{8}{9} & -\frac{4}{9} \\ \frac{8}{9} & \frac{1}{9} & \frac{4}{9} \\ -\frac{4}{9} & \frac{4}{9} & \frac{7}{9} \end{pmatrix}$ represents a reflection in plane Π
>
> Find the eigenvalues and eigenvectors of $\mathbf{M}$ and hence find the Cartesian equation of the plane Π.

Chapter review 6

 1 $\mathbf{A} = \begin{pmatrix} 1 & 0 & 2 \\ t & 3 & 1 \\ -2 & -1 & 1 \end{pmatrix}$

Given that $\mathbf{A}$ is singular, find the value of t. **(3 marks)**

 2 $\mathbf{M} = \begin{pmatrix} 1 & 0 & 0 \\ x & 2 & 0 \\ 3 & 1 & 1 \end{pmatrix}$

Find $\mathbf{M}^{-1}$ in terms of x. **(4 marks)**

 3 The matrix $\mathbf{M}$ has eigenvalues $\lambda_1 = 5$ and $\lambda_2 = -15$ and $\mathbf{M} = \begin{pmatrix} 1 & 8 \\ 8 & -11 \end{pmatrix}$

   a For each eigenvalue, find a corresponding eigenvector. **(4 marks)**

   b Find a matrix $\mathbf{P}$ such that $\mathbf{P}^T \mathbf{A} \mathbf{P} = \begin{pmatrix} 5 & 0 \\ 0 & -15 \end{pmatrix}$ **(3 marks)**

(P) **4** The matrix $\mathbf{A} = \begin{pmatrix} 5 & 2 \\ 2 & 1 \end{pmatrix}$ and the matrix $\mathbf{B} = \begin{pmatrix} 2 & -1 \\ -4 & 2 \end{pmatrix}$

   a Find $\mathbf{AB}$.

   b Verify that $\mathbf{B}^T\mathbf{A}^T = (\mathbf{AB})^T$

5 A transformation $T: \mathbb{R}^2 \to \mathbb{R}^2$ is represented by the matrix

$\mathbf{A} = \begin{pmatrix} -5 & 8 \\ 3 & -7 \end{pmatrix}$

   a Find the eigenvalues of $\mathbf{A}$.

   b Find Cartesian equations of the two lines passing through the origin which are invariant under T.

 6 Given that 1 is an eigenvalue of the matrix

$\begin{pmatrix} 3 & 1 & 0 \\ 2 & 4 & 0 \\ 1 & 0 & 1 \end{pmatrix}$

   a find a corresponding eigenvector, **(4 marks)**

   b find the other eigenvalues of the matrix. **(3 marks)**

7 The transformation $T: \mathbb{R}^3 \to \mathbb{R}^3$ is represented by the matrix **T** where $\mathbf{T} = \begin{pmatrix} 4 & 3 & 0 \\ 0 & -2 & 1 \\ 3 & 1 & -2 \end{pmatrix}$

The line l_1 is transformed by T to the line l_2. The line l_1 has vector equation

$\mathbf{r} = \begin{pmatrix} 1 \\ 0 \\ 2 \end{pmatrix} + t \begin{pmatrix} 2 \\ -3 \\ 0 \end{pmatrix}$, where t is a real parameter.

Find Cartesian equations of l_2. **(5 marks)**

8 $\mathbf{A} = \begin{pmatrix} 3 & 4 & -4 \\ 4 & 5 & 0 \\ -4 & 0 & 1 \end{pmatrix}$

a Show that 3 is an eigenvalue of **A** and find the other two eigenvalues. **(3 marks)**
b Find an eigenvector corresponding to the eigenvalue 3. **(2 marks)**

Given that the vectors $\begin{pmatrix} 2 \\ 2 \\ -1 \end{pmatrix}$ and $\begin{pmatrix} 2 \\ -1 \\ 2 \end{pmatrix}$ are eigenvectors corresponding to the other two eigenvalues,

c find a matrix **P** such that $\mathbf{P}^T\mathbf{AP}$ is a diagonal matrix. **(4 marks)**

9 $\mathbf{A} = \begin{pmatrix} 2 & -2 & 0 \\ -2 & 1 & 2 \\ 0 & 2 & 5 \end{pmatrix}$

a Show that $\begin{pmatrix} 2 \\ 3 \\ -1 \end{pmatrix}$ and $\begin{pmatrix} 2 \\ -1 \\ 1 \end{pmatrix}$ are eigenvectors of **A**, giving their corresponding eigenvalues. **(6 marks)**

b Given that 6 is the third eigenvalue of **A**, find a corresponding eigenvector. **(2 marks)**
c Hence write down a matrix such that $\mathbf{P}^{-1}\mathbf{AP}$ is a diagonal matrix. **(2 matrix)**

10 a Calculate the inverse of the matrix

$\mathbf{A}(x) = \begin{pmatrix} 1 & x & -1 \\ 3 & 0 & 2 \\ 1 & 1 & 0 \end{pmatrix}$, $x \neq \frac{5}{2}$ **(4 marks)**

The image of the vector $\begin{pmatrix} a \\ b \\ c \end{pmatrix}$ when it is transformed by the matrix $\begin{pmatrix} 1 & 3 & -1 \\ 3 & 0 & 2 \\ 1 & 1 & 0 \end{pmatrix}$ is the

vector $\begin{pmatrix} 4 \\ 3 \\ 5 \end{pmatrix}$

b Find the values of a, b and c. **(5 marks)**

11 a Show that for all values of the constant α, an eigenvalue of the matrix **A** is 1, where

$\mathbf{A} = \begin{pmatrix} \alpha & 0 & 2 \\ 4 & 3 & 0 \\ -2 & -1 & 1 \end{pmatrix}$ **(3 marks)**

An eigenvector of the matrix **A** is $\begin{pmatrix} 2 \\ -2 \\ 1 \end{pmatrix}$ and the corresponding eigenvalue is β ($\beta \neq 1$).

b Find the value of α and the value of β. **(4 marks)**

c For your value of α, find the third eigenvalue of **A**. **(2 marks)**

12 The matrix **A** is defined by $\mathbf{A} = \begin{pmatrix} 1 & -1 & 3 \\ 2 & 1 & u \\ 0 & 1 & 1 \end{pmatrix}$

a Find $\mathbf{A}^{-1}$ in terms of u, stating the condition for which **A** is non-singular. **(4 marks)**

The image vector of $\begin{pmatrix} a \\ b \\ c \end{pmatrix}$ when transformed by the matrix $\mathbf{A} = \begin{pmatrix} 1 & -1 & 3 \\ 2 & 1 & 4 \\ 0 & 1 & 1 \end{pmatrix}$ is $\begin{pmatrix} -2.8 \\ 5.3 \\ 2.3 \end{pmatrix}$

b Find the values of a, b and c. **(4 marks)**

13 $\mathbf{M} = \begin{pmatrix} 3 & 0 & 0 \\ 1 & 1 & 1 \\ 4 & -1 & 3 \end{pmatrix}$

a Show that the matrix **M** has only two distinct eigenvalues. **(4 marks)**

b Find an eigenvector corresponding to each of these eigenvalues. **(4 marks)**

14 The matrix $\mathbf{P} = \begin{pmatrix} \frac{1}{2} & -\frac{1}{2} & \frac{1}{\sqrt{2}} \\ \frac{1}{2} & -\frac{1}{2} & -\frac{1}{\sqrt{2}} \\ \frac{1}{\sqrt{2}} & \frac{1}{\sqrt{2}} & 0 \end{pmatrix}$

a Show that the matrix **P** is orthogonal. **(4 marks)**

The transformation $P: \mathbb{R}^3 \to \mathbb{R}^3$ is represented by the matrix **P**.

The plane Π_1 is transformed by A to the plane Π_2.
The plane Π_2 has Cartesian equation $x + y - \sqrt{2}z = 0$

b Find a Cartesian equation of the plane Π_1. **(3 marks)**

15 a Determine the eigenvalues of the matrix

$$\mathbf{A} = \begin{pmatrix} 3 & -3 & 6 \\ 0 & 2 & -8 \\ 0 & 0 & -2 \end{pmatrix}$$

b Show that $\begin{pmatrix} 3 \\ 1 \\ 0 \end{pmatrix}$ is an eigenvector of **A**.

$$\mathbf{B} = \begin{pmatrix} 7 & -6 & 2 \\ 1 & 2 & 3 \\ 1 & -3 & 2 \end{pmatrix}$$

c Show that $\begin{pmatrix} 3 \\ 1 \\ 0 \end{pmatrix}$ is an eigenvector of **B** and write down the corresponding eigenvalue.

d Hence, or otherwise, write down an eigenvector of the matrix **AB**, and state the corresponding eigenvalue.

E/P 16 $\mathbf{A} = \begin{pmatrix} 1 & 0 & 1 \\ 3 & 1 & 1 \\ 4 & 2 & 7 \end{pmatrix}$

    a Showing your working, find $\mathbf{A}^{-1}$. **(4 marks)**

    The transformation $T: \mathbb{R}^3 \to \mathbb{R}^3$ is represented by the matrix $\mathbf{A}$.

    b Find Cartesian equations of the line which is mapped by T onto the line

$$x = \frac{y}{4} = \frac{z}{3}$$ **(5 marks)**

Challenge

The **trace** (tr) of a matrix is defined as the sum of the elements along the leading diagonal.

Let $\mathbf{A} = \begin{pmatrix} a & b \\ c & d \end{pmatrix}$ and $\mathbf{B} = \begin{pmatrix} e & f \\ g & h \end{pmatrix}$

a Show that $\operatorname{tr}(\mathbf{AB}) = \operatorname{tr}(\mathbf{BA})$

b Hence prove that, if there exists a non-singular matrix $\mathbf{P}$ such that
$\mathbf{P}^{-1}\mathbf{MP} = \begin{pmatrix} p & 0 \\ 0 & q \end{pmatrix}$ then the trace of matrix $\mathbf{P} = p + q$

Summary of key points

1. The **transpose** of a matrix, $\mathbf{A}^\mathrm{T}$, is obtained by interchanging the rows and columns of the matrix $\mathbf{A}$.

2. If $\mathbf{A} = \mathbf{A}^\mathrm{T}$, the matrix $\mathbf{A}$ is **symmetric**.

3. The identity 3×3 matrix is $\mathbf{I} = \begin{pmatrix} 1 & 0 & 0 \\ 0 & 1 & 0 \\ 0 & 0 & 1 \end{pmatrix}$

4. The zero 3×3 matrix is $\mathbf{0} = \begin{pmatrix} 0 & 0 & 0 \\ 0 & 0 & 0 \\ 0 & 0 & 0 \end{pmatrix}$

5. If $\mathbf{A}$ and $\mathbf{B}$ are matrices with dimensions $n \times m$ and $m \times p$, then $(\mathbf{AB})^\mathrm{T} = \mathbf{B}^\mathrm{T}\mathbf{A}^\mathrm{T}$

6. $\begin{vmatrix} a & b & c \\ d & e & f \\ g & h & i \end{vmatrix} = a\begin{vmatrix} e & f \\ h & i \end{vmatrix} - b\begin{vmatrix} d & f \\ g & i \end{vmatrix} + c\begin{vmatrix} d & e \\ g & h \end{vmatrix}$

7. If $\det(\mathbf{A}) = 0$, then $\mathbf{A}$ is a **singular** matrix.

   If $\det(\mathbf{A}) \neq 0$, then $\mathbf{A}$ is a **non-singular** matrix.

8 If the matrix **A** is non-singular then it has an inverse $\mathbf{A}^{-1}$ and $\mathbf{AA}^{-1} = \mathbf{A}^{-1}\mathbf{A} = \mathbf{I}$

9 The **minor** of an element of a 3 × 3 matrix is the determinant of the elements which remain when the row and the column containing the element are crossed out.

10 To find the inverse of a non-singular 3 × 3 matrix **A**:
   1 Find the determinant of **A**, det(**A**).
   2 Form **M**, the matrix of the minors of **A**.
   3 Form **C**, the matrix of cofactors, by changing the signs of some elements of **M** according to the rule of alternating signs.
   4 Write down the transpose, $\mathbf{C}^T$, of the matrix of cofactors.
   5 The inverse of the matrix **A** is given by the formula $\mathbf{A}^{-1} = \dfrac{1}{\det(\mathbf{A})} \mathbf{C}^T$

11 If **A** and **B** are non-singular matrices then $(\mathbf{AB})^{-1} = \mathbf{B}^{-1}\mathbf{A}^{-1}$

12 If the transformation T is represented by the matrix **T** and the transformation U is represented by the matrix **U**, then the matrix **UT** represents the combined transformation of the transformation T followed by the transformation U.

13 An **eigenvector** of a matrix **A** is a non-zero column vector **x** which satisfies the equation $\mathbf{Ax} = \lambda \mathbf{x}$, where λ is a scalar which is the corresponding **eigenvalue** of the matrix.

14 The equation $\det(\mathbf{A} - \lambda \mathbf{I}) = 0$ is the **characteristic equation** of **A**. You solve this equation of a matrix to find the eigenvalues of the matrix.

15 If $\begin{pmatrix} a \\ b \\ c \end{pmatrix}$ is an eigenvector of a matrix, then $\begin{pmatrix} \dfrac{a}{\sqrt{(a^2 + b^2 + c^2)}} \\ \dfrac{a}{\sqrt{(a^2 + b^2 + c^2)}} \\ \dfrac{c}{\sqrt{(a^2 + b^2 + c^2)}} \end{pmatrix}$ is the corresponding **normalised eigenvector**.

16 If **M** is a square matrix such that $\mathbf{MM}^T = \mathbf{I}$, then **M** is called an **orthogonal matrix**.

17 If **M** is an orthogonal matrix, then $\mathbf{M}^{-1} = \mathbf{M}^T$

18 Two eigenvectors $\mathbf{x}_1$ and $\mathbf{x}_2$ are **orthogonal** if their scalar product $\mathbf{x}_1.\mathbf{x}_2 = 0$

19 If **M** is an orthogonal matrix consisting of the normalised column vectors $\mathbf{x}_1$, $\mathbf{x}_2$ and $\mathbf{x}_3$, then $\mathbf{x}_1.\mathbf{x}_2 = \mathbf{x}_2.\mathbf{x}_3 = \mathbf{x}_3.\mathbf{x}_1 = 0$

20 A **diagonal matrix** is a square matrix in which all of the elements which are not on the diagonal from the top left to the bottom right of the matrix are zero. $\begin{pmatrix} a & 0 \\ 0 & b \end{pmatrix}$ and $\begin{pmatrix} d & 0 & 0 \\ 0 & e & 0 \\ 0 & 0 & f \end{pmatrix}$ and are diagonal matrices.

21 To reduce a symmetric matrix **A** to a diagonal matrix **D**:
  1 Find normalised eigenvectors of **A**.
  2 Form an orthogonal matrix **P** with columns consisting of the normalised eigenvectors of **A**.
  3 Write down $\mathbf{P}^T$, the transpose of the matrix **P**.
  4 The diagonal matrix **D** is given by $\mathbf{P}^T\mathbf{AP} = \mathbf{D}$

22 When symmetric matrix **A** is reduced to a diagonal matrix **D**, the elements on the diagonal are the eigenvalues of **A**.

Review exercise

1 Find the magnitude of the vector
$(-\mathbf{i} - \mathbf{j} + \mathbf{k}) \times (-\mathbf{i} + \mathbf{j} - \mathbf{k})$ (3)

← Further Pure 3 Section 5.2

2 Given that $\mathbf{p} = 3\mathbf{i} + \mathbf{k}$ and $\mathbf{q} = \mathbf{i} + 3\mathbf{j} + c\mathbf{k}$, find the value of the constant c for which the vector $(\mathbf{p} \times \mathbf{q}) + \mathbf{p}$ is parallel to the vector $\mathbf{k}$. (4)

← Further Pure 3 Section 5.1

3 Referred to a fixed origin O, the position vectors of three non-linear points A, B and C are $\mathbf{a}$, $\mathbf{b}$ and $\mathbf{c}$ respectively. By considering $\overrightarrow{AB} \times \overrightarrow{AC}$, prove that the area of triangle ABC can be expressed in the form $\frac{1}{2}|\mathbf{a} \times \mathbf{b} + \mathbf{b} \times \mathbf{c} + \mathbf{c} \times \mathbf{a}|$ (5)

← Further Pure 3 Sections 5.1, 5.3

4 The figure shows a right prism with triangular ends ABC and DEF, and parallel edges AD, BE, CF.

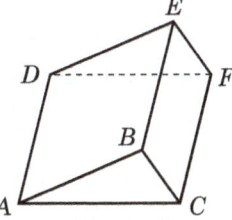

Given that
A is $(2, 7, -1)$, B is $(5, 8, 2)$, C is $(6, 7, 4)$ and D is $(12, 1, -9)$
a find $\overrightarrow{AB} \times \overrightarrow{AC}$ (3)
b find $\overrightarrow{AD}.(\overrightarrow{AB} \times \overrightarrow{AC})$ (3)
c Calculate the volume of the prism. (2)

← Further Pure 3 Sections 5.1, 5.3

5 The plane Π_1 has vector equation
$\mathbf{r} = (5\mathbf{i} + \mathbf{j}) + u(-4\mathbf{i} + \mathbf{j} + 3\mathbf{k}) + v(\mathbf{j} + 2\mathbf{k})$,
where u and v are parameters.
a Find a vector $\mathbf{n}_1$ normal to Π_1. (3)
The plane Π_2 has equation
$3x + y - z = 3$
b Write down a vector $\mathbf{n}_2$ normal to Π_2. (1)

c Show that $4\mathbf{i} + 13\mathbf{j} + 25\mathbf{k}$ is perpendicular to both $\mathbf{n}_1$ and $\mathbf{n}_2$. (3)
Given that the point $(1, 1, 1)$ lies on both Π_1 and Π_2,
d write down an equation of the line of intersection of Π_1 and Π_2 in the form $\mathbf{r} = \mathbf{a} + t\mathbf{b}$, where t is a parameter. (2)

← Further Pure 3 Section 5.5

6 The points A, B and C lie on the plane Π and, relative to a fixed origin O, they have position vectors
$\mathbf{a} = 3\mathbf{i} - \mathbf{j} + 4\mathbf{k}$, $\mathbf{b} = -\mathbf{i} + 2\mathbf{j}$,
$\mathbf{c} = 5\mathbf{i} - 3\mathbf{j} + 7\mathbf{k}$
respectively.
a Find $\overrightarrow{AB} \times \overrightarrow{AC}$. (3)
b Obtain the equation of Π in the form $\mathbf{r}.\mathbf{n} = p$. (3)
The point D has position vector $5\mathbf{i} + 2\mathbf{j} + 3\mathbf{k}$.
c Calculate the volume of the tetrahedron $ABCD$. (4)

← Further Pure 3 Sections 5.1, 5.4 and 5.5

7 The points A and B have position vectors $4\mathbf{i} + \mathbf{j} - 7\mathbf{k}$ and $2\mathbf{i} + 6\mathbf{j} + 2\mathbf{k}$ respectively relative to a fixed origin O.
a Show that angle AOB is a right angle. (3)
b Find a vector equation for the median AM of the triangle OAB. (4)
c Find a vector equation of the plane OAB, giving your answer in the form $\mathbf{r}.\mathbf{n} = p$ (4)

← Further Pure 3 Section 5.4

8 Referred to a fixed origin O, the point A has position vector $a(4\mathbf{i} + \mathbf{j} + 2\mathbf{k})$ and the plane Π has equation

$\mathbf{r}.(\mathbf{i} - 5\mathbf{j} + 3\mathbf{k}) = 5a$

where a is a scalar constant.

a Show that A lies in the plane Π. (5)

The point B has position vector $a(2\mathbf{i} + 11\mathbf{j} - 4\mathbf{k})$.

b Show that $\overrightarrow{BA}$ is perpendicular to the plane Π. (4)

c Calculate, to the nearest one tenth of a degree, $\angle OBA$. (3)

← Further Pure 3 Section 5.5

9 The points A, B, C and D have coordinates $(3, 1, 2)$, $(5, 2, -1)$, $(6, 4, 5)$ and $(-7, 6, -3)$ respectively.

a Find $\overrightarrow{AC} \times \overrightarrow{AD}$. (3)

b Find a vector equation of the line through A which is perpendicular to $\overrightarrow{AC}$ and $\overrightarrow{AD}$. (3)

c Verify that B lies on this line. (1)

d Find the volume of the tetrahedron $ABCD$. (4)

← Further Pure 3 Sections 5.1, 5.4

10 The line l_1 has equation

$\mathbf{r} = \mathbf{i} + 6\mathbf{j} - \mathbf{k} + \lambda(2\mathbf{i} + 3\mathbf{k})$

and the line l_2 has equation

$\mathbf{r} = 3\mathbf{i} + p\mathbf{j} + \mu(\mathbf{i} - 2\mathbf{j} + \mathbf{k})$

where p is a constant.

The plane Π_1 contains l_1 and l_2.

a Find a vector which is normal to Π_1. (3)

b Show that an equation for Π_1 is $6x + y - 4z = 16$. (3)

c Find the value of p. (2)

The plane Π_2 has equation $\mathbf{r}.(\mathbf{i} + 2\mathbf{j} + \mathbf{k}) = 2$.

d Find an equation for the line of intersection of Π_1 and Π_2, giving your answer in the form

$(\mathbf{r} - \mathbf{a}) \times \mathbf{b} = 0$ (4)

← Further Pure 3 Section 5.5

11 The plane Π passes through the points $A(-2, 3, 5)$, $B(1, -3, 1)$ and $C(4, -6, -7)$.

a Find $\overrightarrow{AC} \times \overrightarrow{BC}$ (3)

b Hence, or otherwise, find the equation of the plane Π in the form $\mathbf{r}.\mathbf{n} = p$. (3)

The perpendicular from the point $(25, 5, 7)$ to Π meets the plane at the point F.

c Find the coordinates of F. (2)

← Further Pure 3 Section 5.5

12 The plane Π passes through the points $P(-1, 3, -2)$, $Q(4, -1, -1)$ and $R(3, 0, c)$, where c is a constant.

a Find, in terms of c, $\overrightarrow{RP} \times \overrightarrow{RQ}$ (3)

Given that $\overrightarrow{RP} \times \overrightarrow{RQ} = 3\mathbf{i} + d\mathbf{j} + \mathbf{k}$, where d is a constant,

b find the value of c and show that $d = 4$. (4)

c Find an equation of Π in the form $\mathbf{r}.\mathbf{n} = p$, where p is a constant.

The point S has position vector $\mathbf{i} + 5\mathbf{j} + 10\mathbf{k}$. The point S' is the image of S under reflection in Π.

d Find the position vector of S'.

← Further Pure 3 Sections 5.1, 5.5

13 The points A, B and C lie on the plane Π_1 and, relative to a fixed origin O, they have position vectors respectively.

$\mathbf{a} = \mathbf{i} + 3\mathbf{j} - \mathbf{k}$, $\mathbf{b} = 3\mathbf{i} + 3\mathbf{j} - 4\mathbf{k}$, $\mathbf{c} = 5\mathbf{i} - 2\mathbf{j} - 2\mathbf{k}$

a Find $(\mathbf{b} - \mathbf{a}) \times (\mathbf{c} - \mathbf{a})$ (3)

b Find an equation of Π_1, giving your answer in the form $\mathbf{r}.\mathbf{n} = p$ (4)

The plane Π_2 has Cartesian equation $x + z = 3$ and Π_1 and Π_2 intersect in the line l.

c Find an equation of l in the form $(\mathbf{r} - \mathbf{p}) \times \mathbf{q} = 0$ (3)

The point P is the point on l that is nearest to the origin O.

d Find the coordinates of P. (4)

← Further Pure 3 Section 5.5

14 The points $A(2, 0, -1)$ and $B(4, 3, 1)$ have position vectors **a** and **b** respectively with respect to a fixed origin O.

  a Find $\mathbf{a} \times \mathbf{b}$ (3)

The plane Π_1 contains the points O, A and B.

  b Verify that an equation of Π_1 is
  $x - 2y + 2z = 0$ (2)

The plane Π_2 has equation $\mathbf{r}.\mathbf{n} = d$ where $\mathbf{n} = 3\mathbf{i} + \mathbf{j} - \mathbf{k}$ and d is a constant. Given that B lies on Π_2,

  c find the value of d. (4)

The planes Π_1 and Π_2 intersect in the line L.

  d Find an equation of L in the form $\mathbf{r} = \mathbf{p} + t\mathbf{q}$, where t is a parameter. (2)

  e Find the position vector of the point X on L where OX is perpendicular to L. (2)

  ← Further Pure 3 Sections 5.1, 5.5

15 The points A, B and C have position vectors, relative to a fixed origin O,

$\mathbf{a} = 2\mathbf{i} - \mathbf{j}$, $\mathbf{b} = \mathbf{i} + 2\mathbf{j} + 3\mathbf{k}$
$\mathbf{c} = 2\mathbf{i} + 3\mathbf{j} + 2\mathbf{k}$

respectively. The plane Π passes through A, B and C.

  a Find $\overrightarrow{AB} \times \overrightarrow{AC}$ (3)

  b Show that a Cartesian equation of Π is $3x - y + 2z = 7$ (3)

The line l has equation
$(\mathbf{r} - 5\mathbf{i} - 5\mathbf{j} - 3\mathbf{k}) \times (2\mathbf{i} - \mathbf{j} - 2\mathbf{k}) = 0$.
The line l and the plane Π intersect at the point T.

  c Find the coordinates of T. (2)

  d Show that A, B and T lie on the same straight line. (3)

  ← Further Pure 3 Sections 5.2, 5.5

16 The plane Π passes through the points $A(-1, -1, 1)$, $B(4, 2, 1)$ and $C(2, 1, 0)$.

  a Find a vector equation of the line perpendicular to Π which passes through the point $D(1, 2, 3)$. (2)

  b Find the volume of the tetrahedron $ABCD$. (4)

  c Obtain the equation of Π in the form $\mathbf{r}.\mathbf{n} = p$ (3)

The perpendicular from D to the plane Π meets Π at the point E.

  d Find the coordinates of E. (2)

  e Show that $DE = \dfrac{11\sqrt{35}}{35}$ (3)

The point D' is the reflection of D in Π.

  f Find the coordinates of D'. (2)

  ← Further Pure 3 Section 5.5

17 The points A, B and C have position vectors $(\mathbf{j} + 2\mathbf{k})$, $(2\mathbf{i} + 3\mathbf{j} + \mathbf{k})$ and $(\mathbf{i} + \mathbf{j} + 3\mathbf{k})$, respectively, relative to the origin O. The plane Π contains the points A, B and C.

  a Find a vector which is perpendicular to Π. (2)

  b Find the area of triangle ABC. (3)

  c Find a vector equation of Π in the form $\mathbf{r}.\mathbf{n} = p$. (3)

  d Hence, or otherwise, obtain a Cartesian equation of Π. (2)

  e Find the distance of the origin O from Π. (3)

The point D has position vector $(3\mathbf{i} + 4\mathbf{j} + \mathbf{k})$. The distance of D from Π is $\dfrac{1}{\sqrt{17}}$.

  f Using this distance, or otherwise, calculate the acute angle between the line AD and Π, giving your answer in degrees to one decimal place. (4)

  ← Further Pure 3 Sections 5.4, 5.5

18 Relative to a fixed origin O the lines l_1 and l_2 have equations

$l_1 : \mathbf{r} = -\mathbf{i} + 2\mathbf{j} - 4\mathbf{k} + s(-2\mathbf{i} + \mathbf{j} + 3\mathbf{k})$,
$l_2 : \mathbf{r} = -\mathbf{j} + 7\mathbf{k} + t(-\mathbf{i} + \mathbf{j} - \mathbf{k})$

where s and t are variable parameters.

  a Show that the lines intersect and are perpendicular to each other. (3)

b Find a vector equation of the straight line l_3 which passes through the point of intersection of l_1 and l_2 and the point with position vector $4\mathbf{i} + \lambda\mathbf{j} - 3\mathbf{k}$, where λ is a real number. (3)

The line l_3 makes an angle θ with the plane containing l_1 and l_2.

c Find $\sin\theta$ in terms of λ. (3)

Given that l_1, l_2 and l_3 are coplanar,

d find the value of λ. (3)

← Further Pure 3 Sections 5.2, 5.5

E/P 19 Referred to a fixed origin O, the planes Π_1 and Π_2 have equations $\mathbf{r}.(2\mathbf{i} - \mathbf{j} + 2\mathbf{k}) = 9$ and $\mathbf{r}.(4\mathbf{i} + 3\mathbf{j} - \mathbf{k}) = 8$ respectively.

a Determine the shortest distance from O to the line of intersection of Π_1 and Π_2. (5)

b Find, in vector form, an equation of the plane Π_3 which is perpendicular to Π_1 and Π_2 and passes through the point with position vector $2\mathbf{j} + \mathbf{k}$. (3)

c Find the position vector of the point that lies in Π_1, Π_2 and Π_3. (2)

← Further Pure 3 Section 5.5

E/P 20 Vector equations of the two straight lines l and m are respectively
$$\mathbf{r} = \mathbf{j} + 3\mathbf{k} + t(2\mathbf{i} + \mathbf{j} - \mathbf{k})$$
$$\mathbf{r} = \mathbf{i} + \mathbf{j} - \mathbf{k} + u(-2\mathbf{i} + \mathbf{j} + \mathbf{k})$$

a Show that these lines do not intersect.

The point A with parameter t_1 lies on l and the point B with parameter u_1 lies on m. (3)

b Write down the vector $\overrightarrow{AB}$ in terms of $\mathbf{i}, \mathbf{j}, \mathbf{k}, t_1$ and u_1. (2)

Given that the line AB is perpendicular to both l and m,

c find the values of t_1 and u_1 and show that, in this case, the length of AB is $\dfrac{7}{\sqrt{5}}$. (4)

← Further Pure 3 Sections 5.1, 5.2, 5.3

E/P 21 $\mathbf{A} = \begin{pmatrix} 1 & 1 & 2 \\ 0 & 1 & 1 \\ 0 & 0 & 1 \end{pmatrix}$

Prove by induction, that for all positive integers n,

$$\mathbf{A}^n = \begin{pmatrix} 1 & n & \frac{1}{2}(n^2 + 3n) \\ 0 & 1 & n \\ 0 & 0 & 1 \end{pmatrix}$$ (8)

← Further Pure 3 Section 6.1

E/P 22 $\mathbf{A} = \begin{pmatrix} k & 1 & -2 \\ 0 & -1 & k \\ 9 & 1 & 0 \end{pmatrix}$

where k is a real constant.

a Find the values of k for which $\mathbf{A}$ is singular. (4)

Given that $\mathbf{A}$ is non-singular,

b find, in terms of k, $\mathbf{A}^{-1}$. (4)

← Further Pure 3 Section 6.2

E/P 23 The matrix $\mathbf{M}$ is given by
$$\mathbf{M} = \begin{pmatrix} 1 & 4 & -1 \\ 3 & 0 & p \\ a & b & c \end{pmatrix}$$
where p, a, b and c are constants and $a > 0$.

Given that $\mathbf{MM}^T = k\mathbf{I}$ for some constant k, find

a the value of p, (3)

b the value of k, (3)

c the values of a, b and c, (3)

d $\det \mathbf{M}$. (2)

← Further Pure 3 Section 6.3

E/P 24 **a** Given that $\mathbf{A} = \begin{pmatrix} 1 & 1 & 2 \\ 0 & 2 & 1 \\ 1 & 0 & 2 \end{pmatrix}$ find $\mathbf{A}^2$. (3)

b Using $\mathbf{A}^3 = \begin{pmatrix} 10 & 9 & 23 \\ 5 & 9 & 14 \\ 9 & 5 & 19 \end{pmatrix}$ show that
$$\mathbf{A}^3 - 5\mathbf{A}^2 + 6\mathbf{A} - \mathbf{I} = 0$$ (4)

c Deduce that
$$\mathbf{A}(\mathbf{A} - 2\mathbf{I})(\mathbf{A} - 3\mathbf{I}) = \mathbf{I}$$ (2)

d Hence find $\mathbf{A}^{-1}$. (3)

← Further Pure 3 Sections 6.2, 6.3

25 Given that $\mathbf{A} = \begin{pmatrix} 1 & 0 & 0 \\ 0 & 2 & 1 \\ 0 & 0 & 1 \end{pmatrix}$ use matrix multiplication to find

  a $\mathbf{A}^2$ (3)
  b $\mathbf{A}^3$. (3)
  c Prove by induction that
  $$\mathbf{A}^n = \begin{pmatrix} 1 & 0 & 0 \\ 0 & 2^n & 2^n - 1 \\ 0 & 0 & 1 \end{pmatrix} n \geq 1 \quad (6)$$
  d Find the inverse of $\mathbf{A}^n$. (3)
  ← Further Pure 3 Section 6.2

26 $\mathbf{A} = \begin{pmatrix} 3 & 1 & -1 \\ 1 & 1 & 1 \\ 5 & 3 & u \end{pmatrix} u \neq 1$

  a Show that $\det \mathbf{A} = 2(u - 1)$. (4)
  b Find the inverse of $\mathbf{A}$.
  The image of the vector $\begin{pmatrix} a \\ b \\ c \end{pmatrix}$ when transformed by the matrix $\begin{pmatrix} 3 & 1 & -1 \\ 1 & 1 & 1 \\ 5 & 3 & 6 \end{pmatrix}$
  is $\begin{pmatrix} 3 \\ 1 \\ 6 \end{pmatrix}$ (3)
  c Find the values of a, b and c. (4)
  ← Further Pure 3 Section 6.1

27 The transformation R is represented by the matrix $\mathbf{M}$, where
$$\mathbf{M} = \begin{pmatrix} 3 & a & 0 \\ 2 & b & 0 \\ c & 0 & 1 \end{pmatrix}$$
and where a, b and c are constants.
Given that $\mathbf{M} = \mathbf{M}^{-1}$,
  a find the values of a, b and c, (4)
  b evaluate the determinant of $\mathbf{M}$, (3)
  c find an equation satisfied by all the points which remain invariant under R. (3)
  ← Further Pure 3 Section 6.2

28 The transformation $T: \mathbb{R}^3 \to \mathbb{R}^3$ is represented by the matrix $\mathbf{M}$.
The vector $\begin{pmatrix} 2 \\ -1 \\ 0 \end{pmatrix}$ is transformed by T to $\begin{pmatrix} -5 \\ -1 \\ 0 \end{pmatrix}$, the vector $\begin{pmatrix} 0 \\ -1 \\ 2 \end{pmatrix}$ is transformed to $\begin{pmatrix} -1 \\ 9 \\ 0 \end{pmatrix}$ and the vector $\begin{pmatrix} \alpha \\ 0 \\ 1 \end{pmatrix}$ is transformed to $\begin{pmatrix} -\alpha + 1 \\ 5 \\ 2\alpha + 2 \end{pmatrix}$ where α ($\alpha \neq -1$) is a constant.

  a Find $\mathbf{M}$. (6)
  The plane Π_1 has equation
  $$\mathbf{r} = \begin{pmatrix} 3 \\ 0 \\ 1 \end{pmatrix} + \lambda \begin{pmatrix} 2 \\ -1 \\ 0 \end{pmatrix} + \mu \begin{pmatrix} 0 \\ -1 \\ 2 \end{pmatrix}$$
  where λ and μ are parameters, and T transforms Π_1 to the plane Π_2.
  b Find a Cartesian equation of Π_2. (4)
  ← Further Pure 3 Sections 6.2, 6.3, 6.4

29 The transformation $S: \mathbb{R}^3 \to \mathbb{R}^3$ maps the point $\begin{pmatrix} x \\ y \\ z \end{pmatrix}$ onto the point $\begin{pmatrix} a \\ b \\ c \end{pmatrix}$ where
$$\begin{aligned} a &= x + y - z \\ b &= y + z \\ c &= z, \end{aligned}$$
The matrix of this transformation is $\mathbf{A}$.
  a By solving the given equations for x, y and z in terms of a, b and c, or otherwise, write down the matrix $\mathbf{A}^{-1}$. (4)
  The transformation $T: \mathbb{R}^3 \to \mathbb{R}^3$ has matrix
  $$\mathbf{B} = \begin{pmatrix} 1 & -2 & 2 \\ 2 & -1 & -2 \\ 2 & 2 & 1 \end{pmatrix}$$
  b Given that $\mathbf{BB}^\mathsf{T} = k\mathbf{I}$, find the value of k. (5)
  U is the composite transformation consisting of T followed by S.

c Find the point whose image under U

is $\begin{pmatrix} 1 \\ 0 \\ 1 \end{pmatrix}$ (3)

← Further Pure 3 Sections 6.4, 6.5

E/P 30 $\mathbf{M} = \begin{pmatrix} 4 & -5 \\ 6 & -9 \end{pmatrix}$

a Find the eigenvalues of **M**. (4)

A transformation $T: \mathbb{R}^2 \to \mathbb{R}^2$ is represented by the matrix **M**. There is a line through the origin for which every point on the line is mapped onto itself under T.

b Find the Cartesian equation of this line. (3)

← Further Pure 3 Section 6.6

E/P 31 A transformation $T: \mathbb{R}^2 \to \mathbb{R}^2$ is represented by the matrix

$\mathbf{A} = \begin{pmatrix} k & 2 \\ 6 & -1 \end{pmatrix}$, where k is a constant.

For the case $k = -4$,

a find the image under T of the line with equation $y = 2x + 1$ (3)

For the case $k = 2$, find

b the two eigenvalues of **A**, (4)

c a Cartesian equation of the two lines passing through the origin which are invariant under T. (4)

← Further Pure 3 Section 6.5

E/P 32 The eigenvalues of the matrix **M**, where

$\mathbf{M} = \begin{pmatrix} 4 & -2 \\ 1 & 1 \end{pmatrix}$

are λ_1 and λ_2, where $\lambda_1 < \lambda_2$

a Find the value of λ_1 and the value of λ_2. (3)

b Find $\mathbf{M}^{-1}$. (3)

c Verify that the eigenvalues of $\mathbf{M}^{-1}$ are λ_1^{-1} and λ_2^{-1}. (2)

A transformation $T: \mathbb{R}^2 \to \mathbb{R}^2$ is represented by the matrix **M**. There are two lines, passing through the origin, each of which is mapped onto itself under the transformation T.

d Find Cartesian equations for each of these lines. (3)

← Further Pure 3 Section 6.6

E/P 33 Find the eigenvalues and corresponding eigenvectors for the matrix

$\begin{pmatrix} 2 & -3 & 1 \\ 3 & 1 & 3 \\ -5 & 2 & -4 \end{pmatrix}$ (7)

← Further Pure 3 Section 6.6

E/P 34 Given that $\begin{pmatrix} 0 \\ 1 \\ -1 \end{pmatrix}$ is an eigenvector of the matrix **A** where

$\mathbf{A} = \begin{pmatrix} 3 & 4 & p \\ -1 & q & -4 \\ 1 & 1 & 3 \end{pmatrix}$

a find the eigenvalue of **A** corresponding to $\begin{pmatrix} 0 \\ 1 \\ -1 \end{pmatrix}$ (3)

b find the value of p and the value of q.

The image of the vector $\begin{pmatrix} l \\ m \\ n \end{pmatrix}$ when transformed by **A** is $\begin{pmatrix} 10 \\ -4 \\ 3 \end{pmatrix}$ (3)

c Using the values of p and q from part **b**, find the values of the constants l, m and n. (4)

← Further Pure 3 Section 6.6

E/P 35 $\mathbf{A} = \begin{pmatrix} 5 & 1 & -2 \\ -1 & 6 & 1 \\ 0 & 1 & 3 \end{pmatrix}$

a Show that 3 is an eigenvalue of **A**. (2)

b Find the other two eigenvalues of **A**. (3)

c Find also a normalised eigenvector corresponding to the eigenvalue 3. (4)

← Further Pure 3 Section 6.6

36 $A = \begin{pmatrix} 3 & 2 & 4 \\ 2 & 0 & 2 \\ 4 & 2 & k \end{pmatrix}$

a Show that $\det A = 20 - 4k$ (3)

b Find A^{-1}. (3)

Given that $k = 3$ and that $\begin{pmatrix} 0 \\ 2 \\ -1 \end{pmatrix}$ is an eigenvector of A,

c find the corresponding eigenvalue.

Given that the only other distinct eigenvalue of A is 8, (3)

d find a corresponding eigenvector. (3)

← Further Pure 3 Section 6.6

37 $A = \begin{pmatrix} 1 & 0 & 4 \\ 0 & 5 & 4 \\ 4 & 4 & 3 \end{pmatrix}$

a Verify that $\begin{pmatrix} 2 \\ -2 \\ 1 \end{pmatrix}$ is an eigenvector of A and find the corresponding eigenvalue. (4)

b Show that 9 is another eigenvalue of A and find the corresponding eigenvector. (4)

c Given that the third eigenvector of A is $\begin{pmatrix} 2 \\ 1 \\ -2 \end{pmatrix}$, find a matrix P and a diagonal matrix D such that $P^T A P = D$ (4)

← Further Pure 3 Section 6.6

38 $A = \begin{pmatrix} 6 & 2 & -3 \\ 2 & 0 & 0 \\ -3 & 0 & 2 \end{pmatrix}$

Given that $\lambda = -1$ and $\lambda = 8$ are two eigenvalues of A,

a find the third eigenvalue of A. (2)

b Find the normalised eigenvector corresponding to the eigenvalue $\lambda = 8$ (3)

Given that $\begin{pmatrix} \frac{1}{\sqrt{14}} \\ \frac{2}{\sqrt{14}} \\ \frac{3}{\sqrt{14}} \end{pmatrix}$ and $\begin{pmatrix} \frac{1}{\sqrt{6}} \\ \frac{-2}{\sqrt{6}} \\ \frac{1}{\sqrt{6}} \end{pmatrix}$ are normalised eigenvectors corresponding to the other two eigenvalues,

c find a matrix P such that $P^T A P$ is a diagonal matrix. (5)

d Find $P^T A P$. (3)

← Further Pure 3 Section 6.6

39 $M = \begin{pmatrix} 1 & 0 & 1 \\ 0 & 2 & 0 \\ 4 & 3 & 1 \end{pmatrix}$

a Find the eigenvalues and corresponding eigenvectors of M. (5)

The transformation $T: \mathbb{R}^3 \to \mathbb{R}^3$ is represented by the matrix M.

b Find Cartesian equations of the image of the line
$$\frac{x}{2} = y = \frac{z}{-1}$$
under this transformation. (4)

← Further Pure 3 Section 6.6

40 a Show that 9 is an eigenvalue of the matrix $\begin{pmatrix} 6 & -2 & 2 \\ -2 & 5 & 0 \\ 2 & 0 & 7 \end{pmatrix}$ (2)

b Find the other two eigenvalues of the matrix. (2)

c Find also normalised eigenvectors x_1, x_2 and x_3 corresponding to each of these eigenvalues. (4)

d Verify that the matrix P with columns x_1, x_2 and x_3 is an orthogonal matrix. (3)

← Further Pure 3 Sections 6.6, 6.7

Challenge

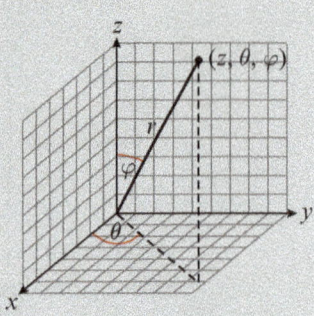

1 **Spherical polar coordinates** are defined by the distance from the origin, r, the 'azimuthal angle' (measured anti-clockwise from the x-axis in the xy-plane), θ, and the 'polar angle' (measured from the positive z-axis), φ.

A line L passes through the origin and the point with spherical polar coordinates $\left(3, \frac{\pi}{4}, \frac{\pi}{3}\right)$

a Find, in their simplest form, the direction cosines of L.

b Find, in terms of θ and φ, expressions for the direction cosines of the line which passes through the origin and the point with spherical coordinates (r, θ, φ).

2 The **trace** (tr) of a matrix is defined as the sum of the elements along the leading diagonal.

Let $\mathbf{A} = \begin{pmatrix} a & b \\ c & d \end{pmatrix}$ and $\mathbf{B} = \begin{pmatrix} e & f \\ g & h \end{pmatrix}$

a Show that $\operatorname{tr}(\mathbf{AB}) = \operatorname{tr}(\mathbf{BA})$

b Hence prove that, if there exists a non-singular matrix $\mathbf{P}$ such that $\mathbf{P}^{-1}\mathbf{MP} = \begin{pmatrix} p & 0 \\ 0 & q \end{pmatrix}$, then the trace of matrix $\mathbf{M}$ is equal to $p + q$

Exam practice

Mathematics
International Advanced Subsidiary/ Advanced Level Further Pure 3

Time: 1 hour 30 minutes
You must have: Mathematical Formulae and Statistical Tables, Calculator
Answer ALL questions

1. The line $x = 6$ is a directrix of the ellipse with equation
$$\frac{x^2}{p^2} + \frac{y^2}{q^2} = 1$$
The point $(3, 0)$ is the corresponding focus.

   a Show that the value of p is $3\sqrt{2}$ (3)

   b Find the value of q. (2)

2. a Starting from the definitions of $\sinh x$ and $\cosh x$ in terms of exponentials, show that
$$\cosh 2x = 1 + \sinh^2 x$$ (3)

   b Solve the equation
$$\cosh 2x - 2\sinh x - 16 = 0$$ (5)

3. $I_n = \int_0^3 \frac{x^n}{\sqrt{(9 - x^2)}} \, dx \quad n \geq 0$

   a Find an expression for $\int \frac{x}{\sqrt{9 - x^2}} \, dx, \quad 0 \leq x \leq 3$ (2)

   b Hence, or otherwise, show that $I_n = \frac{9(n-1)}{n} I_{n-2} \quad n \geq 2$ (5)

   c Find I_4 in the form $k\pi$, where k is a rational number. (4)

4. a Differentiate $x \operatorname{arsinh} 2x$ with respect to x (3)

   b Hence, or otherwise, find the exact value of $\int_0^{\sqrt{2}} x \operatorname{arsinh} 2x \, dx$

   Give your answer in the form $A \ln B + C$ where A, B and C are real numbers. (7)

5. The curve with parametric equations $x = \cosh 2\theta, \; y = 4 \sinh \theta \quad 0 \leq \theta \leq 1$
is rotated through 2π radians about the x-axis.
Find the exact area of the surface generated. (7)

6 The position vectors of the points A, B and C relative to an origin O are

$\mathbf{i} - 2\mathbf{j} - 2\mathbf{k}$, $7\mathbf{i} - 3\mathbf{k}$ and $4\mathbf{i} + 4\mathbf{j}$ respectively

Find

a $\overrightarrow{AC} \times \overrightarrow{BC}$ (4)

b the area of triangle ABC (2)

c an equation of the plane ABC in the form $\mathbf{r.n} = p$ (2)

7 i Find without using a calculator

$$\int_0^5 \frac{1}{\sqrt{15 + 2x - x^2}} \, dx$$

Give your answer as a multiple of π (4)

ii a Show that

$$5\cosh x - 4\sinh x = \frac{e^{2x} + 9}{2e^{2x}}$$ (4)

b Hence using algebraic integration, find

$$\int \frac{1}{\cosh x - 4\sinh x} \, dx$$ (3)

8 $\mathbf{M} = \begin{pmatrix} 1 & 0 & 3 \\ 0 & -2 & 1 \\ k & 0 & 1 \end{pmatrix}$ where k is a constant

Given that $\begin{pmatrix} 6 \\ 1 \\ 6 \end{pmatrix}$ is an eigenvector of $\mathbf{M}$

a find the eigenvalue of $\mathbf{M}$ corresponding to $\begin{pmatrix} 6 \\ 1 \\ 6 \end{pmatrix}$ (2)

b show that $k = 3$ (2)

c show that $\mathbf{M}$ has exactly two eigenvectors. (4)

d A transformation $T: \mathbb{R}^3 \to \mathbb{R}^3$ is represented by $\mathbf{M}$.

The transformation T maps the line l_1 with Cartesian equations

$$\frac{x-2}{1} = \frac{y}{-3} = \frac{z+1}{4}, \text{ on to the line } l_2$$

Taking $k = 3$, find Cartesian equations of l_2 (5)

TOTAL FOR PAPER: 75 MARKS

GLOSSARY

algebraic representing mathematical information with letters and numbers as symbols

arc a smooth curve joining two points

arcsin the **inverse** sine function

arccos the **inverse** cosine function

arctan the **inverse** tangent function

arcosh the **inverse hyperbolic** cosine function

arsinh the inverse **hyperbolic** sine function

artanh the inverse **hyperbolic** tangent function

asymptote a line that a curve approaches but never quite reaches

calculus the mathematical study of continuous change

cartesian coordinates a unique point in a plane specified by a pair of numerical **coordinates**

chord a straight line segment whose endpoints lie on a curve or a circle

conic section the curve obtained when a plane **intersects** with a cone. The three types of **conic section** are a **hyperbola**, a **parabola** and an **ellipse**

constant a term that does not include a **variable**. In the expression $3x^3 - 5x + 4$, the constant term is 4

coordinates a set of values that show an exact position. In a two-dimensional grid, the first number represents a point on the x-axis and the second number represents a point on the y-axis

cosecant the trigonometric **function** defined as $\dfrac{1}{\sin x}$

cosine the trigonometric function that is equal to the ratio of the side adjacent to an acute angle (in a right-angle triangle) to the hypotenuse

cosech (hyperbolic cosecant) is defined as $\dfrac{1}{\operatorname{sech} x}$

cosh (hyperbolic cosine) is defined as $\dfrac{e^x + e^{-y}}{2}$

cotangent the trigonometric **function** defined as $\dfrac{1}{\tan x}$

coth (hyperbolic cotangent) is defined as $\dfrac{1}{\tanh x}$

derivative a way to represent the rate of change. In other words, $\dfrac{dy}{dx}$ is the first derivative and $\dfrac{d^2y}{dx^2}$ is the second derivative

determinant the value calculated from the elements of a **matrix**

For the matrix $\begin{pmatrix} a & b \\ c & c \end{pmatrix}$ the determinant is calculated as $\dfrac{1}{ad - bc}$

differentiation the **instantaneous** rate of change of a function with respect to one of its variables

directrix (of a **conic section**) a line which together with the **focus**, defines a **conic section** as the **locus** of points whose distance from the focus is proportional to the horizontal distance from the **directrix**

domain the set of values that a function exists for. For example, $f(x) = \sqrt{x}$ only exists for $x \geqslant 0$

eccentricity a parameter associated with a conic section. It is denoted by the letter e. It can be thought of as the measure in which the conic section deviates from the circle (which has no eccentricity)

eigenvalue a scalar (number) associated with a matrix or a system of equations.

eigenvector (of a **matrix**) a non-zero column vector **x** which satisfies the equation $\mathbf{Ax} = \lambda\mathbf{x}$ where λ is a **scalar** and is known as the **eigenvalue**

ellipse the set of all points where the sum of the distance from two fixed points is **constant**

equidistant at an equal distance from two or more places

equation a statement that values of two mathematical expressions are equal. Solving an **equation** consists of determining the value(s) of the variable

exponential something is said to increase or decrease exponentially if its rate of change is expressed using exponents. A graph of such a rate is a curve that continually becomes steeper or shallower. For example:

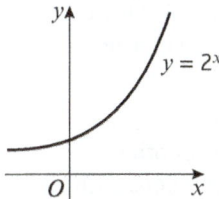

finite a value that is limited or of fixed size; not **infinite**

focus lies on the axis of **symmetry** of a **conic section**

function – the relationship between a set of inputs and a set of possible outputs, where each input is related to exactly one output

gradient the slope of a line

horizontal flat and level rather than up and down

hyperbola (rectangular) a symmetrical curve which has **parametric** equations $x = ct$, $y = \dfrac{c}{t}$, $t \in \mathbb{R}$ or in Cartesian form $xy = c^2$ where c is a positive constant

hyperbolic functions are a set of six functions that are related to the hyperbola

GLOSSARY

identity an equality that is always true regardless of the values of the variables
For example $3(2x - y) = 6x - 3y$

infinite without limit; not **finite**

instantaneous happening immediately

integer a whole number. The symbol for an integer is $\mathbb{Z}$

integration the **inverse** of **differentiation**. It is used to find a **function** given its derivative, the area under a curve, the volume of rotation of a curve, the length of an **arc** and the surface area generated by the revolution of a curve

intersect two or more curves are said to **intersect** if they cross each other

intersection the point at which two or more lines or curves cross (**intersect**)

invariant (**point** or **line**) a fixed point or line that does not move under a **transformation**

inverse the opposite. For example, **differentiation** is the **inverse** of **integration** and vice versa

linear where the **variables** have the power 1. For example, $y = 2x + 3$ is **linear**, but $y = x^2$ is not. A linear function can be represented by a straight line

linear transformation a **transformation** moving all points (x, y) in a plane according to some rule

locus (**of a set of points**) – the set of points that satisfies a given condition or a rule.
For example, the locus of points where the sum of the squares of the **coordinates** is a constant, is a circle whose centre is the origin, and the value of the **constant** is the square of the radius

logarithm the **inverse** to an **exponential**. It is the power to which a number must be raised to achieve another value.
For example; $\log_3 8 = 2$ because $2^3 = 8$

matrix a rectangular array of numbers or other mathematical **functions** for which operations such as addition and multiplication are defined.
For example, $\begin{pmatrix} a & b & c \\ d & e & f \end{pmatrix}$ is a 2 × 3 matrix; 2 rows and 3 columns

midpoint a point which lies directly between two other points

natural logarithm a **logarithm** to the base of **e**, where **e** is an irrational number that is approximately equal to 2.718281....

normal a line that is **perpendicular** to another line. For example, the normal to a curve is perpendicular to the **tangent** to the curve

non-singular matrix a **matrix** is non-singular if a determinant can be found. That is, the determinant ≠ 0

parabola the **locus** of points such that the distance to the **focus** equals the distance to the **directrix**. The parametric equations of a parabola are $x = at^2$, $y = 2at$, $t \in \mathbb{R}$ or in Cartesian form $y^2 = 4ax$, where a is a positive **constant**

parallelepiped a solid three-dimensional shape in which all faces are parallelograms

parametric equations equations that express a set of quantities as **functions** of independent variables, known as 'parameters'
For example, $x = ct$, $y = \frac{c}{t}$, $t \in \mathbb{R}$ are the parametric equations of a **hyperbola**.

perpendicular means at right angles. A line meeting another at 90°

quadrant the axes of a two-dimensional axes is divided into four quadrants.

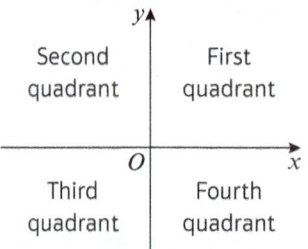

radian a measure of angles used in calculus and trigonometry. One **radian** is the angle subtended at the centre of a circle by an arc that is equal in length to the radius of the circle

reciprocal (**functions**) a function of the form $f(x) = \frac{a}{bx + c} + d$ which forms a curve of two branches, the ends of each approach asymptotes

reduction formula a method of reducing an integral into a simpler form

roots the roots of an equation are the set of all possible solutions. A quadratic equation has up to 2 roots

scalar a real number that is not a **vector** or a **matrix**

scale factor a number which multiplies some quantity. The ratio of any two corresponding lengths in two **similar** geometrical figures is called a **scale factor**

scalar triple product the value of the **scalar triple product** is the volume of a **parallelepiped**

secant the trigonometric **function** defined as $\frac{1}{\cos x}$

sech (**hyperbolic secant**) defined as $\frac{1}{\cosh x}$

sine the trigonometric **function** that is equal to the ratio of the side opposite to an acute angle (in a right-angle triangle) to the hypotenuse

sinh (**hyperbolic sine**) is a **function** defined as $\sinh x = \frac{e^x - e^{-x}}{2}$

singular matrix a **matrix** in which the **determinant** = 0, and therefore the **inverse** of the matrix does not exist.

For example, $\mathbf{A} = \begin{pmatrix} 4 & 2 \\ 2 & 1 \end{pmatrix}$

det $\mathbf{A} = 4 \times 1 - 2 \times 2 = 0$ and $\frac{1}{0}$ is not defined so **A** is singular

symmetry when a shape looks the same following a **transformation** such as reflection or rotation

tangent the
(i) trigonometric function that is equal to the ratio of the side opposite to an acute angle (in a right-angle triangle) to the adjacent.
(ii) line that touches a curve at a point without crossing over and matching the gradient of the curve at that point.

tanh (**hyperbolic tangent**) defined as $\tanh x = \dfrac{\sinh x}{\cosh x}$

transformation a **linear** mapping that is either a reflection, rotation or stretch

transpose (of a **matrix**) where the columns and rows in a **matrix** are interchanged. The transpose of matrix **A** is denoted by $\mathbf{A}^T$

vector a quantity that has both a magnitude and a direction. Geometrically, a vector is a directed line segment whose length is the magnitude of the vector with an arrow indicating the direction

variable a quantity that is able to be changed, i.e. not **constant**

vector product (cross product) a vector that is **perpendicular** to two vectors and is therefore normal to the plane that contains them. The vector product of the vectors a and b is defined as;

$a \times b = |a||b| \sin \theta \, \mathbf{n}$ where n is a unit vector perpendicular to both a and b

vertex (**plural vertices**) where two lines meet at an angle, especially in a polygon

ANSWERS

CHAPTER 1
Prior knowledge check
1. $\ln\left(\dfrac{1+\sqrt{3}}{2}\right)$
2. $\dfrac{1}{\cos^2 x} - \tan^2 x \equiv \sec^2 x - \tan^2 x \equiv (1 + \tan^2 x) - \tan^2 x \equiv 1$

Exercise 1A
1. **a** 27.29 (2 d.p.) **b** 1.13 (2 d.p.)
   c −0.96 (2 d.p.) **d** 0.01 (2 d.p.)
2. **a** $\dfrac{e - e^{-1}}{2}$ **b** $\dfrac{e^4 + e^{-4}}{2}$
   c $\dfrac{e - 1}{e + 1}$ **d** $\dfrac{2}{e^{-1} + e}$
3. **a** $\dfrac{3}{4}$ **b** $\dfrac{5}{3}$
   c $\dfrac{3}{5}$ **d** $\dfrac{2\pi}{\pi^2 - 1}$
4. $x = 1.32$ (2 d.p.)
   $x = -1.32$ (2 d.p.)
5. $x = 0.88$ (2 d.p.)
6. $x = -0.55$ (2 d.p.)
7. $x = 0.10$ (2 d.p.)
8. $x = 2.77$ (2 d.p.)
   $x = -2.77$ (2 d.p.)

Exercise 1B
1.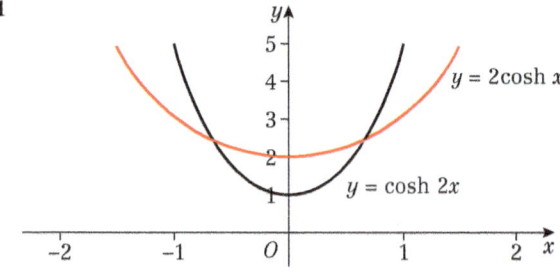

2. **a**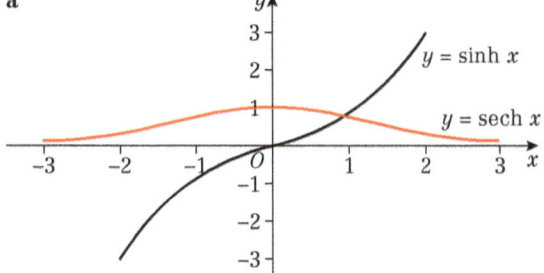

3. **a** $f(x) \in \mathbb{R}$ (All real numbers.)
   b $f(x) \geq 1$
   c $-1 < f(x) < 1$
      $|f(x)| < 1$
   d $0 < f(x) \leq 1$
   e $f(x) \in \mathbb{R}, f(x) \neq 0$
      (All real numbers except zero.)
   f $f(x) < -1, f(x) > 1$
      $|f(x)| > 1$

4. **a**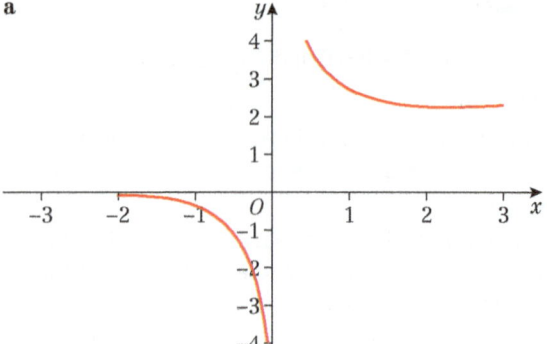

   b $x = 0$
   $y = 2$
   $y = 0$

5.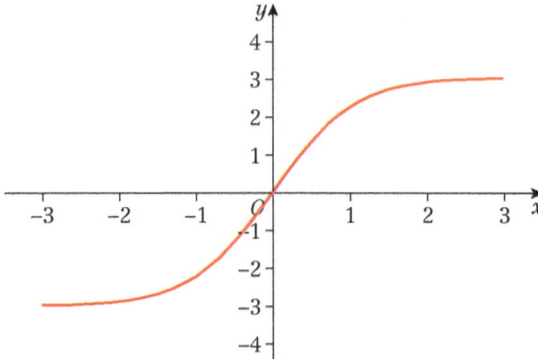

   b $y = -3$
   $y = 3$

Challenge
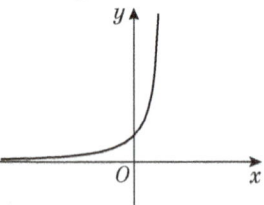

Exercise 1C
1.

$y = \operatorname{artanh} x$

Online — Worked solutions are available in SolutionBank.

ANSWERS

2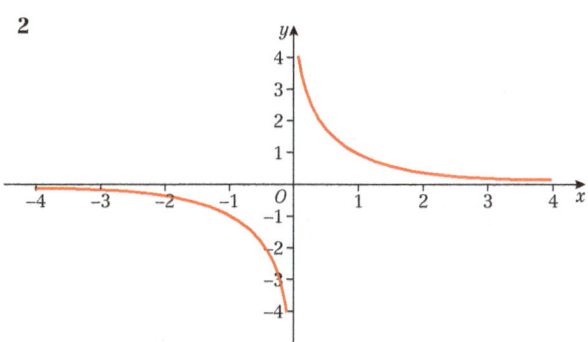

$y = \text{arcosech } x, x \neq 0$

3 Let $y = \text{artanh } x$. Then $x = \tanh y = \dfrac{e^{2y} - 1}{e^{2y} + 1}$ and $|x| < 1$.

So $(e^{2y} + 1)x = e^{2y} - 1 \Rightarrow e^{2y}(1 - x) = 1 + x$

$\Rightarrow e^{2y} = \dfrac{1 + x}{1 - x} \Rightarrow y = \dfrac{1}{2}\ln\left(\dfrac{1 + x}{1 - x}\right)$ for $|x| < 1$.

4 **a** $\ln(2 + \sqrt{5})$ **b** $\ln(3 + 2\sqrt{2})$ **c** $\dfrac{1}{2}\ln 3$

5 **a** $\ln(\sqrt{2} + \sqrt{3})$ **b** $\ln(2 + \sqrt{5})$ **c** $\dfrac{1}{2}\ln\left(\dfrac{11}{9}\right)$

6 **a** $\ln(-3 + \sqrt{10})$ **b** $\ln\left(\dfrac{3 + \sqrt{5}}{2}\right)$ **c** $\dfrac{1}{2}\ln(2 + \sqrt{3})$

7 $\dfrac{1}{2}\ln\left(\dfrac{1 + x}{1 - x}\right) + \dfrac{1}{2}\ln\left(\dfrac{1 + y}{1 - y}\right) = \ln\sqrt{3}$

$\Rightarrow \ln\left(\sqrt{\dfrac{1 + x}{1 - x}} \times \sqrt{\dfrac{1 + y}{1 - y}}\right) = \ln\sqrt{3}$

$\Rightarrow (1 + x)(1 + y) = 3(1 - x)(1 - y) \Rightarrow xy - 2x - 2y + 1 = 0$

$\Rightarrow y(x - 2) = 2x - 1 \Rightarrow y = \dfrac{2x - 1}{x - 2}$

Exercise 1D

1 **a** RHS $\equiv 2 \sinh A \cosh A$

$\equiv 2\left(\dfrac{e^A - e^{-A}}{2}\right)\left(\dfrac{e^A + e^{-A}}{2}\right)$

$\equiv \dfrac{1}{2}(e^{2A} - 1 + 1 - e^{-2A})$

$\equiv \dfrac{e^{2A} - e^{-2A}}{2}$

$\equiv \sinh 2A \equiv$ LHS

b RHS $\equiv \cosh A \cosh B - \sinh A \sinh B$

$\equiv \left(\dfrac{e^A + e^{-A}}{2}\right)\left(\dfrac{e^B + e^{-B}}{2}\right) - \left(\dfrac{e^A - e^{-A}}{2}\right)\left(\dfrac{e^B - e^{-B}}{2}\right)$

$\equiv \dfrac{e^{A+B} + e^{-A+B} + e^{A-B} + e^{-A-B}}{4} - \dfrac{e^{A+B} - e^{-A+B} - e^{A-B} + e^{-A-B}}{4}$

$\equiv \dfrac{2(e^{-A+B} + e^{A-B})}{4}$

$\equiv \dfrac{e^{A-B} + e^{-(A-B)}}{2}$

$\equiv \cosh(A - B) \equiv$ LHS

c RHS $\equiv 4 \cosh^3 A - 3 \cosh A$

$\equiv 4\left(\dfrac{e^A + e^{-A}}{2}\right)^3 - 3\left(\dfrac{e^A + e^{-A}}{2}\right)$

$(e^A + e^{-A})^3 \equiv e^{3A} + 3e^{2A}e^{-A} + 3e^A e^{-2A} + e^{-3A}$

$\equiv e^{3A} + 3e^A + 3e^{-A} + e^{-3A}$

RHS $\equiv \dfrac{e^{3A} + 3e^A + 3e^{-A} + e^{-3A}}{2} - \dfrac{3(e^A + e^{-A})}{2}$

$\equiv \dfrac{e^{3A} + e^{-3A}}{2} \equiv \cosh 3A \equiv$ LHS

d RHS $\equiv 2 \sinh\left(\dfrac{A - B}{2}\right)\cosh\left(\dfrac{A + B}{2}\right)$

$\equiv 2\left(\dfrac{e^{\frac{A-B}{2}} - e^{\frac{-A+B}{2}}}{2}\right)\left(\dfrac{e^{\frac{A+B}{2}} + e^{\frac{-A-B}{2}}}{2}\right)$

$\equiv \dfrac{1}{2}\left(e^{\frac{A-B}{2} + \frac{A+B}{2}} - e^{\frac{-A+B}{2} + \frac{A+B}{2}} + e^{\frac{A-B}{2} + \frac{-A-B}{2}} - e^{\frac{-A+B}{2} + \frac{-A-B}{2}}\right)$

$\equiv \dfrac{1}{2}(e^A - e^B + e^{-B} - e^{-A})$

$\equiv \dfrac{1}{2}(e^A - e^{-A}) - \dfrac{1}{2}(e^B - e^{-B})$

$\equiv \sinh A - \sinh B$

$\equiv$ LHS

2 **a** $\sinh(A - B) \equiv \sinh A \cosh B - \cosh A \sinh B$

b $\sinh 3A \equiv 3 \sinh A + 4 \sinh^3 A$

c $\cosh A + \cosh B \equiv 2 \cosh\left(\dfrac{A + B}{2}\right)\cosh\left(\dfrac{A - B}{2}\right)$

d $\cosh 2A \equiv \dfrac{1 + \tanh^2 A}{1 - \tanh^2 A}$

e $\cosh 2A \equiv \cosh^4 A - \sinh^4 A$

3 **a** $\sinh x \equiv \pm\sqrt{3}$ **b** $\tanh x \equiv \pm\dfrac{\sqrt{3}}{2}$ **c** $\cosh 2x \equiv 7$

4 **a** $\cosh x \equiv \sqrt{2}$ **b** $\sinh 2x \equiv -2\sqrt{2}$ **c** $\tanh 2x \equiv -\dfrac{2\sqrt{2}}{3}$

5 **a** $x = \ln\left(\dfrac{1}{7}\right), x = 0$ **b** $x = \ln 3$

c $x = \ln\left(\dfrac{7}{2}\right), x = \ln 4$ **d** $x = \ln\left(\dfrac{5}{3}\right)$

e $x = \ln\left(\dfrac{-3 + \sqrt{13}}{2}\right), x = \ln(4 + \sqrt{17})$

f $x = \ln(4 + \sqrt{15})$ **g** $x = 0, x = \ln\left(\dfrac{7 \pm 3\sqrt{5}}{2}\right)$

h $x = \ln\left(\dfrac{5}{2}\right), x = \ln 3$ **i** $x = \ln(1 + \sqrt{2})$

6 **a** RHS $\equiv 2\left(\dfrac{1}{4}(e^{2x} + 2 + e^{-2x})\right) - 1 \equiv \dfrac{1}{2}(e^{2x} + e^{-2x}) \equiv$ LHS

b $\ln(3 \pm 2\sqrt{2})$

7 $\ln\left(\dfrac{7}{2} \pm \dfrac{3}{2}\sqrt{5}\right)$

8 He has not applied Osborn's rule in line 1 – correct identity should be $\text{sech}^2 x = 1 - \tanh^2 x$ since implied $\sin^2$ term; he has split the denominator of the fraction in line 2 which is invalid; he has taken the reciprocal of both terms in line 3 – this is mathematically incorrect.

Correct proof: $\dfrac{1 + \tanh^2 x}{1 - \tanh^2 x} \equiv \dfrac{2 - \text{sech}^2 x}{\text{sech}^2 x} \equiv \dfrac{2}{\text{sech}^2 x} - 1$

$\equiv 2 \cosh^2 x - 1$

9 **a** $8 \cosh(x + 0.693)$

b 8

c $0.148, -1.534$

Chapter review 1

1 **a** $\dfrac{4}{3}$ **b** $\dfrac{13}{5}$ **c** $-\dfrac{15}{17}$

2 $y = \dfrac{12 - 13x}{12x - 13}$

3 RHS $= \sinh A \cosh B - \cosh A \sinh B$

$= \left(\dfrac{e^A - e^{-A}}{2}\right)\left(\dfrac{e^B + e^{-B}}{2}\right) - \left(\dfrac{e^A + e^{-A}}{2}\right)\left(\dfrac{e^B - e^{-B}}{2}\right)$

$= \dfrac{e^{A+B} - e^{-A+B} + e^{A-B} - e^{-A-B}}{4} - \dfrac{e^{A+B} + e^{-A+B} - e^{A-B} - e^{-A-B}}{4}$

$= \dfrac{2(e^{A-B} - e^{-A+B})}{4} = \dfrac{e^{A-B} - e^{-(A-B)}}{2}$

$= \sinh(A - B) =$ LHS

4 RHS $= \dfrac{2 \tanh \dfrac{1}{2}x}{1 - \tanh^2 \dfrac{1}{2}x}$

$2 \tanh \dfrac{1}{2}x = \dfrac{2(e^x - 1)}{e^x + 1}$

$1 - \tanh^2 \dfrac{1}{2}x = 1 - \left(\dfrac{e^x - 1}{e^x + 1}\right)^2 = \dfrac{(e^x + 1)^2 - (e^x - 1)^2}{(e^x + 1)^2}$

$= \dfrac{4e^x}{(e^x + 1)^2}$

So RHS $= \dfrac{2(e^x - 1)}{e^x + 1} \times \dfrac{(e^x + 1)^2}{4e^x} = \dfrac{(e^x - 1)(e^x + 1)}{2e^x}$

$= \dfrac{e^{2x} - 1}{2e^x} = \dfrac{e^x - e^{-x}}{2} = \sinh x = $ LHS

5 $x = \ln\left(\dfrac{1}{2}\right), x = \ln 7$

6 $x = \ln\left(\dfrac{5}{3}\right)$

7 $x = \ln\left(\dfrac{-14 + \sqrt{205}}{3}\right)$

$x = \ln(1 + \sqrt{2})$

8 a

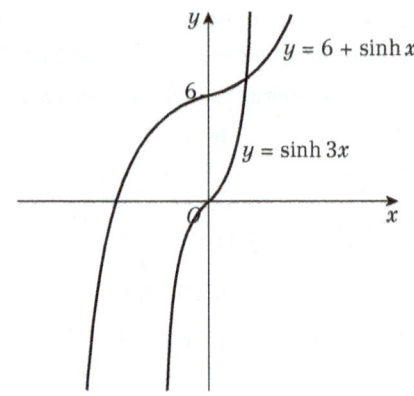

b $6 + \sinh x = 3\sinh x + 4\sinh^3 x$
$\Rightarrow 2\sinh^3 x + \sinh x - 3 = 0$
$\Rightarrow \sinh x = 1 \Rightarrow x = \ln(1 + \sqrt{2})$
$\sinh(3\ln(1 + \sqrt{2})) = 7$,
so $(\ln(1 + \sqrt{2}), 7)$ is the point of intersection.

9 a $R = 12, \alpha = 0.405$ **b** 12

10 a $4\sinh(x + 0.693)$
  b $x = 0.75$ (2 d.p.)
  c 0.75 (2 d.p.)

Challenge

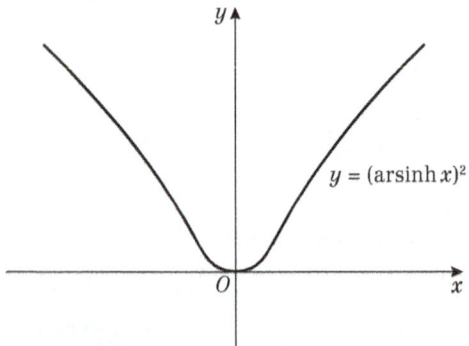

CHAPTER 2
Prior knowledge check
1 $\dfrac{1}{2}$

2 $x = \pm\dfrac{a}{\sqrt{1 + k^2}}$

3 $y^2 = 4ax$

Exercise 2A
1 i a

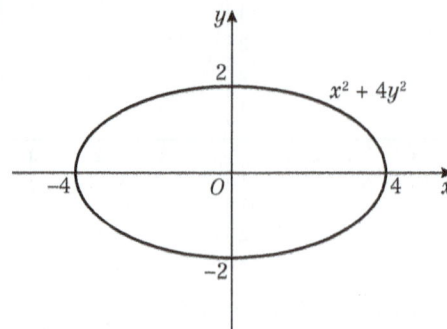

  b $x = 4\cos\theta, y = 2\sin\theta$

ii a

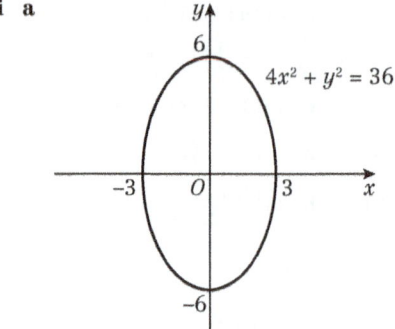

  b $x = 3\cos\theta, y = 6\sin\theta$

iii a

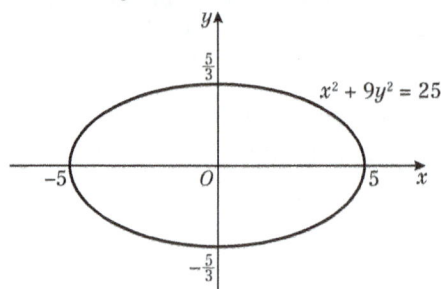

  b $x = 5\cos\theta, y = \dfrac{5}{3}\sin\theta$

2 i a

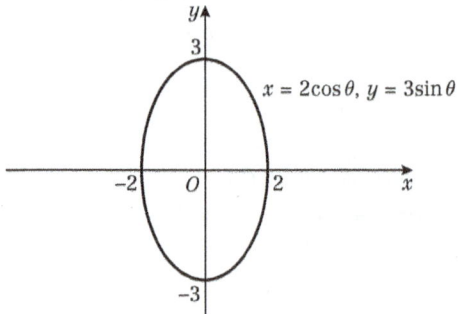

  b $\dfrac{x^2}{4} + \dfrac{y^2}{9} = 1$

ii a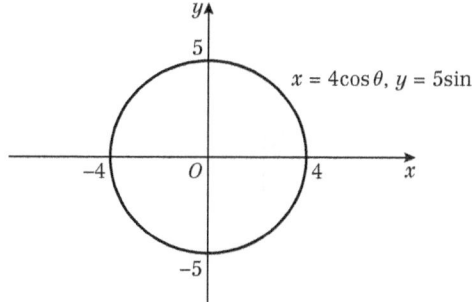

b $\dfrac{x^2}{16} + \dfrac{y^2}{25} = 1$

iii a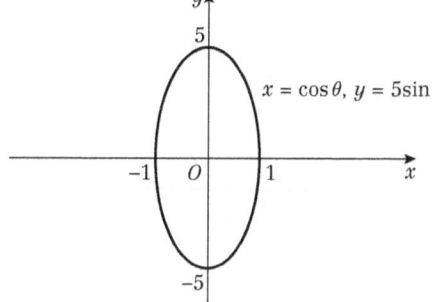

b $x^2 + \dfrac{y^2}{25} = 1$

iv a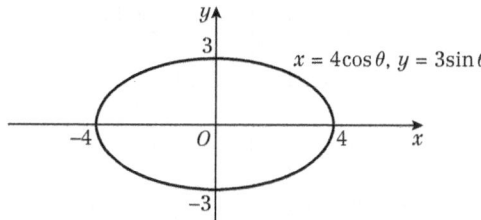

b $\dfrac{x^2}{16} + \dfrac{y^2}{9} = 1$

3 a $(b\cos\theta, a\sin\theta)$

b Ellipse $\dfrac{x^2}{b^2} + \dfrac{y^2}{a^2} = 1$

c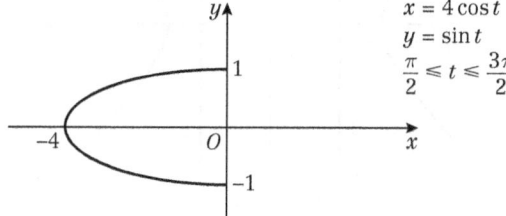

Challenge

$\begin{pmatrix} \frac{1}{\sqrt{2}} & -\frac{1}{\sqrt{2}} \\ \frac{1}{\sqrt{2}} & \frac{1}{\sqrt{2}} \end{pmatrix} \begin{pmatrix} a\cos t \\ b\sin t \end{pmatrix} = \begin{pmatrix} \frac{a}{\sqrt{2}}\cos t - \frac{b}{\sqrt{2}}\sin t \\ \frac{a}{\sqrt{2}}\cos t + \frac{b}{\sqrt{2}}\sin t \end{pmatrix}$

Show $(x+y)^2 = 2a^2\cos^2 t$

Show $(x-y)^2 = 2b^2\sin^2 t$

Substitute into $\dfrac{(x+y)^2}{2a^2} + \dfrac{(x-y)^2}{2b^2} = 1$ and simplify.

Exercise 2B

1 a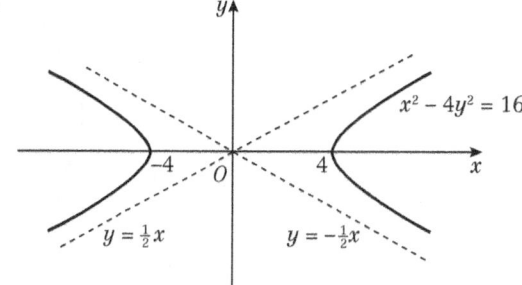

b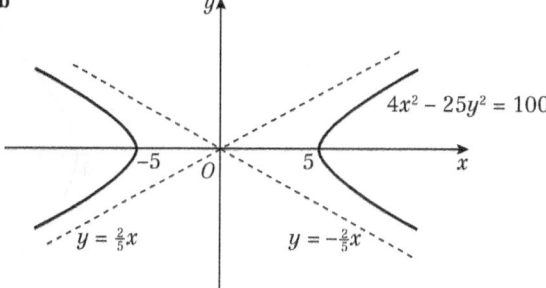

c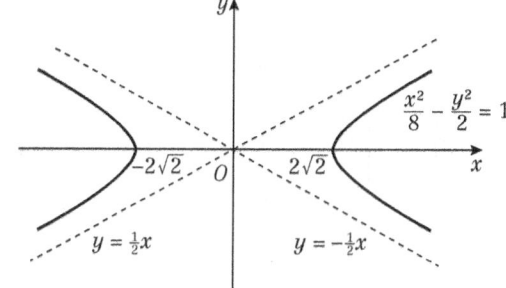

2 i a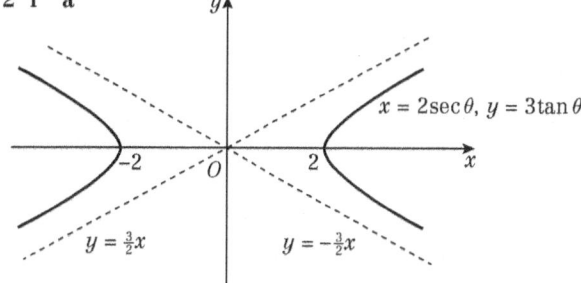

b $\dfrac{x^2}{4} - \dfrac{y^2}{9} = 1$

ii a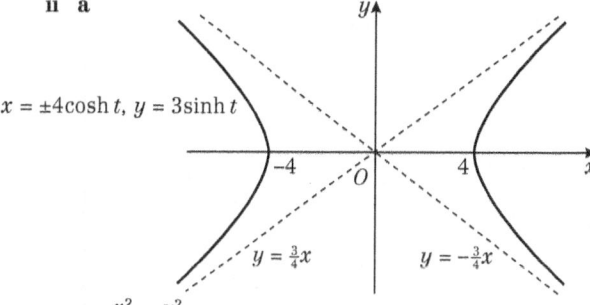

b $\dfrac{x^2}{16} - \dfrac{y^2}{9} = 1$

iii a

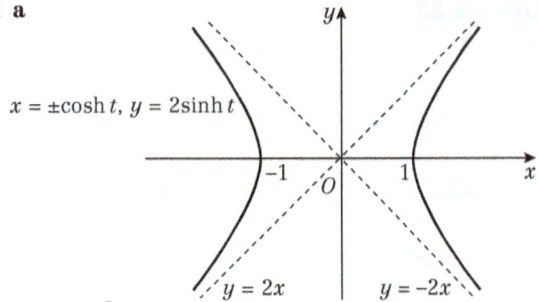

$x = \pm\cosh t$, $y = 2\sinh t$

b $x^2 - \dfrac{y^2}{4} = 1$

iv a

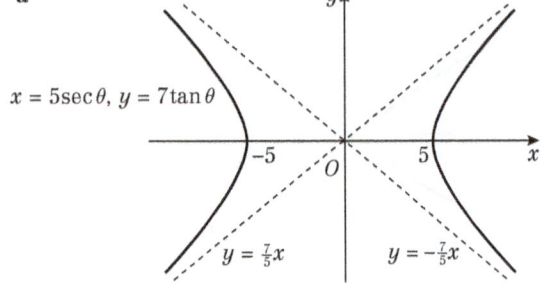

$x = 5\sec\theta$, $y = 7\tan\theta$

b $\dfrac{x^2}{25} - \dfrac{y^2}{49} = 1$

Challenge

$\begin{pmatrix} \frac{1}{\sqrt{2}} & -\frac{1}{\sqrt{2}} \\ \frac{1}{\sqrt{2}} & \frac{1}{\sqrt{2}} \end{pmatrix}\begin{pmatrix} ct \\ \frac{c}{t} \end{pmatrix} = \begin{pmatrix} \frac{ct}{\sqrt{2}} - \frac{c}{t\sqrt{2}} \\ \frac{ct}{\sqrt{2}} + \frac{c}{t\sqrt{2}} \end{pmatrix}$, so $x^2 = \dfrac{c^2t^2}{2} - c^2 + \dfrac{c^2}{2t^2}$ and

$y^2 = \dfrac{c^2t^2}{2} + c^2 + \dfrac{c^2}{2t^2}$. Therefore $y^2 - x^2 = 2c^2$,

so $a^2 = 2c^2 \Rightarrow a = \pm c\sqrt{2}$, so $y^2 - x^2 = a^2$

Exercise 2C

1 **a** $e = \dfrac{2}{3}$

   b $e = \dfrac{\sqrt{7}}{4}$

   c $e = \dfrac{1}{\sqrt{2}}$

2 **a** Foci = $(\pm 1, 0)$; directrices $x = \pm 4$

   b Foci = $(\pm 3, 0)$; directrices $x = \pm\dfrac{16}{3}$

   c Foci = $(0, \pm 2)$; directrices $y = \pm\dfrac{9}{2}$

3 **a** The foci are on the x-axis, so $a > b$.

   b **i** $e = \dfrac{1}{2}$ **ii** $a = 6$, $b = 3\sqrt{3}$

   c

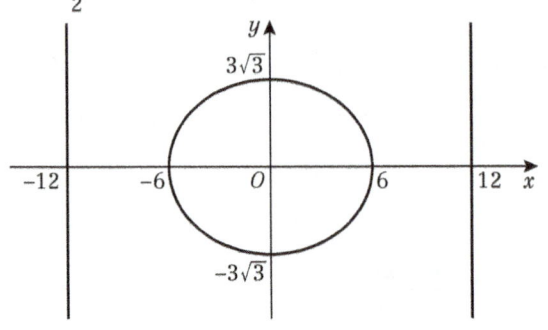

4 **a** The foci are on the y-axis, so $b > a$.

   b **i** $e = \dfrac{1}{2}$ **ii** $a = 2\sqrt{3}$, $b = 4$

   c

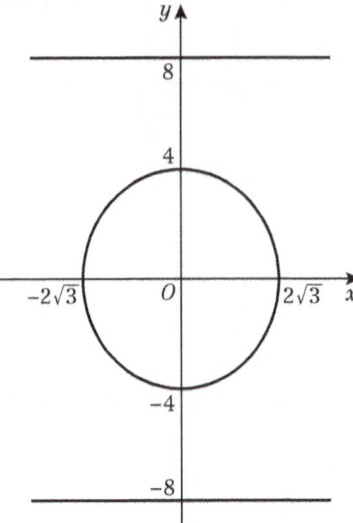

5 **a** $e = \dfrac{2\sqrt{10}}{5}$ **b** $e = \dfrac{4}{3}$ **c** $e = \dfrac{5}{3}$

6 **a**

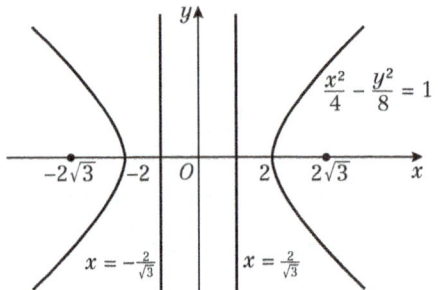

   b

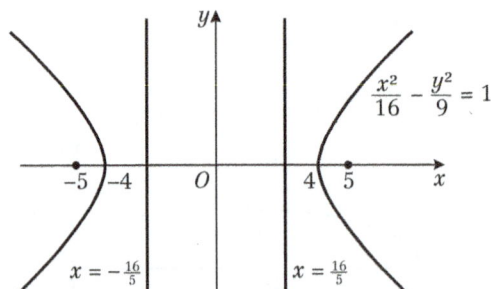

   c

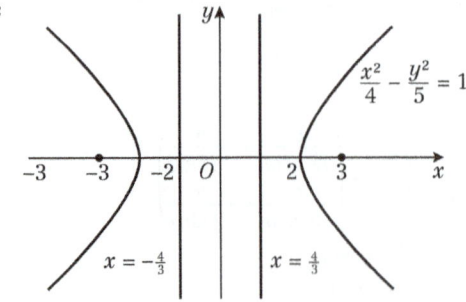

7 a i $e = \dfrac{5}{\sqrt{24}}$; foci are $\left(\pm\dfrac{5}{\sqrt{24}} \times \sqrt{24}, 0\right) = (\pm 5, 0)$

  ii $e = 5$; foci are $(\pm 5 \times 1, 0) = (\pm 5, 0)$

  iii $e = \dfrac{5}{4}$; foci are $\left(\dfrac{5}{4} \times 4, 0\right) = (\pm 5, 0)$

  iv $e = \dfrac{5}{3}$; foci are $\left(\dfrac{5}{3} \times 3, 0\right) = (\pm 5, 0)$

b

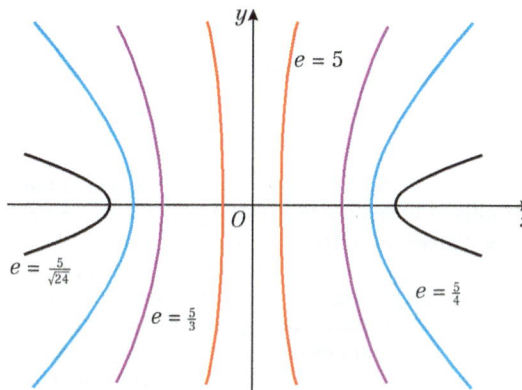

8 Use the fact that $(ae, 0)$ is on the chord.
$\dfrac{(ae)^2}{a^2} + \dfrac{y^2}{b^2} = 1$ simplifies to $y = \pm\dfrac{b^2}{a}$.
Therefore the length of the chord is $\dfrac{2b^2}{a}$

9 a $\dfrac{4}{5}$ **b** $\dfrac{x^2}{36} + \dfrac{y^2}{100} = 1$

10 Let P have coordinates (x, y).
$PA^2 = (x + 3\sqrt{3})^2 + y^2 = x^2 + 6x\sqrt{3} + 27 + \left(9 - \dfrac{x^2}{4}\right)$
$= \dfrac{3}{4}(x + 4\sqrt{3})^2$
$x + 4\sqrt{3} > 0 \Rightarrow PA = \dfrac{\sqrt{3}}{2}x + 6$
Similarly, $PB = 6 - \dfrac{\sqrt{3}}{2}x$
So $PA + PB = 12$.

11

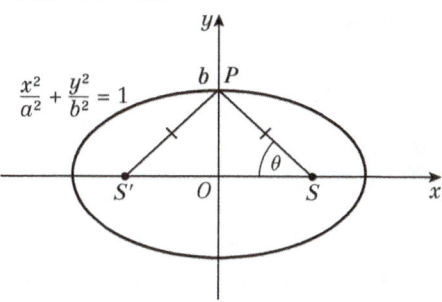

Consider $\triangle POS$

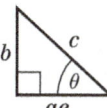

$c^2 = b^2 + a^2e^2$, but $b^2 = a^2(1 - e^2)$
$\Rightarrow c^2 = a^2 - a^2e^2 + a^2e^2 = a^2$
$\Rightarrow c = a$
So $\cos\theta = \dfrac{ae}{a} = e$
If you use the result that $SP + S'P = 2a$ then since $S'P = SP$ it is clear $SP = a$
Hence $\cos\theta = \dfrac{ae}{a} = e$

12

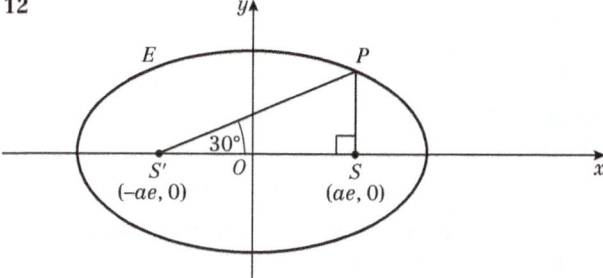

PS is y where $\dfrac{a^2e^2}{a^2} + \dfrac{y^2}{b^2} = 1$
$y^2 = b^2(1 - e^2)$
$y = b\sqrt{1 - e^2}$
$SS' = 2ae$
$\tan 30° = \dfrac{1}{\sqrt{3}} = \dfrac{y}{2ae} = \dfrac{b\sqrt{1 - e^2}}{2ae}$
But $b^2 = a^2(1 - e^2)$
$\Rightarrow \quad \dfrac{1}{\sqrt{3}} = \dfrac{a\sqrt{1 - e^2}\sqrt{1 - e^2}}{2ae}$
$\Rightarrow \quad \dfrac{2e}{\sqrt{3}} = 1 - e^2$
$\Rightarrow e^2 + \dfrac{2}{\sqrt{3}}e - 1 = 0$
$\Rightarrow e = \dfrac{1}{\sqrt{3}}, (e > 0)$

Exercise 2D

1 a Tangent: $x\cos\theta + 2y\sin\theta = 2$
Normal: $2x\sin\theta - y\cos\theta = 3\sin\theta\cos\theta$
b Tangent: $3x\cos\theta + 5y\sin\theta = 15$
Normal: $5x\sin\theta - 3y\cos\theta = 16\sin\theta\cos\theta$

2 a Tangent: $6y + \sqrt{5}x = 9$
Normal: $3\sqrt{5}y = 18x - 16\sqrt{5}$
b Tangent: $2\sqrt{3}y - x = 8$
Normal: $y + 2\sqrt{3}x = -3\sqrt{3}$

3 $\dfrac{dy}{dx} = -\dfrac{b\cos t}{a\sin t}$
So tangent is $y - b\sin t = -\dfrac{b\cos t}{a\sin t}(x - a\cos t)$
$\Rightarrow bx\cos t + ay\sin t = ab(\sin^2 t + \cos^2 t) = ab$

4 a $y = x + \sqrt{5}$ meets the ellipse when
$\dfrac{x^2}{4} + (x + \sqrt{5})^2 = 1$
$\Rightarrow 5x^2 + 8\sqrt{5}x + 16 = 0$
This has discriminant $(8\sqrt{5})^2 - 4 \times 5 \times 16 = 0$
So the line meets the ellipse at only one point, therefore is a tangent to the ellipse.
b $\left(-\dfrac{4}{5}\sqrt{5}, \dfrac{1}{5}\sqrt{5}\right)$

5 a $2y\cos\theta - 3x\sin\theta = -5\sin\theta\cos\theta$
b P is $(3, 0), (-3, 0), \left(-\dfrac{3}{2}, \sqrt{3}\right)$ or $\left(-\dfrac{3}{2}, -\sqrt{3}\right)$

6 $c = \pm 2\sqrt{2}$

7 $m = \pm 2$

8 a $m = 2$ **b** $\left(-\dfrac{3}{2}, 1\right)$ **c** $\left(0, \dfrac{1}{4}\right)$ **d** $\dfrac{45}{16}$

9 a $\dfrac{dy}{dx} = \dfrac{\frac{dy}{d\theta}}{\frac{dx}{d\theta}} = \dfrac{2\cos\theta}{-3\sin\theta} = -\dfrac{2}{3}\cot\theta$

b $\dfrac{\left(\frac{9}{5}\right)^2}{9} + \dfrac{\left(-\frac{8}{5}\right)^2}{4} = \dfrac{9}{25} + \dfrac{16}{25} = 1$, so $Q\left(\dfrac{9}{5}, -\dfrac{8}{5}\right)$ lies on E.

c $\frac{1}{2}$

d $\tan\theta = \frac{1}{3}$; $\left(\frac{9}{\sqrt{10}}, \frac{2}{\sqrt{10}}\right)$ and $\left(-\frac{9}{\sqrt{10}}, -\frac{2}{\sqrt{10}}\right)$

10 $m = \pm\sqrt{2}$, $c = \pm 8$

11 $3x\sin\theta\cos\theta - 2y = 24\sin\theta$

12 a $m = \frac{3\cos\theta}{-5\sin\theta}$, $(x_1, y_1) = (5\cos\theta, 3\sin\theta)$

Substitute into $y - y_1 = m(x - x_1)$ and simplify.

b $3y\sin\theta\cos\theta - 9\cos\theta = 5x\sin^2\theta$

c At $(-4, 0)$, $-9\cos\theta = -20\sin^2\theta$
Use $\sin^2\theta + \cos^2\theta \equiv 1$ to obtain
$20\cos^2\theta + 9\cos\theta - 20 = 0$ and therefore $\cos\theta = \frac{4}{5}$

13 a $\frac{dy}{dx} = \frac{-2\cos t}{\sin t}$ and substitute into $y - y_1 = m(x - x_1)$
using $m = \frac{-2\cos t}{\sin t}$ and $(x_1, y_1) = (2\cos t, 4\sin t)$

b Find $l_2 : 2y\cos t = x\sin t$ and equate/substitute l_1 and l_2

Exercise 2E

1 a Tangent: $8y = 3x - 4$
 Normal: $3y + 8x = 108$
 b Tangent: $3y = 2x - 6$
 Normal: $2y + 3x = 48$
 c Tangent: $5y = 2x - 5$
 Normal: $2y + 5x = 56$

2 a Tangent: $5y\sinh t + 10 = 2x\cosh t$
 Normal: $2y\cosh t + 5x\sinh t = 29\cosh t\sinh t$
 b Tangent: $y\tan t + 3 = 3x\sec t$
 Normal: $3y\sec t + x\tan t = 10\sec t\tan t$

3 $\frac{dy}{dx} = -\frac{b\sec t}{a\tan t}$
So tangent is $y - b\tan t = -\frac{b\sec t}{a\tan t}(x - a\sec t)$
$\Rightarrow bx\sec t - ay\tan t = ab(\sec^2 t - \tan^2 t) = ab$

4 $\frac{dy}{dx} = \frac{b\cosh t}{a\sinh t}$, so gradient of normal is $-\frac{a\sinh t}{b\cosh t}$
So equation of normal is
$y - b\sinh t = -\frac{a\sinh t}{b\cosh t}(x - a\cosh t)$
$\Rightarrow ax\sinh t + by\cosh t = (a^2 + b^2)\sinh t\cosh t$

5 a $\left(0, -\frac{3}{\sinh t}\right)$
 b $\left(0, \frac{25}{3}\sinh t\right)$
 c $\frac{2}{3}|(25\sinh^2 t + 9)\coth t|$

6 P and Q are $(4, 3\sqrt{3})$ and $(4, -3\sqrt{3})$

7 $c = \pm 6$

8 $m = \pm\frac{13}{7}$

9 $m = \pm 4$ and $c = \pm 7$

10 a $c = 3$
 b $\left(\frac{25}{3}, -\frac{16}{3}\right)$

11 a Find normal gradient $= -\frac{a\sinh t}{b\cosh t}$ and substitute into
$y - y_1 = m(x - x_1)$
 b $\left(\left(\frac{a^2 + b^2}{a}\right)\cosh t, 0\right)$
 c $\left(a, \frac{(a^2 + b^2)\sinh t\cosh t - a^2\sinh t}{b\cosh t}\right)$

12 a Substitute $m = \frac{5}{7\sin\theta}$ and
$(x_1, y_1) = (7\sec\theta, 5\tan\theta)$ into $y - y_1 = m(x - x_1)$.

b l_1 has gradient $\frac{5}{7\sin\theta}$, so equation of l_2 is $y = -\frac{7\sin\theta}{5}$
So at Q, $-\frac{49}{5}(\sin^2\theta)x = 5x - 35\cos\theta$
$\Rightarrow x = \frac{175\cos\theta}{25 + 49\sin^2\theta}$, $y = \frac{-245\sin\theta\cos\theta}{25 + 49\sin^2\theta}$

13 $\frac{dy}{dx} = \frac{x}{4y}$
$y - y_1 = m(x - x_1) \Rightarrow xx_1 - 4yy_1 = 16$, $xx_2 - 4yy_2 = 16$
Equate and substitute (m, n) to obtain $\frac{y_2 - y_1}{x_2 - x_1} = \frac{m}{4n}$
This is the gradient of the line joining (x_1, y_1) and (x_2, y_2)
$y - y_1 = \frac{m}{4n}(x - x_1) \Rightarrow mx - 4ny = 16$

14 Substitute $(6, 4)$ into the general equation of the tangent to get $3\sec\theta - 4\tan\theta = 2 \Rightarrow 2\cos\theta + 4\sin\theta = 3$
$\Rightarrow \sqrt{20}\cos(\theta + 1.107...) = 3$
$\Rightarrow \theta + 1.107... = ..., 0.835..., 5.447..., 7.118..., ...$
This gives two values of θ in the range $[0, 2\pi)$, so there are two tangents to the hyperbola passing through $(6, 4)$.

15 a The asymptotes of H are $y = x$ and $y = -x$.
Let $A = (a, a)$ and $B = (b, -b)$, so the midpoint of
AB is $\left(\frac{a + b}{2}, \frac{a - b}{2}\right)$.
Now we compute a and b for the generic point P on H
$P = (X, Y)$.
Differentiating H we get $2x - 2y\frac{dy}{dx} = 0 \Rightarrow \frac{dy}{dx} = \frac{x}{y}$
Gradient of the tangent at P is $\frac{X}{Y}$
So the tangent has equation $y - Y = \frac{X}{Y}(x - X)$
At A: $a - Y = \frac{X}{Y}(a - X) \Rightarrow a = X + Y$
At B: $b - Y = \frac{X}{Y}(-b - X) \Rightarrow b = Y - X$
So $X = \frac{a + b}{2}$ and $Y = \frac{a - b}{2}$

b $|OA| = \sqrt{2}|a|$ and $|OB| = \sqrt{2}|b|$
So $|OA| \times |OB| = 2|ab| = 2|X^2 - Y^2| = 2|1| = 2$
which is constant.

Exercise 2F

1 a $(apq, a(p + q))$
 b Chord PQ has gradient
$\frac{2ap - 2aq}{ap^2 - aq^2} = \frac{2a(p - q)}{a(p - q)(p + q)} = \frac{2}{(p + q)}$
Equation of chord PQ is: $y - 2ap = \frac{2}{p + q}(x - ap^2)$
$\Rightarrow y(p + q) - 2ap^2 - 2apq = 2x - 2ap^2$
$\Rightarrow y(p + q) = 2x + 2apq$
Chord passes through $(a, 0) \Rightarrow 0 = 2a + 2apq$
or $pq = -1$
Locus of R is $x = -a$
 c $y = a$

2 a $\frac{2x}{a^2} - \frac{2y}{b^2} \times \frac{dy}{dx} = 0 \Rightarrow \frac{dy}{dx} = \frac{b^2 x}{a^2 y}$
So gradient of tangent at P is $\frac{b^2 a\sec t}{a^2 b\tan t} = \frac{b}{a\sin t}$
Equation of tangent is $y - b\tan t = \frac{b}{a\sin t}(x - a\sec t)$
$\Rightarrow bx\sec t - ay\tan t = ab(\sec^2 t - \tan^2 t) = ab$

b A is where $y = 0 \Rightarrow x = \dfrac{ab}{b\sec t} = a\cos t$,
i.e. $A(a\cos t, 0)$.
B is where $x = 0 \Rightarrow y = \dfrac{ab}{-a\tan t} = -b\cot t$,
i.e. $B(0, -b\cot t)$.
Midpoint of AB is $\left(\dfrac{a}{2}\cos t, -\dfrac{b}{2}\cot t\right)$
$x = \dfrac{a}{2}\cos t \Rightarrow \sec t = \dfrac{a}{2x}$
$y = -\dfrac{b}{2}\cot t \Rightarrow \tan t = -\dfrac{b}{2y}$
Use $\sec^2 t \equiv 1 + \tan^2 t$
$\Rightarrow \dfrac{a^2}{4x^2} = 1 + \dfrac{b^2}{4y^2}$ which gives locus.

3 a $\dfrac{2x}{a^2} - \dfrac{2y}{b^2} \times \dfrac{dy}{dx} = 0 \Rightarrow \dfrac{dy}{dx} = \dfrac{b^2 x}{a^2 y}$
So gradient of normal at P is $-\dfrac{a^2 b \tan t}{b^2 a \sec t} = -\dfrac{a}{b}\sin t$
Equation of tangent is $y - b\tan t = -\dfrac{a}{b}\sin t(x - a\sec t)$
$\Rightarrow ax\sin t + by = (a^2 + b^2)\tan t$

b $y = 0 \Rightarrow x = \left(\dfrac{a^2 + b^2}{a}\right)\sec t \Rightarrow A$ is $\left(\dfrac{a^2 + b^2}{a}\sec t, 0\right)$
$x = 0 \Rightarrow y = \left(\dfrac{a^2 + b^2}{b}\right)\tan t \Rightarrow B$ is $\left(0, \dfrac{a^2 + b^2}{b}\tan t\right)$
Midpoint of AB is $\left(\dfrac{a^2 + b^2}{2a}\sec t, \dfrac{a^2 + b^2}{2b}\tan t\right)$
$x = \dfrac{a^2 + b^2}{2a}\sec t \Rightarrow \sec t = \dfrac{2ax}{a^2 + b^2}$
$y = \dfrac{a^2 + b^2}{2b}\tan t \Rightarrow \tan t = \dfrac{2by}{a^2 + b^2}$
Use $\sec^2 t \equiv 1 + \tan^2 t$:
$4a^2 x^2 = (a^2 + b^2)^2 + 4b^2 y^2$

4 a Find $\dfrac{dy}{dx} = \dfrac{3\cos t}{-5\sin t}$ and substitute into
$y - y_1 = m(x - x_1)$ using $m = \dfrac{5\sin\theta}{3\cos\theta}$ and
$(x_1, y_1) = (5\cos\theta, 3\sin\theta)$

b Find midpoint $\left(\dfrac{8}{5}\cos\theta, -\dfrac{8}{3}\sin\theta\right)$ and use
$\sin^2\theta + \cos^2\theta \equiv 1$

5 a Tangent at P is $x + p^2 y = 2cp$
Tangent at Q is $x + q^2 y = 2cq$
Tangents intersect when $(p^2 - q^2)y = 2c(p - q)$
So $R = \left(\dfrac{2cpq}{p+q}, \dfrac{2c}{p+q}\right)$

b Gradient of PQ is $\dfrac{\dfrac{c}{q} - \dfrac{c}{p}}{cq - cp} = -\dfrac{1}{pq}$
So equation of PQ is
$y - \dfrac{c}{p} = -\dfrac{1}{pq}(x - cp) \Rightarrow ypq + x = c(p + q)$

c i $y = -2x, x \neq 0$ **ii** $y = 2c^2, x < 0$ **iii** $x = 2c^2$

6 a $\dfrac{1}{t}$

b $y - 2at = \dfrac{1}{t}(x - at^2) \Rightarrow ty - 2at^2 = x - at^2$
$\Rightarrow x - ty + at^2 = 0$

c T is $(0, at)$. Perpendicular bisector of OT is $y = \dfrac{at}{2}$
Perpendicular bisector of OP is $y - at = -\dfrac{t}{2}\left(x - \dfrac{at^2}{2}\right)$
Centre of circle is where perpendicular bisectors intersect: $\dfrac{at}{2} - at = -\dfrac{t}{2}\left(x - \dfrac{at^2}{2}\right)$
Therefore centre of circle is $\left(\dfrac{at^2}{2} + a, \dfrac{at}{2}\right)$.

7 $y = \dfrac{1}{2}, x < 0$

8 a $\dfrac{x^2}{2^2} + \dfrac{(2y - 6)^2}{4^2} = 1$

b Simplifies to $x^2 + (y - 3)^2 = 4$ which is a circle of centre $(0, 3)$ and radius 2.

Challenge

A is (x_1, y_1) and B is (x_2, y_2).
Then $k = \dfrac{y_2 - y_1}{x_2 - x_1}$ and the midpoint is $(x, y) = \left(\dfrac{x_1 + x_2}{2}, \dfrac{y_1 + y_2}{2}\right)$.
$\left.\begin{array}{l}b^2 x_1^2 + a^2 y_1^2 = a^2 b^2 \\ b^2 x_2^2 + a^2 y_2^2 = a^2 b^2\end{array}\right\} \Rightarrow a^2(y_2^2 - y_1^2) = b^2(x_2^2 - x_1^2)$
$\Rightarrow a^2(y_1 + y_2)(y_2 - y_1) = b^2(x_1 + x_2)(x_2 - x_1)$
$\Rightarrow -\dfrac{ka^2}{b^2}(y_1 + y_2) = (x_1 + x_2) \Rightarrow -\dfrac{ka^2}{b^2}y = x \Rightarrow ka^2 y + b^2 x = 0$

Chapter review 2

1 a $\dfrac{x^2}{16} + \dfrac{y^2}{81} = 1$

b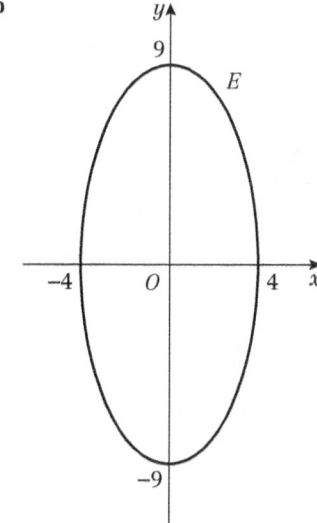

c $4x\sin\theta - 9y\cos\theta = -65\cos\theta\sin\theta$

2 a $\dfrac{x^2}{4} - \dfrac{y^2}{25} = 1$

b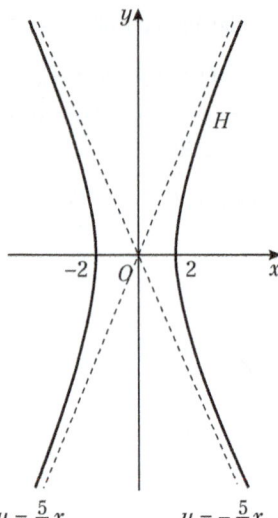

$y = \frac{5}{2}x$, $y = -\frac{5}{2}x$

c $2y\sinh t + 10 = 5x\cosh t$

3 a $\dfrac{x^2}{a^2} - \dfrac{y^2}{a^2m^2} = 1$

b A is $\left(\dfrac{a^2 + b^2}{a}\sec t, 0\right)$ and B is $\left(0, \dfrac{a^2 + b^2}{b}\tan t\right)$.

So midpoint is $(x, y) = \left(\dfrac{a^2 + b^2}{2a}\sec t, \dfrac{a^2 + b^2}{2b}\tan t\right)$.

Using $\sec^2 t \equiv 1 + \tan^2 t$, $\dfrac{4a^2x^2}{(a^2+b^2)^2} = \dfrac{4b^2y^2}{(a^2+b^2)^2} + 1$

So the locus of the midpoint is $4a^2x^2 = (a^2 + b^2)^2 + 4b^2y^2$.

4 a Gradient of chord $= \dfrac{\frac{c}{p} - \frac{c}{q}}{cp - cq} = \dfrac{c(q-p)}{pqc(p-q)} = -\dfrac{1}{pq}$

b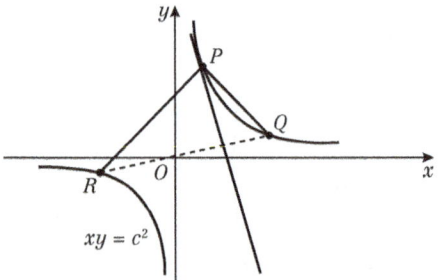

Gradient of $PQ = -\dfrac{1}{pq}$ and gradient of $PR = -\dfrac{1}{pr}$

So $-1 = p^2qr$ (1)

Gradient of tangent at P is $-\dfrac{1}{p^2}$ and gradient of chord $RQ = -\dfrac{1}{qr}$

So $\left(-\dfrac{1}{qr}\right)\left(-\dfrac{1}{p^2}\right) = \dfrac{1}{p^2qr}$

But from (1), $p^2qr = -1$

Therefore tangent at P is perpendicular to chord QR.

5 a $y = ct^{-1}, x = ct \Rightarrow \dfrac{dy}{dx} = \dfrac{-ct^{-2}}{c} = -\dfrac{1}{t^2}$

Equation of tangent is: $y - \dfrac{c}{t} = -\dfrac{1}{t^2}(x - ct)$

$\Rightarrow yt^2 - ct = -x + ct$ or $t^2y + x = 2ct$

b $\left(-\dfrac{4}{3}, -12\right)$ and $\left(12, \dfrac{4}{3}\right)$

6 a Let P have coordinates (x, y).

$PA^2 = (x+4)^2 + y^2 = x^2 + 8x + 16 + \left(9 - \dfrac{9}{25}x^2\right)$

$= \left(\dfrac{4}{5}x + 5\right)^2$

$\dfrac{4}{5}x + 5 > 0 \Rightarrow PA = \dfrac{4}{5}x + 5$

Similarly, $PB = 5 - \dfrac{4}{5}x$, so $PA + PB = 10$.

b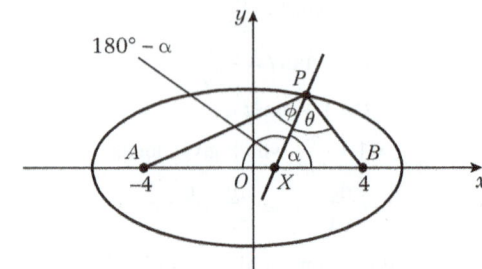

Normal at P is $5x\sin t - 3y\cos t = 16\cos t \sin t$

X is when $y = 0$, i.e. $x = \dfrac{16}{5}\cos t$

$PB^2 = (5\cos t - 4)^2 + (3\sin t)^2 = (4\cos t - 5)^2$

$\Rightarrow PB = 5 - 4\cos t$ and $PA = 10 - PB = 5 + 4\cos t$

$AX = 4 + \dfrac{16}{5}\cos t$, $BX = 4 - \dfrac{16}{5}\cos t$

Consider sine rule on $\triangle PAX$:

$\sin\phi = \dfrac{\sin(180° - \alpha)AX}{AP} = \dfrac{\sin\alpha\left(4 + \frac{16}{5}\cos t\right)}{5 + 4\cos t}$

$= \dfrac{4}{5}\sin\alpha$

Consider sine rule on $\triangle PBX$:

$\sin\theta = \dfrac{\sin\alpha BX}{PB} = \dfrac{\sin\alpha\left(4 - \frac{16}{5}\cos t\right)}{5 - 4\cos t}$

$= \dfrac{4}{5}\sin\alpha$

So $\sin\phi = \sin\theta$ and, since both angles are acute, $\theta = \phi$

Therefore normal bisects APB.

7 a $y = ct^{-1}, x = ct \Rightarrow \dfrac{dy}{dx} = \dfrac{-ct^{-2}}{c} = -\dfrac{1}{t^2}$

Equation of tangent is: $y - \dfrac{c}{t} = -\dfrac{1}{t^2}(x - ct)$

$yt^2 - ct = -x + ct$ or $t^2y + x = 2ct$

b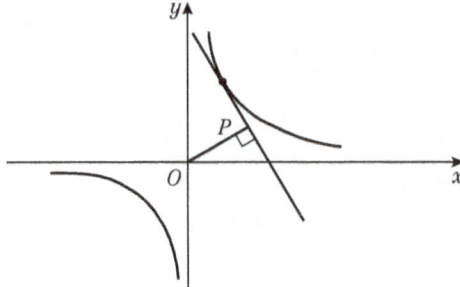

Gradient of tangent is $-\dfrac{1}{t^2}$

Gradient of OP is t^2

Equation of OP is $y = t^2x$

Equation of tangent is $t^2y = 2ct - x$

Solving, $t^4x = 2ct - x$

$\Rightarrow \quad x = \dfrac{2ct}{1+t^4}, y = \dfrac{2ct^3}{1+t^4}$

$x^2 + y^2 = \dfrac{4c^2t^2 + 4c^2t^6}{(1+t^4)^2} = \dfrac{4c^2t^2}{1+t^4}$

$\Rightarrow (x^2+y^2)^2 = \dfrac{16c^4t^4}{(1+t^4)^2}$
$xy = \dfrac{4c^2t^4}{(1+t^4)^2}$ $\bigg\} \Rightarrow (x^2+y^2)^2 = 4c^2xy$

8 a OP has gradient $\dfrac{2ap}{ap^2} = \dfrac{2}{p}$ and OQ has gradient $\dfrac{2}{q}$

Since OP and OQ are perpendicular, $\dfrac{2}{p} \times \dfrac{2}{q} = -1$, so $pq = -4$

b $y + xq = aq^3 + 2aq$

c Normal at P is $y + xp = ap^3 + 2ap$
Solve equations simultaneously to get
$x = a(q^2 + p^2 + qp + 2), y = apq(q + p)$
$pq = -4 \Rightarrow R$ is $(ap^2 + aq^2 - 2, -4pq(p+q))$

d $x = a((p+q)^2 - 2pq - 2) = a((p+q)^2 + 6)$
$y = 4a(p+q) \Rightarrow p + q = \dfrac{y}{4a}$

$\Rightarrow \quad x = a\left(\dfrac{y^2}{16a^2} + 6\right)$

$\Rightarrow x - 6a = \dfrac{y^2}{16a} \Rightarrow y^2 = 16ax - 96a^2$

9 $y = mx + c$ and $\dfrac{x^2}{a^2} + \dfrac{y^2}{b^2} = 1$
$\Rightarrow b^2x^2 + a^2(mx + c)^2 = a^2b^2$
$\Rightarrow x^2(b^2 + a^2m^2) + 2a^2mcx + a^2(c^2 - b^2) = 0$
For a tangent the discriminant is 0:
$4a^4m^2c^2 = 4(b^2 + a^2m^2)a^2(c^2 - b^2)$
$\Rightarrow \quad c = \pm\sqrt{a^2m^2 + b^2}$
So the lines $y = mx \pm \sqrt{a^2m^2 + b^2}$ are tangents.

10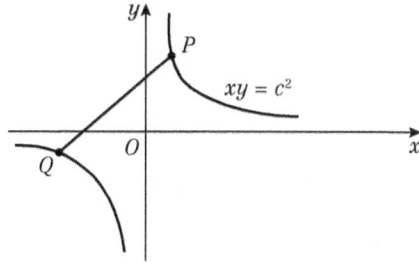

Chord PQ has gradient $\dfrac{\frac{c}{p} - \frac{c}{q}}{cp - cq} = \dfrac{c(q-p)}{pqc(p-q)} = -\dfrac{1}{pq}$

If gradient = 1, then $pq = -1$.
Tangent at P is $p^2y + x = 2cp$
Tangent at Q is $q^2y + x = 2cq$
Intersection: $(p^2 - q^2)y = 2c(p - q) \Rightarrow y = \dfrac{2c}{p+q}$

$\Rightarrow x = 2cp - \dfrac{2cp^2}{p+q} = \dfrac{2cpq}{p+q}$

So R is $\left(\dfrac{2cpq}{p+q}, \dfrac{2c}{p+q}\right)$

But $pq = -1$ so R is $\left(x = \dfrac{-2c}{p+q}, y = \dfrac{2c}{p+q}\right)$
The locus of R is the line $y = -x$

11 a Find $\dfrac{dy}{dx} = \dfrac{4\cos\theta}{-6\sin\theta} = \dfrac{2\cos\theta}{-3\sin\theta}$ and substitute into
$y - y_1 = m(x - x_1)$ using $m = \dfrac{2\cos\theta}{-3\sin\theta}$ and
$(x_1, y_1) = (6\cos\theta, 4\sin\theta)$

b Find midpoint $\left(\dfrac{3}{\cos\theta}, \dfrac{2}{\sin\theta}\right)$ and use $\sin^2\theta + \cos^2\theta \equiv 1$

12 a $m = \dfrac{5\cos\theta}{-13\sin\theta}, (x_1, y_1) = (13\cos\theta, 5\sin\theta)$
Substitute into $y - y_1 = m(x - x_1)$ and simplify.

b $5y\sin\theta\cos\theta - 25\cos\theta = 13x\sin^2\theta$

c $(-ae, 0) = (-12, 0)$ as $a = 13$, $b = 5$ and $e = \dfrac{12}{13}$
Given line passes through this point,
$-25\cos\theta = -156\sin^2\theta$
Use $\sin^2\theta + \cos^2\theta \equiv 1$ to obtain
$156\cos^2\theta + 25\cos\theta - 156 = 0$ and therefore
$\cos\theta = \dfrac{12}{13} = e$

13 a Find normal gradient $= -\dfrac{\sin\theta}{2}$ and substitute into
$y - y_1 = m(x - x_1)$

b $A(20\sec\theta, 0), B(1, 10\tan\theta)$ and midpoint of
AB is $(10\sec\theta, 5\tan\theta)$
Use $\tan^2\theta + 1 \equiv \sec^2\theta$ to obtain $\dfrac{x^2}{100} - \dfrac{y^2}{25} = 1$

14 a $\dfrac{dy}{dx} = \dfrac{b\cos t}{-a\sin t}$
So gradient of normal at $(a\cos t, b\sin t)$ is $\dfrac{a\sin t}{b\cos t}$
Equation of normal is $y - b\sin t = \dfrac{a\sin t}{b\cos t}(x - a\cos t)$
$\Rightarrow ax\sin t - by\cos t = (a^2 - b^2)\cos t \sin t$

b $y = 0 \Rightarrow x = \left(\dfrac{a^2 - b^2}{a}\right)\cos t \Rightarrow M$ is $\left(\dfrac{a^2 - b^2}{a}\cos t, 0\right)$
$x = 0 \Rightarrow y = -\left(\dfrac{a^2 - b^2}{b}\right)\sin t \Rightarrow N$ is $\left(0, -\dfrac{a^2 - b^2}{b}\sin t\right)$
Midpoint of MN is $\left(\dfrac{a^2 - b^2}{2a}\cos t, -\dfrac{a^2 - b^2}{2b}\sin t\right)$
$x = \dfrac{a^2 - b^2}{2a}\cos t \Rightarrow \cos t = \dfrac{2ax}{a^2 - b^2}$
$y = -\dfrac{a^2 - b^2}{2b}\sin t \Rightarrow \sin t = \dfrac{2by}{a^2 - b^2}$
$\sin^2 t + \cos^2 t \equiv 1 \Rightarrow 4b^2y^2 + 4a^2x^2 = (a^2 - b^2)^2$

15 $a = 5, b = 3 \Rightarrow e = \frac{4}{5}$, so foci are $(\pm 4, 0)$.
Let P have coordinates (x, y).
$PS^2 = (x + 4)^2 + y^2 = x^2 + 8x + 16 + \left(9 - \dfrac{9}{25}x^2\right)$
$= \left(\dfrac{4}{5}x + 5\right)^2$
$\dfrac{4}{5}x + 5 > 0 \Rightarrow PS = \dfrac{4}{5}x + 5$
Similarly, $PS' = 5 - \dfrac{4}{5}x$, so $PS + PS' = 10$.

16 $\dfrac{108}{5}$

Challenge
$QS = ePS \Leftrightarrow QS^2 = e^2PS^2$
$QS^2 = a^2e^4\cos^2\theta - 2a^2e^3\cos\theta + a^2e^2$
$PS^2 = a^2\cos^2\theta - 2a^2e\cos\theta + a^2e^2 + b^2\sin^2\theta$
Use rearrangements of $b^2 = a^2(1 - e^2)$ to simplify.

CHAPTER 3
Prior knowledge check

1. $y = \sin^2 x \Rightarrow \dfrac{dy}{dx} = \sin x \cos x + \sin x \cos x$
   $= 2\sin x \cos x = \sin 2x$

2. $y = \dfrac{\sin x}{\cos x} \Rightarrow \dfrac{dy}{dx} = \dfrac{\cos x \cos x - \sin x (-\sin x)}{\cos^2 x}$
   $= \dfrac{\cos^2 x + \sin^2 x}{\cos^2 x} = \sec^2 x$

3. $\dfrac{dy}{dx} = 2x\cos 3x - 3x^2 \sin 3x$

Exercise 3A

1. $2\cosh 2x$
2. $5\sinh 5x$
3. $2\text{sech}^2 2x$
4. $3\cosh 3x$
5. $-4\text{cosech}^2 4x$
6. $-2\tanh 2x \,\text{sech}\, 2x$
7. $e^{-x}(\cosh x - \sinh x)$
8. $\cosh 3x + 3x \sinh 3x$
9. $\dfrac{x\cosh x - \sinh x}{3x^2}$
10. $x(2\cosh 3x + 3x \sinh 3x)$
11. $2\cosh 2x \cosh 3x + 3\sinh 2x \sinh 3x$
12. $\tanh x$
13. $3x^2 \cosh x^3$
14. $4\cosh 2x \sinh 2x$
15. $\sinh x \, e^{\cosh x}$
16. $-\coth x \,\text{cosech}\, x$
18. $y = 12.13$
19. $\dfrac{dy}{dx} = 3\sinh 3x \sinh x + \cosh 3x \cosh x$
    $\dfrac{d^2y}{dx^2} = 2(5\cosh 3x \sinh x + 3\sinh 3x \cosh x)$
20. Tangent: $4y \sinh q - x \cosh q = -16$ or $x \cosh q - 4y \sinh q = 16$
    Normal: $4x \sinh q - y \cosh q = 63 \sinh q \cosh q$

Exercise 3B

1. **a** $\dfrac{2}{\sqrt{4x^2 - 1}}$ **b** $\dfrac{1}{\sqrt{(x+1)^2 + 1}}$
   c $\dfrac{3}{1 - 9x^2}$ **d** $\dfrac{-1}{x(1-x^2)^{\frac{1}{2}}}$
   e $\dfrac{2x}{\sqrt{x^4 - 1}}$ **f** $\dfrac{3}{\sqrt{9x^2 - 1}}$
   g $2x\,\text{arcosh}\, x + \dfrac{x^2}{\sqrt{x^2 - 1}}$
   h $y = \text{arsinh}\,\dfrac{x}{2}$
   Let $t = \dfrac{x}{2}$ $y = \text{arsinh}\, t$
   $\dfrac{dt}{dx} = \dfrac{1}{2}$ $\dfrac{dy}{dt} = \dfrac{1}{\sqrt{t^2 + 1}}$
   $\dfrac{dy}{dx} = \dfrac{1}{2\sqrt{\left(\dfrac{x}{2}\right)^2 + 1}}$
   $= \dfrac{1}{\sqrt{x^2 + 4}}$
   i $y = e^{x^3} \text{arsinh}\, x$
   $\dfrac{dy}{dx} = 3x^2 e^{x^3} \text{arsinh}\, x + \dfrac{e^{x^3}}{\sqrt{x^2 + 1}}$
   j $y = \text{arsinh}\, x \,\text{arcosh}\, x$
   $\dfrac{dy}{dx} = \dfrac{1}{\sqrt{x^2 + 1}} \text{arcosh}\, x + \dfrac{1}{\sqrt{x^2 - 1}} \text{arsinh}\, x$
   k $y = \text{arcosh}\, x \,\text{sech}\, x$
   $\dfrac{dy}{dx} = \dfrac{1}{\sqrt{x^2 + 1}} \text{sech}\, x - \text{arcosh}\, x \tanh x \,\text{sech}\, x$
   $= \text{sech}\, x \left(\dfrac{1}{\sqrt{x^2 + 1}} - \text{arcosh}\, x \tanh x \right)$

2. **a** $y = \text{arcosh}\, x \Rightarrow \cosh y = x$
   $\sinh y \cdot \dfrac{dy}{dx} = 1 \Rightarrow \dfrac{dy}{dx} = \dfrac{1}{\sinh y} = \dfrac{1}{\sqrt{\cosh^2 y - 1}} = \dfrac{1}{\sqrt{x^2 - 1}}$
   b $y = \text{artanh}\, x \Rightarrow \tanh y = x$
   $\text{sech}^2 y \cdot \dfrac{dy}{dx} = 1 \Rightarrow \dfrac{dy}{dx} = \dfrac{1}{\text{sech}^2 y} = \dfrac{1}{1 - \tanh^2 y} = \dfrac{1}{1 - x^2}$

3. $y = \text{artanh}\left(\dfrac{e^x}{2}\right) \Rightarrow \tanh\left(\dfrac{e^x}{2}\right) = x$
   $\dfrac{dy}{dx} = \dfrac{1}{1 - \left(\dfrac{e^{2x}}{4}\right)} \cdot \dfrac{e^x}{2} = \dfrac{4}{4 - e^{2x}} \cdot \dfrac{e^x}{2} \Rightarrow (4 - e^{2x})\dfrac{dy}{dx} = 2e^x$

Exercise 3C

2. **a** $-\dfrac{2}{\sqrt{1 - 4x^2}}$
   b Let $y = \arctan \dfrac{x}{2}$
   Let $t = \dfrac{x}{2}$ $y = \arctan t$
   $\dfrac{dt}{dx} = \dfrac{1}{2}$ $\dfrac{dy}{dt} = \dfrac{1}{1 + t^2}$
   $\dfrac{dy}{dx} = \dfrac{1}{1 + t^2} \cdot \dfrac{1}{2} = \dfrac{1}{2\left(1 + \dfrac{x^2}{4}\right)} = \dfrac{2}{4 + x^2}$ or $\dfrac{2}{x^2 + 4}$
   c $\dfrac{3}{\sqrt{1 - 9x^2}}$ **d** $-\dfrac{1}{1 + x^2}$
   e $\dfrac{1}{x\sqrt{x^2 - 1}}$ **f** $-\dfrac{1}{x\sqrt{x^2 - 1}}$
   g $-\dfrac{1}{(x-1)\sqrt{1-2x}}$ **h** $-\dfrac{2x}{\sqrt{1 - x^4}}$
   i $e^x\left(\arccos x - \dfrac{1}{\sqrt{1-x^2}}\right)$ **j** $\dfrac{\cos x}{\sqrt{1-x^2}} - \sin x \arcsin x$
   k $x\left(2\arccos x - \dfrac{x}{\sqrt{1-x^2}}\right)$ **l** $\dfrac{e^{\arctan x}}{1+x^2}$

3. $\dfrac{1}{1 + x^2 (\arctan x)^2}\left(\arctan x + \dfrac{x}{1+x^2}\right)$

4. $y = \arcsin x \Rightarrow \sin y = x$
   $\cos y \dfrac{dy}{dx} = 1 \Rightarrow \dfrac{dy}{dx} = \dfrac{1}{\cos y} = \dfrac{1}{\sqrt{1-x^2}}$
   $\dfrac{d^2y}{dx^2} = \dfrac{\sqrt{1-x^2} \times 0 - 1 \times \left(-\dfrac{1}{2}\right)(-2x)(1-x^2)^{-\frac{1}{2}}}{1-x^2}$
   $\Rightarrow \dfrac{d^2y}{dx^2} = \dfrac{x\dfrac{dy}{dx}}{1-x^2} \Rightarrow (1 - x^2)\dfrac{d^2y}{dx^2} = x\dfrac{dy}{dx}$

5. $\sqrt{3}y - \dfrac{\pi\sqrt{3}}{6} = 4x - 1$

Challenge

$\dfrac{dy}{dx} = -\cos x \sinh x - \sin x \cosh x$
$\dfrac{d^2y}{dx^2} = -2\sin x \sinh x$

$\dfrac{d^3y}{dx^3} = -2\cos x \sinh x - 2\sin x \cosh x$

$\dfrac{d^4y}{dx^4} = -2y + 2\sin x \sinh x - 2\sin x \sinh x - 2y = -4y$

Chapter review 3

1 $2\sinh 2x$

2 **a** $\dfrac{3}{\sqrt{9x^2+1}}$ **b** $\dfrac{2x}{\sqrt{x^4+1}}$
   c $\dfrac{1}{\sqrt{x^2-4}}$ **d** $2x\left(\text{arcosh } 2x + \dfrac{x}{\sqrt{4x^2-1}}\right)$

3 $y = \arctan x \Rightarrow \tan y = x$
   $\sec^2 y \cdot \dfrac{dy}{dx} = 1 \Rightarrow \dfrac{dy}{dx} = \dfrac{1}{\sec^2 y} = \dfrac{1}{1+\tan^2 y} = \dfrac{1}{1+x^2}$

4 $y = (\text{arsinh } x)^2$
   $\dfrac{dy}{dx} = \dfrac{\text{arsinh } x}{\sqrt{1+x^2}} + \dfrac{\text{arsinh } x}{\sqrt{1+x^2}} = \dfrac{2\,\text{arsinh } x}{\sqrt{1+x^2}}$

   $\dfrac{d^2y}{dx^2} = \dfrac{2 - x\,\text{arsinh } x}{\dfrac{\sqrt{1+x^2}}{1+x^2}}$

   $= \dfrac{2 - x\dfrac{dy}{dx}}{1+x^2} \Rightarrow (1+x^2)\dfrac{d^2y}{dx^2} + x\dfrac{dy}{dx} - 2 = 0$

5 **a** $\dfrac{dy}{dx} = 5\sinh x - 3\cosh x$
   b $(\ln 2, 4)$

6 $y = \arcsin x^2$
   $\dfrac{dy}{dx} = \dfrac{2\arcsin x}{\sqrt{1-x^2}}$

   $\dfrac{d^2y}{dx^2} = \dfrac{\dfrac{2\sqrt{1-x^2}}{\sqrt{1-x^2}} + \dfrac{2\arcsin x(x)}{(1-x^2)^{\frac{1}{2}}}}{1-x^2} = \dfrac{2 + 2x\dfrac{dy}{dx}}{1-x^2}$

   $\Rightarrow (1-x^2)\dfrac{d^2y}{dx^2} - 2x\dfrac{dy}{dx} - 2 = 0$

7 $x = \text{arcosh}(\sinh 2x)$
   Let $t = \sinh 2x$ $x = \text{arcosh } t$
   $\dfrac{dt}{dx} = 2\cosh 2x$ $\dfrac{dx}{dt} = \dfrac{1}{\sqrt{t^2-1}}$
   $\dfrac{dy}{dx} = \dfrac{1}{\sqrt{t^2-1}} \times 2\cosh 2x$
   $= \dfrac{2\cosh 2x}{\sqrt{\sinh^2 2x - 1}}$

8 $y = x - \arctan x$
   $\dfrac{dy}{dx} = 1 - \dfrac{1}{1+x^2}$
   $\dfrac{d^2y}{dx^2} = \dfrac{(0-2x)}{(1+x^2)^2}$
   $= \dfrac{2x}{(1+x^2)^2}$
   $= 2x\left(1 - \left(1 - \dfrac{1}{1+x^2}\right)\right)^2$
   $\dfrac{d^2y}{dx^2} = 2x\left(1 - \dfrac{dy}{dx}\right)^2$

9 $\dfrac{1}{x^2+1}$

10 $p = \sqrt{2} \pm 1$

11 tangent
    $ay \sinh q - xb \cosh q + ab = 0$
    normal
    $ax \sinh q + by \cosh q - \sinh q \cosh q (a^2 + b^2) = 0$

CHAPTER 4
Prior knowledge check

1 $\cos^2 x = \dfrac{\cos 2x}{2} - \dfrac{1}{2} \Rightarrow \int\left(\dfrac{\cos 2x}{2} - \dfrac{1}{2}\right)dx = \dfrac{\sin 2x}{4} - \dfrac{x}{2} + c$

2 10.03

3 $\dfrac{1}{2}e^x(\sin x - \cos x) + c$

Exercise 4A

1 **a** $\cosh x + 3\sinh x + C$
   b $5\tanh x + C$
   c $-\coth x + C$
   d $\sinh x - \tanh x + C$
   e $-\text{sech } x + C$
   f $-3\,\text{cosech } x + C$
   g $\tanh x - \text{sech } x + C$
   h $\tanh x + \coth x + C$

2 **a** $\dfrac{1}{2}\cosh 2x + c$
   b $3\sinh\left(\dfrac{x}{3}\right) + C$
   c $\dfrac{1}{2}\tanh(2x-1) + C$
   d $-\dfrac{1}{5}\coth 5x + C$
   e $-\dfrac{1}{2}\text{cosech } 2x + c$
   f $-\dfrac{1}{\left(\dfrac{1}{\sqrt{2}}\right)}\text{sech}\left(\dfrac{x}{\sqrt{2}}\right) + C = -\sqrt{2}\,\text{sech}\left(\dfrac{x}{\sqrt{2}}\right) + C$
   g $\cosh 5x - \sinh 4x + 6\tanh\left(\dfrac{x}{2}\right) + C$

3 **a** $\arctan x + C$ **b** $\text{arsinh } x + C$
   c $\ln|1+x| + C$ **d** $\ln(1+x^2) + C$
   e $\arcsin x + C$ **f** $\text{arcosh } x + C$
   g $3\sqrt{x^2-1} + C$ **h** $-\dfrac{3}{(1+x)} + C$

4 **a** $-2\sqrt{(1-x^2)} + \arcsin x + C$
   b $\text{arcosh } x + \sqrt{(x^2-1)} + C$
   c $\sqrt{(1+x^2)} - 3\,\text{arsinh } x + C$

5 **a** $\dfrac{x^2}{1+x^2} = \dfrac{1+x^2-1}{1+x^2} = \dfrac{1+x^2}{1+x^2} - \dfrac{1}{1+x^2} = 1 - \dfrac{1}{1+x^2}$
   b $x - \arctan x + C$

Exercise 4B

1 **a** $\dfrac{1}{4}\sinh 4x + C$
   b $\dfrac{1}{4}\ln \cosh 4x + C$
   c $\dfrac{1}{6}\tanh 6x + C$
   d $-\dfrac{1}{7}\text{cosech}^7 x + C$
   e $\dfrac{1}{3}(\cosh 2x)^{\frac{3}{2}} + C$
   f $-\dfrac{1}{30}\text{sech}^{10} 3x + C$

2 a $\frac{1}{3}\ln(2+3\cosh x)+C$

b $\tanh x + \frac{1}{2}\tanh^2 x + 2$ or $\tanh x - \frac{1}{2}\text{sech}^2 x + C$

c $5x + 2\ln\cosh x + C$

3 a $\int \coth x\, dx = \int \frac{\cosh x}{\sinh x}\, dx = \ln\sinh x + C$

$\int \coth 2x\, dx = \frac{1}{2}\ln\sinh 2x + C$

$\int_1^2 \coth 2x\, dx = \left[\frac{1}{2}\ln\sinh 2x\right]_1^2 = \frac{1}{2}(\ln\sinh 4 - \ln\sinh 2)$

b $= \frac{1}{2}\ln\left(\frac{\sinh 4}{\sinh 2}\right)$

$= \frac{1}{2}\ln\left(\frac{e^4 - e^{-4}}{e^2 - e^{-2}}\right)$

$= \frac{1}{2}\ln(e^2 + e^{-2})$

$= \ln\sqrt{e^2 + \frac{1}{e^2}}$

4 a $\frac{1}{3}x\cosh 3x - \frac{1}{9}\sinh 3x + C$

b $x\tanh x - \ln\cosh x + C$

5 a $\frac{1}{4}e^{2x} + \frac{1}{2}x + C$

b $\frac{1}{2}e^x + \frac{1}{10}e^{-5x} + C$

c $\frac{1}{16}e^{4x} - \frac{1}{16}e^{-4x} + \frac{1}{8}e^{2x} - \frac{1}{8}e^{-2x} + C$

or $\frac{1}{8}\sinh 4x + \frac{1}{4}\sinh 2x + C$

6 $\int \cosh^2 3x\, dx = \frac{1}{4}\int (e^{3x} + e^{-3x})^2\, dx$

$= \frac{1}{4}\int (e^{6x} + 2 + e^{-6x})\, dx$

$= \frac{1}{24}e^{6x} - \frac{1}{24}e^{-6x} + \frac{1}{2}x + C$

$= \frac{1}{12}\sinh 6x + \frac{1}{2}x + C$ which agrees with the result in Example **5b**

7 $1 - \frac{1}{e}$

8 a $\frac{1}{4}\sinh 2x - \frac{1}{2}x + C$

b $x + 2\,\text{sech}\,x + C$

c $x - \frac{1}{3}\coth 3x + C$

d $-\frac{1}{8}x + \frac{1}{32}\sinh 4x + C$

e $\sinh x + \frac{2}{3}\sinh^3 x + \frac{1}{5}\sinh^5 x + C$

f $\frac{1}{2}\ln\cosh 2x - \frac{1}{4}\tanh^2 2x + C$

9 $\int_0^{\ln 2} \cosh^2\left(\frac{x}{2}\right) dx = \int_0^{\ln 2}\left(\frac{1+\cosh x}{2}\right) dx$

$= \frac{1}{2}[x + \sinh x]_0^{\ln 2}$

$= \frac{1}{2}\left[\ln 2 + \left[\frac{e^{\ln 2} - e^{-\ln 2}}{2}\right]\right]_0^{\ln 2}$

$= \frac{1}{2}\left[\ln 2 + \frac{3}{4}\right]$

$= \frac{1}{8}[4\ln 2 + 3]$

$= \frac{1}{8}[\ln 16 + 3]$

11 a $4\arctan(e^x) + C$

b $\arctan(e^{2x}) + C$

c $4\arctan\left(e^{\frac{x}{2}}\right) + C$

12 a $\frac{1}{8}\sinh(2x^2) + \frac{x^2}{4} + C$

b $\frac{1}{2}\tanh(x^2) + C$

Exercise 4C

1 Using $x = a\tan A$, $dx = a\sec^2 A\, dA$

$\int \frac{1}{a^2 + x^2}\, dx = \int \frac{1}{a^2 + a^2\tan^2\theta}\, a\sec^2\theta\, d\theta$

$= \int \frac{a\sec^2\theta}{a^2\sec^2\theta}\, d\theta$

$= \frac{1}{a}\int d\theta$

$= \frac{\theta}{a} + C$

$= \frac{1}{a}\arctan\left(\frac{x}{a}\right) + C$

2 Using $x = \cos A$, $dx = -\sin A\, dA$

$\int \frac{1}{\sqrt{1-x^2}}\, dx = \int \frac{1}{\sqrt{1-\cos^2 A}}\,(-\sin A)\, dA$

$= -\int dA$

$= -A + C$

$= -\arccos x + C$

3 a $3\arcsin\left(\frac{x}{2}\right) + C$ **b** $\text{arcosh}\left(\frac{x}{3}\right) + C$

c $\frac{4\sqrt{5}}{5}\arctan\left(\frac{x}{\sqrt{5}}\right) + C$ **d** $\frac{1}{2}\text{arsinh}\left(\frac{2x}{5}\right) + C$

4 a $\arcsin\left(\frac{x}{5}\right) + C$ **b** $3\,\text{arsinh}\left(\frac{x}{3}\right) + C$

c $\text{arcosh}\left(\frac{x}{\sqrt{2}}\right) + C$ **d** $\frac{1}{2}\arctan\left(\frac{x}{4}\right) + C$

5 a $\frac{1}{2}\text{arcosh}\left(\frac{x}{\sqrt{3}}\right) + C$ **b** $\frac{\sqrt{3}}{6}\arctan\frac{\sqrt{3}x}{2} + C$

c $\frac{1}{3}\text{arsinh}\left(\frac{3x}{4}\right) + C$ **d** $\frac{1}{2}\arcsin\left(\frac{2x}{\sqrt{3}}\right) + C$

6 a 0.927 (3 s.f.) **b** 0.977 (3 s.f.)

c 0.719 (3 s.f.)

7 a $\ln\{1 + \sqrt{2}\}$ **b** $\ln\left(\frac{4}{3}\right)$

c $\left(\frac{\pi}{12}\right)$

8 a 2.27 (s.f.) **b** 4.13 (s.f.)

9 a Cartesian equation of circle is $x^2 + y^2 = r^2$

Area of C can be written as $4\int_0^r y\, dx = 4\int_0^r \sqrt{r^2 - x^2}\, dx$

b Use substitution $x = r\sin A$ so $dx = r\cos A$

$4\int_0^r \sqrt{r^2 - x^2}\, dx = 4\int_0^{\frac{\pi}{2}} \sqrt{r^2 - r^2\sin^2 A}\, r\cos A\, dA$

$= 2r^2\int_0^{\frac{\pi}{2}} (1 + \cos 2A)\, dA$

$= 2r^2\left[A + \frac{\sin 2A}{2}\right]_0^{\frac{\pi}{2}}$

$= 2r^2\left[\frac{\pi}{2}\right]$

$= \pi r^2$

10 a $\frac{x}{9} - \frac{2}{27}\arctan\frac{3x}{2} + C$

b $\sqrt{x}\sqrt{1+x} - \text{arsinh}(\sqrt{x}) + C$

11 a $\sqrt{x^2 - 4} - 2\,\text{arcosh}\left(\frac{x}{2}\right) + C$

b $-2\sqrt{2-x^2} - \arcsin\left(\frac{x}{\sqrt{2}}\right) + C$

c $\frac{2\sqrt{3}}{3}\arctan(\sqrt{3}x) + \frac{1}{2}\ln(1+3x^2) + C$

12 $2\ln x + \frac{4}{\sqrt{5}}\arctan\left(\frac{x}{\sqrt{5}}\right) - \frac{1}{2}\ln(x^2+5) + C$

13 $\dfrac{2}{(x+1)(x^2+1)} = \dfrac{A}{x+1} + \dfrac{Bx+C}{x^2+1}$

$\Rightarrow 2 = A(x^2+1) + (Bx+C)(x+1)$

$x = -1 \Rightarrow 2 = 2A \Rightarrow A = 1$

Coefficient of $x^2 \Rightarrow A + B = 0 \Rightarrow B = -1$

Coefficent of $x \Rightarrow B + C = 0 \Rightarrow C = 1$

$\displaystyle\int_0^1 \dfrac{2}{(x+1)(x^2+1)}\,dx = \int_0^1 \left(\dfrac{1}{x+1}\right)dx + \int_0^1 \left(\dfrac{1-x}{x^2+1}\right)dx$

$= \displaystyle\int_0^1 \left(\dfrac{1}{x+1}\right)dx + \int_0^1 \left(\dfrac{1}{x+1}\right)dx + \int_0^1 \left(\dfrac{1-x}{x^2+1}\right)dx$

$= [\ln(x+1)]_0^1 + [\arctan x]_0^1 - \left[\dfrac{1}{2}\ln(x^2+1)\right]_0^1$

$= \ln 2 + \arctan 1 - \dfrac{1}{2}\ln 2$

$= \dfrac{1}{4}(\pi + 2\ln 2)$

14 0.824 (3 s.f.)

15 With $x = \dfrac{1}{2}\sin A$, $dx = \dfrac{1}{2}\cos A\,dA$

$1 - 4x^2 = 1 - \sin^2 A = \cos^2 A$ and so $\dfrac{x^2}{\sqrt{1-4x^2}} = \dfrac{\sin^2 A}{4\cos A}$

$\displaystyle\int_0^{\frac{1}{4}} \dfrac{x^2}{\sqrt{1-4x^2}}\,dx = \int_0^{\frac{\pi}{6}} \dfrac{\sin^2 A}{4\cos A} \times \dfrac{1}{2}\cos A\,dA$

$= \dfrac{1}{8}\displaystyle\int_0^{\frac{\pi}{6}} \sin^2 A\,dA$

$= \dfrac{1}{8}\displaystyle\int_0^{\frac{\pi}{6}} (1 - \cos 2A)\,dA$

$= \dfrac{1}{8}\left[A - \dfrac{\sin 2A}{2}\right]_0^{\frac{\pi}{6}}$

$= \dfrac{1}{16}\left[\dfrac{\pi}{6} - \dfrac{\sqrt{3}}{4}\right]$

$= \dfrac{1}{192}(2\pi - 3\sqrt{3})$

16 **a** With $x = 2\cosh u$, $dx = 2\sinh u\,du$

$\displaystyle\int \sqrt{x^2-4}\,dx = \int 2\sqrt{\cosh^2 u - 1} \times 2\sinh u\,du$

$= 4\displaystyle\int \sinh^2 u\,du$

$= 2\displaystyle\int (\cosh 2u - 1)\,du$

$= 2\left(\dfrac{\sinh 2u}{2} - u\right) + C$

$= 2\sinh u \cosh u - 2u + C$

$= 2\left(\dfrac{\sqrt{x^2-4}}{2}\right)\left(\dfrac{x}{2}\right) - 2\operatorname{arcosh}\left(\dfrac{x}{2}\right) + C$

$= 2x\sqrt{x^2-4} - 2\operatorname{arcosh}\left(\dfrac{x}{2}\right) + C$

b 12.9 (s.f.)

17 **a** $2\cosh x - \sinh x = 2\left(\dfrac{e^x + e^{-x}}{2}\right) - \left(\dfrac{e^x - e^{-x}}{2}\right) = \dfrac{e^x + 3e^{-x}}{2}$

$\displaystyle\int \dfrac{1}{2\cosh x - \sinh x}\,dx = \int \dfrac{2}{e^x + 3e^{-x}}\,dx$

$= \displaystyle\int \dfrac{2e^x}{e^{2x} + 3}\,dx$

b $\dfrac{2}{\sqrt{3}}\arctan\left(\dfrac{e^x}{\sqrt{3}}\right) + C$

18 0.360 (3 s.f.)

19 **a** **i** Using partial fractions

$\dfrac{1}{a^2 - x^2} = \dfrac{1}{2a}\left\{\dfrac{1}{a-x} + \dfrac{1}{a+x}\right\}$

So $\displaystyle\int \dfrac{dx}{a^2 - x^2} = \dfrac{1}{2a}\int \left\{\dfrac{1}{a-x} + \dfrac{1}{a+x}\right\}dx$

$= \dfrac{1}{2a}[-\ln|a-x| + \ln|a+x|] + C$

$= \dfrac{1}{2a}\ln\left|a + \dfrac{x}{a} - x\right| + C$

ii $\dfrac{1}{a}\operatorname{artanh}\left(\dfrac{x}{a}\right) + D$

b $\dfrac{1}{2}\ln\left|\dfrac{a+x}{a-x}\right|$

20 **a** $\operatorname{arcsec} x + C$

b $\sqrt{x^2 - 1} - \operatorname{arcsec} x + C$

Exercise 4D

1 **a** $\arcsin\left(\dfrac{x+2}{3}\right) + C$ **b** $\operatorname{arcosh}\left(\dfrac{x-2}{4}\right) + C$

  c $\operatorname{arsinh}(x+3) + C$ **d** $\operatorname{arcosh}(x-1) + C$

  e $\dfrac{\sqrt{10}}{10}\arctan\left(\dfrac{\sqrt{2}(x+1)}{\sqrt{5}}\right) + C$

  f $\dfrac{1}{2}\arcsin\left(\dfrac{2x+3}{3}\right) + C$

  g $\dfrac{1}{\sqrt{2}}\arcsin\left(\dfrac{x+3}{4}\right) + C$

  h $\dfrac{1}{3}\operatorname{arcosh}\left(\dfrac{9x-4}{\sqrt{7}}\right) + C$

2 **a** $\dfrac{1}{2}\operatorname{arsinh}(2x-3) + C$

  b $\dfrac{1}{2}\operatorname{arcosh}\left(\dfrac{2x-3}{\sqrt{5}}\right) + C$

3 **a** 0.400 (3 s.f.) **b** 0.325 (3 s.f.) **c** 0.597 (3 s.f.)

4 **a** $\ln\{2 + \sqrt{5}\}$ **b** $\dfrac{\pi}{3\sqrt{3}}$

5 $3x^2 - 6x + 7 = 3\left(x^2 - 2x + \dfrac{7}{3}\right) = 3\left(x - 1^2 + \dfrac{4}{3}\right)$

$\displaystyle\int \dfrac{1}{\sqrt{3x^2 - 6x + 7}}\,dx = \dfrac{1}{\sqrt{3}}\int \dfrac{1}{\sqrt{(x-1)^2 + \left(\dfrac{2}{\sqrt{3}}\right)^2}}\,dx$

Let $u = x - 1 \Rightarrow du = dx$

$\Rightarrow \displaystyle\int_1^3 \dfrac{1}{\sqrt{3x^2 - 6x + 7}}\,dx = \dfrac{1}{\sqrt{3}}\int_0^2 \dfrac{1}{\sqrt{u^2 + \left(\dfrac{2}{\sqrt{3}}\right)^2}}\,du$

$= \dfrac{1}{\sqrt{3}}\left[\operatorname{arsinh}\dfrac{\sqrt{3}u}{2}\right]_0^2$

$= \dfrac{1}{\sqrt{3}}\operatorname{arsinh}\sqrt{3}$

$= \dfrac{1}{\sqrt{3}}\ln(\sqrt{3} + \sqrt{3+1})$

$= \dfrac{1}{\sqrt{3}}\ln(\sqrt{3} + 2)$

6 **a** $\operatorname{arsinh}(x+2) + C$

  b $\arcsin\left(\dfrac{x-2}{3}\right) + C$

7 $\dfrac{\pi\sqrt{3}}{90}$

8 $\ln(4 - \sqrt{7})$

9 $x = 1 + \sinh u$, $dx = \cosh u\,du$

$x^2 - 2x + 2 = \sinh^2 u + 2\sinh u + 1 - 2\sinh u + 1 + 2$

$\qquad\qquad = \sinh^2 u + 1 = \cosh^2 u$

$\displaystyle\int \dfrac{1}{(x^2 - 2x + 2)^{\frac{3}{2}}}\,dx = \int \dfrac{1}{\cosh^3 u}\cosh u\,du$

$= \displaystyle\int \operatorname{sech}^2 u\,du$

$= \tanh u + C$

$= x - \dfrac{1}{\sqrt{x^2 - 2x + 2}} + C$

10 Using the substitution $x = 2\sin\theta - 1$, $dx = 2\cos\theta\, d\theta$
and $3 - 2x - x^2 = 3 - 2(2\sin\theta - 1)$
$\qquad\qquad\qquad\quad - (4\sin^2\theta - 4\sin\theta + 1)$
$\qquad\qquad\qquad = 4 - 4\sin^2\theta$
$\qquad\qquad\qquad = 4\cos^2\theta$

So $\int \dfrac{x}{\sqrt{3-2x-x^2}}\, dx = \int \dfrac{2\sin\theta - 1}{2\cos\theta} \times 2\cos\theta\, d\theta$
$\qquad\qquad\qquad\qquad = \int (2\sin\theta - 1)\, d\theta$
$\qquad\qquad\qquad\qquad = -2\cos\theta - \theta + C$
$\qquad\qquad\qquad\qquad = -2\sqrt{1 - \left(\dfrac{x+1}{2}\right)^2} - \theta + C$
$\qquad\qquad\qquad\qquad = -\sqrt{3 - 2x - x^2} - \arcsin\left(\dfrac{x+1}{2}\right) + C$

Challenge

1 a $\dfrac{1}{8}\sinh(2x^2) + \dfrac{x^2}{4} + C$ **b** $\dfrac{1}{2}\tanh(x^2) + C$

Exercise 4E

1 a $I = \int 1 \cdot \operatorname{arsinh} x\, dx$

Let $u = \operatorname{arsinh} x$, $\dfrac{dv}{dx} = 1$

$\dfrac{du}{dx} = \dfrac{1}{\sqrt{x^2 + 1}}$, $v = x$

$I = -x\operatorname{arsinh} x - \int \dfrac{x}{\sqrt{x^2}} + 1\, dx$

$\quad = -x\operatorname{arsinh} x - \sqrt{x^2 + 1} - C$

b 0.467 (3 s.f.)

c $\dfrac{1}{2}[(2x+1)\operatorname{arsinh}(2x+1) - \sqrt{4x^2 + 4x + 2}] + C$

2 Let $u = \arctan 3x$, $\dfrac{dv}{dx} = 1$

$\dfrac{du}{dx} = \dfrac{1}{1 + (3x)^2}$, $v = x$

$\int \arctan 3x\, dx = -x\arctan 3x - \int \dfrac{3x}{1 + 9x^2}\, dx$

$\qquad\qquad\qquad = -x\arctan 3x - \dfrac{1}{6}\int \dfrac{18x}{1 + 9x^2}\, dx$

$\qquad\qquad\qquad = -x\arctan 3x - \dfrac{1}{6}\ln(1 + 9x^2) + C$

3 a Let $u = \operatorname{arcosh} x$, $\dfrac{dv}{dx} = 1$

$\dfrac{du}{dx} = \dfrac{1}{\sqrt{x^2 - 1}}$, $v = x$

$\int \operatorname{arcosh} x\, dx = x\operatorname{arcosh} x - \int \dfrac{1}{\sqrt{x^2 - 1}}\, dx$

$\qquad\qquad\qquad = x\operatorname{arcosh} x - \sqrt{x^2 - 1} + C$

b $\int_1^2 \operatorname{arcosh} x = [2\operatorname{arcosh} 2 - \sqrt{3}] - [\operatorname{arcosh} 1]$

$\qquad\qquad = [2\operatorname{arcosh} 2 - \sqrt{3}]$

As $\operatorname{arcosh} x = \ln(x + \sqrt{x^2 - 1})$

$\int_2^2 \operatorname{arcosh} x = [2\ln(2 + \sqrt{3}) - \sqrt{3}]$

$\qquad\qquad = [\ln(2 + \sqrt{3})^2 - \sqrt{3}]$

$\qquad\qquad = [\ln(7 + 4\sqrt{3}) - \sqrt{3}]$

4 a $I = \int 1 \times \arctan x\, dx$

Let $u = \arctan x$, $\dfrac{dv}{dx} = 1$

so $\dfrac{du}{dx} = \dfrac{1}{x^2} + 1$, $v = x$

$I = x\arctan x - \int \dfrac{x}{x^2 + 1}\, dx$

$\quad = x\arctan x - \dfrac{1}{2}\ln(x^2 + 1) + C$

b $\int_1^{\sqrt{3}} \arctan x = \left[x\arctan x - \dfrac{1}{2}\ln(x^2 + 1)\right]_1^{\sqrt{3}}$

$\quad = \left[\sqrt{3}\arctan\sqrt{3} - \dfrac{1}{2}\ln 4\right] - \left[-\arctan(-1) - \dfrac{1}{2}\ln 2\right]$

$\quad = \dfrac{\sqrt{3}\pi}{3} - \ln 2 + \left(-\dfrac{\pi}{4}\right) + \ln 2$

$\quad = \dfrac{(4\sqrt{3} - 3)\pi}{12} - \dfrac{\ln 2}{2}$

c 4.23 (3 s.f.)

5 a $\dfrac{\sqrt{2}}{8}\pi - 1 + \dfrac{\sqrt{2}}{2}$ **b** $\dfrac{\pi - 2}{4}$

6 $u = \operatorname{arcsec} x$, $\dfrac{dv}{dx} = 1$

$\dfrac{du}{dx} = \dfrac{1}{x\sqrt{x^2 - 1}}$, $v = x$

$\int \operatorname{arcsec} x\, dx = x\operatorname{arcsec} x - \int \dfrac{x}{x\sqrt{x^2 - 1}}\, dx$

$\qquad\qquad = x\operatorname{arcsec} x - \operatorname{arcosh} x + C$

$\qquad\qquad = x\operatorname{arcsec} x - \ln(x + \sqrt{x^2 - 1}) + C$

Exercise 4F

1 a $u = x^n$, $\dfrac{dv}{dx} = e^{\frac{x}{2}}$

$\dfrac{du}{dx} = nx^{n-1}$, $v = 2e^{\frac{x}{2}}$

$I_n = 2x^n e^{\frac{x}{2}} - \int 2nx^{n-1} e^{\frac{x}{2}}\, dx$

$I_n = 2x^n e^{\frac{x}{2}} - 2nx^{n-1} e^{\frac{x}{2}}\, dx$

$I_n = 2x^n e^{\frac{x}{2}} - 2nI_{n-1}$

b $2x^3 e^{\frac{x}{2}} - 12x^2 e^{\frac{x}{2}} + 48xe^{\frac{x}{2}} - 96e^{\frac{x}{2}} + C$

2 a $u = \ln x^n$, $\dfrac{dv}{dx} = x$

$\dfrac{du}{dx} = \dfrac{n(\ln x)^{n-1}}{x}$, $v = \dfrac{x^2}{2}$

$\int_1^e x(\ln x)^n\, dx = \left[\dfrac{x^2(\ln x)^n}{2}\right]_1^e - \int_1^e \dfrac{nx^2(\ln x)^{n-1}}{2x}\, dx$

$\qquad = \left[\dfrac{e^2}{2} - 0\right] - \dfrac{n}{2}\int_1^e x(\ln x)^{n-1}\, dx$

$\qquad = \dfrac{e^2}{2} - \dfrac{n}{2}I_{n-1}$

b $\int_1^e x(\ln x)^4\, dx = I_4$

Substituting 4, 3, 2 and 1 respectively in the reduction formula

$I_4 = \dfrac{e^2}{2} - \dfrac{4}{2}I_3$

$I_4 = \dfrac{e^2}{2} - 2\left(\dfrac{e^2}{2} - \dfrac{3}{2}I_2\right)$

$I_4 = \dfrac{e^2}{2} - e^2 + 3\left(\dfrac{e^2}{2} - \dfrac{2}{2}I_1\right)$

$I_4 = \dfrac{e^2}{2} - e^2 + 3\left(\dfrac{e^2}{2} - \dfrac{3e^2}{2} - \dfrac{1}{2}I_0\right)$,

where $I_0 = \int_1^e x\, dx = \left[\dfrac{x^2}{2}\right]_1^e = \dfrac{e^2}{2} - \dfrac{1}{2}$

$\therefore \int_1^e x\ln x^4\, dx = \dfrac{e^2}{2} - e^2 + \dfrac{3e^2}{2} - \dfrac{3e^2}{2} + \dfrac{3}{2}\left(\dfrac{e^2}{2} - \dfrac{1}{2}\right)$

$\qquad = \dfrac{e^2 - 3}{4}$

3 $\dfrac{16}{7}$

4 a $u = x^n$, $\dfrac{dv}{dx} = e^{-x}$

$\dfrac{du}{dx} = nx^{n-1}$, $v = -e^{-x}$

$\int x^n e^{-x}\, dx = -x^n e^{-x} - \int -nx^{n-1} e^{-x}\, dx = -x^n e^{-x} + nI_{n-1}$

b $-x^3 e^{-x} - 3x^2 e^{-x} - 6xe^{-x} - 6e^{-x} + K$

c $24 - 65e^{-1}$ or $\dfrac{24e - 65}{e}$

5 a $I_n = \int \tanh^n x \, dx = \int \tanh^{n-2} x \, \tanh^2 x \, dx$

$[1 - \text{sech}^2 x = \tanh^2 x]$

$I_n = \int \tanh^{n-2} x \, (1 - \text{sech}^2 x) \, dx$

$= \int \tanh^{n-2} x \, dx - \int \tanh^{n-2} x \, \text{sech}^2 x \, dx$

So $I_n = I_{n-2} - \dfrac{1}{n-1} \tanh^{n-1} x \quad n \neq 1$

b $\ln \cosh x - \dfrac{1}{2} \tanh^2 x - \dfrac{1}{4} \tanh^4 x + C$

c Since $\int \tanh^n x \, dx = \int \tanh^{n-2} x \, dx - \dfrac{1}{n-1} \tanh^{n-1} x$

then $\int_0^{\ln 2} \tanh^n x \, dx = \int_0^{\ln 2} \tanh^{n-2} x \, dx$

$- \left[\dfrac{1}{n-1} \tanh^{n-1} x \right]_0^{\ln 2}$

Now $\tanh(\ln 2) = \dfrac{e^{\ln 2} - e^{-\ln 2}}{e^{\ln 2} + e^{-\ln 2}} = \dfrac{2 - \frac{1}{2}}{2 + \frac{1}{2}} = \dfrac{3}{5}$

So $\int_0^{\ln 2} \tanh^4 x \, dx = \int_0^{\ln 2} \tanh^2 x \, dx - \dfrac{1}{3}\left(\dfrac{3}{5}\right)^3$

$\left(\text{with } n = 4, \tanh(\ln 2) = \dfrac{3}{5}\right)$

$= \left[\int_0^{\ln 2} \tanh^0 x \, dx - 1 \times \left(\dfrac{3}{5}\right) \right] - \dfrac{1}{3} \times \dfrac{27}{125}$

$= \ln 2 - \dfrac{3}{5} - \dfrac{9}{125}$

$= \ln 2 - \dfrac{84}{125}$

6 a $\int \tan^4 x \, dx = \dfrac{1}{3} \tan^3 x - \int \tan^2 x \, dx$

$= \dfrac{1}{3} \tan^3 x - \left(\tan x - \int \tan^0 x \, dx \right)$

$= \dfrac{1}{3} \tan^3 x - \tan x + \int 1 \, dx$

$= \dfrac{1}{3} \tan^3 x - \tan x - x + C$

b $\ln \sqrt{2} - \dfrac{1}{4}$

c $I_n = \int_0^{\frac{\pi}{3}} \tan^6 x \, dx$

$= \left[\dfrac{1}{n-1} \tan^{n-1} x \right]_0^{\frac{\pi}{3}} - I_{n-2} = \dfrac{(\sqrt{3})^{n-1}}{n-1} - I_{n-2}$

So $I_6 = \dfrac{(\sqrt{3})^5}{5} - I_4 = \dfrac{(\sqrt{3})^5}{5} - \left(\dfrac{(\sqrt{3})^3}{3} - I_2 \right) = \dfrac{(\sqrt{3})^5}{5}$

$- \dfrac{(\sqrt{3})^3}{3} + \left(\dfrac{(\sqrt{3})}{1} - I_0 \right)$

As $I_0 = \int_0^{\frac{\pi}{3}} 1 \, dx = \dfrac{\pi}{3}$, $\int_0^{\frac{\pi}{3}} \tan^6 x \, dx$

$= \dfrac{(9\sqrt{3})}{5} - \dfrac{(3\sqrt{3})}{3} + \sqrt{3} - \dfrac{\pi}{3} = \dfrac{9\sqrt{3}}{5} - \dfrac{\pi}{3}$

7 a $I_n = \int_1^a (\ln x)^n \, dx = \int_1^a 1 \, (\ln x)^n \, dx$

$u = (\ln x)^n \quad \dfrac{dv}{dx} = 1$

$\dfrac{du}{dx} = n \dfrac{(\ln x)^{n-1}}{x} \quad v = x$

$\int_1^a 1 \, (\ln x)^n \, dx = [x (\ln x)^n]_1^a - \int_1^a \dfrac{n (\ln x)^{n-1}}{x} x \, dx$

$= [a(\ln a)^n - 0] - n \int_1^a (\ln x)^{n-1} \, dx$

So $I_n = a(\ln a)^n - n I_{n-1}$

b $2(\ln 2)^3 - 6(\ln 2)^2 + 12(\ln 2) - 6$

c $a = e$, $I_n = \int_1^e (\ln x)^n \, dx = e(\ln e)^n - nI_{n-1} = e - nI_{n-1}$

$I_6 = \int_1^e (\ln x)^6 \, dx = e - 6I_5$

$= e - 6(e - 5I_4)$

$= e - 6e + 30(e - 4I_3)$

$= e - 6e + 30e - 120(e - 3I_2)$

$= e - 6e + 30e - 120e + 360(e - 2I_1)$

$= e - 6e + 30e - 120e + 360e - 720(e - I_0)$

$I_0 = \int_1^e 1 \, dx = [x]_1^e = e - 1$

$\int_1^e (\ln x)^6 \, dx = e - 6e + 30e - 120e + 360e$

$- 720e + 720(e - 1)$

$= 265e - 720$

$= 553e - 144$

8 a $\dfrac{16}{35}$ **b** $\dfrac{\pi}{32}$ **c** $\dfrac{8}{105}$ **d** $\dfrac{35\pi}{768}$

9 a $I_{n+1} = \int \dfrac{\sin^{2n+2} x}{\cos x} \, dx$

So $I_n - I_{n+1} = \int \dfrac{\sin^{2n} x - \sin^{2n+2} x}{\cos x} \, dx$

$= \int \dfrac{\sin^{2n} x (1 - \sin^2 x)}{\cos x} \, dx \quad [1 - \sin^2 x = \cos^2 x]$

$= \int \sin^{2n} x \cos x \, dx$

So $I_n - I_{n+1} = \dfrac{\sin^{2n+1} x}{2n+1}$

Or $I_{n+1} = I_n - \dfrac{\sin^{2n+1} x}{2n+1}$

b $\int \dfrac{\sin^4 x}{\cos x} \, dx = I_2$

Substituting $n = 1$ in reduction formula gives

$I_2 = I_1 - \dfrac{\sin^3 x}{3} = \left(I_0 - \dfrac{\sin x}{1} \right) - \dfrac{\sin^3 x}{3}$

$I_0 = \int \dfrac{1}{\cos x} \, dx = \int \sec x \, dx = \ln |(\sec x + \tan x)| + C$

So $\int \dfrac{\sin^4 x}{\cos x} \, dx = \ln |(\sec x + \tan x)| - \sin x$

$- \dfrac{\sin^3 x}{3} + C$

Applying the given limits gives

$\int_0^{\frac{\pi}{4}} \dfrac{\sin^4 x}{\cos x} \, dx = \left[\ln|(\sec x + \tan x)| \, 2\sin x - \dfrac{\sin^3 x}{3} \right]_0^{\frac{\pi}{4}}$

$= \ln(1 + \sqrt{2}) - \dfrac{\sqrt{2}}{2} - \dfrac{\left(\frac{\sqrt{2}}{2}\right)^3}{3}$

$= \ln(1 + \sqrt{2}) - \dfrac{\sqrt{2}}{2} - \dfrac{\sqrt{2}}{12}$

$= \ln(1 + \sqrt{2}) - \dfrac{7\sqrt{2}}{12}$

10 a $u = (1 - x^3)^n \quad \dfrac{dv}{dx} = x$

$\dfrac{du}{dx} = n(1 - x^3)^{n-1} (-3x^2) \quad v = \dfrac{x^2}{2}$

$\int_0^1 x(1 - x^3)^n \, dx = \left[\dfrac{x^2}{2} (1 - x^3)^n \right]_0^1 - \int_0^1 -3nx^2 (1 - x^3)^{n-1} \dfrac{x^2}{2} \, dx$

$= [0 - 0] + \dfrac{3n}{2} \int_0^1 x^4 (1 - x^3)^{n-1} \, dx$ provided $n \geq 0$

Writing $x^4 = x \cdot x^3 = x(1 - (1 - x^3))$ and $I_n = \int_0^1 x(1 - x^3)^n \, dx$

we have $I_n = \dfrac{3n}{2} \int_0^1 x(1 - (1 - x^3))(1 - x^3)^{n-1} \, dx$

$= \dfrac{3n}{2} \int_0^1 x(1 - x^3)_{n-1} \, dx - \dfrac{3n}{2} \int_0^1 x(1 - x^3)^n \, dx$

$= \dfrac{3n}{2} I_{n-1} - \dfrac{3n}{2} I_n$

$\Rightarrow (3n + 2) I_n = 3n I_{n-1}$, so $I_n = \dfrac{3n}{3n+2} I_{n-1}$, $n \geq 1$

b $\dfrac{243}{1540}$

11 a $u = (a^2 - x^2)^n \quad \dfrac{dv}{dx} = 1$

$\dfrac{du}{dx} = -2nx(a^2 - x^2)^{n-1} \quad v = x$

$\int_0^a (a^2 - x^2)^n \, dx = [x(a^2 - x^2)^n]_0^a - \int_0^a x(-2nx(a^2 - x^2)^{n-1}) \, dx$

$= [0 - 0] + 2n \int_0^a x^2 (a^2 - x^2)^{n-1} \, dx$

$= 2n \int_0^a x^2 (a^2 - x^2)^{n-1} \, dx \ (n > 0)$

Writing x^2 as $\{a^2 - (a^2 - x^2)\}$ and defining

$I_n = \int_0^a (a^2 - x^2)^n \, dx$

we have $I_n = 2n \int_0^a \{a^2(a^2 - x^2)^{n-1} - (a^2 - x^2)^n\} \, dx$

$= 2na^2 I_{n-1} - 2n I_n$

So $(2n + 1)I_n = 2na^2 I_{n-1} \Rightarrow I_n = \dfrac{2na^2}{(2n+1)} I_{n-1}$

b i $\dfrac{128}{315}$ **ii** $\dfrac{34992}{35}$ **iii** π

c Use substitution $x = 2\sin\theta$

12 a Integrating by parts with $u = x^n$ and $\dfrac{dv}{dx} = \sqrt{4 - x}$

$\dfrac{du}{dx} = nx^{n-1}, \quad v = -\dfrac{2}{3}(4 - x)^{\frac{3}{2}}$

So $\int_0^4 x^n \sqrt{4 - x} \, dx = \left[-\dfrac{2}{3} x^n (4 - x)^{\frac{3}{2}}\right]_0^4$

$- \int_0^4 -\dfrac{2}{3} nx^{n-1}(4 - x)^{\frac{3}{2}} \, dx$

$= [0 - 0] + \dfrac{2}{3} n \int_0^4 x^{n-1}(4-x)^{\frac{3}{2}} \, dx \ (n > 0)$

$= \dfrac{2}{3} n \int_0^4 x^{n-1}((4-x)\sqrt{4-x}) \, dx$

$= \dfrac{2}{3} n \int_0^4 x^{n-1} 4\sqrt{4-x} \, dx + \dfrac{2}{3} n \int_0^4 x^{n-1}(-x\sqrt{4-x}) \, dx$

$= \dfrac{8}{3} n \int_0^4 x^{n-1}\sqrt{4-x} \, dx - \dfrac{2}{3} n \int_0^4 x^n \sqrt{4-x} \, dx$

So $I_n = \dfrac{8}{3} n I_{n-1} - \dfrac{2}{3} n I_n$

$\Rightarrow (2n + 3)I_n = 8n I_{n-1} \Rightarrow I_n = \dfrac{8n}{2n+3} I^{n-1}, n \geq 1$

b 52.0 (3 s.f.)

13 a $I_n = \int \cos^n x \, dx = \int \cos^{n-1} x \cos x \, dx$

Integrating by parts with $u = \cos^{n-1} x$ and $\dfrac{dv}{dx} = \cos x$

$\dfrac{du}{dx} = (n-1)\cos^{n-2} x(-\sin x), v = \sin x$

So $I_n = \int \cos^n x \, dx = \cos^{n-1} x \sin x$

$- \int -(n-1) \cos^{n-2} x \sin^2 x \, dx$

$= \cos^{n-1} x \sin x + (n-1) \int \cos^{n-2} x (1 - \cos^2 x) \, dx$

$= \cos^{n-1} x \sin x + (n-1) \int \cos^{n-2} x \, dx$

$- (n-1) \int \cos^n x \, dx$

Giving $I_n = \cos^{n-1} x \sin x + (n-1) I_{n-2} - (n-1) I_n$

So $n I_n = \cos^{n-1} x \sin x + (n-1) I_{n-2}$

b $n J_n = (n-1) J_{n-2}$

c i $\dfrac{3\pi}{4}$ **ii** $\dfrac{35\pi}{64}$

d If n is odd, J_n always reduces to a multiple of J_1, but $J_1 = \int_0^{2\pi} \cos x \, dx = [\sin x]_0^{2\pi} = 0$.

(You could also consider the graphical representation.)

14 a Integrating by parts with $u = x^{n-1}$ and $\dfrac{dv}{dx} = \sqrt{1-x^2}$

$\dfrac{du}{dx} = (n-1)x^{n-2}, \quad v = -\dfrac{1}{3}(1-x^2)^{\frac{3}{2}}$

So $I_n = \int_0^1 x^{n-1} \{x\sqrt{1-x^2}\} \, dx$

$= \left[-\dfrac{1}{3} x^{n-1}(1-x^2)^{\frac{3}{2}}\right]_0^1 + \dfrac{(n-1)}{3} \int_0^{\frac{\pi}{2}} x^{n-2}(1-x^2)^{\frac{3}{2}} \, dx$

$= \dfrac{(n-1)}{3} \int_0^{\frac{\pi}{2}} x^{n-2}(1-x^2)^{\frac{3}{2}} dx$ as

$\left[-\dfrac{1}{3}x^{n-1}(1-x^2)^{\frac{3}{2}}\right]_0^1 = 0$

$= \dfrac{(n-1)}{3} \int_0^{\frac{\pi}{2}} x^{n-2}(1-x^2)\sqrt{1-x^2} \, dx$

$= \dfrac{(n-1)}{3} \int_0^{\frac{\pi}{2}} \{x^{n-2}\sqrt{1-x^2} - x^n \sqrt{1-x^2}\} \, dx$

So $I_n = \dfrac{(n-1)}{3} I_{n-2} - \dfrac{(n-1)}{3} I_n$

$\Rightarrow (3 + (n-1)) I_n = (n-1) I_{n-2}$

$\Rightarrow (n+) I_n = (n-1) I_{n-2}$

b $\dfrac{16}{315}$

15 a Integrating by parts with $u = x^n$ and $\dfrac{dv}{dx} = \cosh x$

$\dfrac{du}{dx} = nx^{n-1}, \quad v = \sinh x$

So $\int \cosh x \, dx = x^n \sinh x - \int nx^{n-1} \sinh x \, dx$

Integrating by parts again with $u = x^{n-1}$ and

$\dfrac{dv}{dx} = \sinh x$

$\dfrac{du}{dx} = (n-1)x^{n-2}, \quad v = \cosh x$

So $I_n = x^n \sinh x - n\{x^{n-1} \cosh x - \int (n-1)x^{n-2} \cosh x \, dx\}$

$= x^n \sinh x - nx^{n-1} \cosh x + n(n-1) I_{n-2}, \quad n \geq 2$

b $(x^4 + 12x^2 + 24) \sinh x - (4x^3 - 24x) \cosh x + C$

c $6 - e - 8e^{-1}$ or $\dfrac{6e - e^2 - 8}{e}$

16 a $I_{n-2} = \int \dfrac{\sin(n-2)x}{\sin x} \, dx$

So $I_n - I_{n-2} = \int \dfrac{\sin nx - \sin(n-2)x}{\sin x} \, dx$

$= \int \dfrac{2\cos\left\{\dfrac{n+(n-2)}{2}\right\} x \sin\left\{\dfrac{n-(n-2)}{2}\right\} x}{\sin x} \, dx$

$= \int \dfrac{2\cos(n-1)x \sin x}{\sin x} \, dx$

$= \int 2\cos(n-1)x \, dx$

$= \dfrac{2\sin(n-1)x}{n-1}, n \geq 2$

b i $2 \sin x + \dfrac{2 \sin 3x}{3} + C$ **ii** $\dfrac{\pi}{6} - \dfrac{\sqrt{3}}{2}$

17 a $I_n = \int \sinh^n x \, dx = \int \sinh^{n-1} x \sinh x \, dx$

Integrating by parts with $u = \sinh^{n-1} x$ and $\dfrac{dv}{dx} = \sinh x$

$\dfrac{du}{dx} = (n-1)\sinh^{n-2} x \cosh x, \quad v = \cosh x$

So $I_n = \int \sinh^n x \, dx = \sinh^{n-1} x \cosh x$

$- \int (n-1) \sinh^{n-2} x \cosh^2 x \, dx$

$$= \sinh^{n-1} x \cosh x - (n-1)$$
$$\int \sinh^{n-2} x (1 + \sinh^2 x) \, dx$$
$$= \sinh^{n-1} x \cosh x - (n-1) \int \sinh^{n-2} x \, dx$$
$$- (n-1) \int \sinh^n x \, dx$$

Giving $I_n = \sinh^{n-1} x \cosh x - (n-1)I_{n-2} - (n-1)I_n$

So $nI_n = \sinh^{n-1} x \cosh x - (n-1)I$, $n \geq 2$ *

b i $\dfrac{752}{1215}$

ii $\int \sinh^4 x \, dx = I_4 = \dfrac{1}{4}\sinh^3 x \cosh x - \dfrac{3}{4}I_2$

$= \dfrac{1}{4}\sinh^3 x \cosh x - \dfrac{3}{4}\left(\dfrac{1}{2}\sinh x \cosh x - \dfrac{1}{2}I_0\right)$

$= \dfrac{1}{4}\sinh^3 x \cosh x - \dfrac{3}{8}\sinh x \cosh x + \dfrac{3}{8}\int 1 \, dx$

$= \dfrac{1}{4}\sinh^3 x \cosh x - \dfrac{3}{8}\sinh x \cosh x + \dfrac{3}{8}x + C$

Challenge

a $(a+1)I_n = x^{a+1}(\ln x)^n - nI_{n-1}$

b $\dfrac{2}{27}\sqrt{x^3}(9(\ln x)^3 - 18(\ln x)^2 + 24 \ln x - 16)$

Exercise 4G

1. $\dfrac{56}{3}$ or $18\dfrac{2}{3}$
2. $\ln(2+\sqrt{3})$
3. $\dfrac{3}{2}$
4. $4\dfrac{2}{3}$
5. 6.82 (s.f.)
6. $y = \dfrac{1}{4}(2x^2 - \ln x)$, so $\dfrac{dy}{dx} = x - \dfrac{1}{4x}$

$1 + \left(\dfrac{dy}{dx}\right)^2 = 1 + x^2 - \dfrac{1}{2} + \dfrac{1}{16x^2} = x^2 + \dfrac{1}{2} + \dfrac{1}{16x^2} = \left(x + \dfrac{1}{4x}\right)^2$

So arc length $= \int_1^2 \left(x + \dfrac{1}{4x}\right) dx$

$= \left[\dfrac{x^2}{2} + \dfrac{1}{4}\ln x\right]_1^2$

$= \left[2 + \dfrac{1}{4}\ln 2\right] - \left[\dfrac{1}{2}\right]$

$= \dfrac{1}{4}(6 + \ln 2)$

7. 1.5 They represent the same curve.
8. $\dfrac{1}{2}\ln(4+\sqrt{17}) + 2\sqrt{17}$
9. As $x = r\cos\theta, y = r\sin\theta, \dfrac{dx}{d\theta} = -r\sin\theta, \dfrac{dy}{d\theta} = r\cos\theta$

So $\left(\dfrac{dx}{d\theta}\right)^2 + \left(\dfrac{dy}{d\theta}\right)^2 = r^2(\cos^2\theta + \sin^2\theta) = r^2$

The circumference of the circle $= 4\int_0^{\frac{\pi}{2}} r \, d\theta$

$= 4r [\theta]_0^{\frac{\pi}{2}}$

$= 2\pi r$

10. $3a$

Total length of curve $= 4 \times 3a = 12a$ (symmetry)

11. $2\arctan(e) - \dfrac{\pi}{2}$ or 0.866 (3 s.f.)
12. $4a$
13. $x = t + \sin t, y = 1 - \cos t$

$\dfrac{dx}{dt} = 1 + \cos t, \dfrac{dy}{dt} = \sin t$

So $\left(\dfrac{dx}{dt}\right)^2 + \left(\dfrac{dy}{dt}\right)^2 = ((1 + 2\cos t + \cos^2 t) + (\sin^2 t))$

$= 2(1 + \cos t) = 4\cos^2\left(\dfrac{t}{2}\right)$

Using $s = \int_{t_A}^{t_B} \sqrt{\left(\dfrac{dx}{dt}\right)^2 + \left(\dfrac{dy}{dt}\right)^2} \, dt$

arc length $= \int_0^{\frac{\pi}{3}} \sqrt{4\cos^2\left(\dfrac{t}{2}\right)} \, dt$

$= 2\int_0^{\frac{\pi}{3}} \cos\left(\dfrac{t}{2}\right) dt$

$= 4\left[\sin\left(\dfrac{t}{2}\right)\right]_0^{\frac{\pi}{3}}$

$= 2$

14. $\sqrt{2}[e^{\frac{\pi}{4}} - 1]$ or 1.69 (3 s.f.)

15. $y = \sqrt{3}\sin x$, so $\dfrac{dy}{dx} = \sqrt{3}\cos x$

Using the symmetry of the sine

curve $s = 4\int_0^{\frac{\pi}{2}} \sqrt{1 + \left(\dfrac{dy}{dx}\right)^2} \, dx$

$= 4\int_0^{\frac{\pi}{2}} \sqrt{1 + 3\cos^2 x} \, dx$

b $x = \cos t, y = 2\sin t$

$\dfrac{dx}{dt} = -\sin t, \dfrac{dy}{dt} = 2\cos t$

$\left(\dfrac{dx}{dt}\right)^2 + \left(\dfrac{dy}{dt}\right)^2 = \sin^2 t + 4\cos^2 t$

$= 1 - \cos^2 t + 4\cos^2 t$

$= 1 + 3\cos^2 t$

from the diagram, at A, $t = 0$, at B, $t = \dfrac{\pi}{2}$,

so using the symmetry of the ellipse, the length of

the circumference is $4\int_0^{\frac{\pi}{2}} \sqrt{1 + 3\cos^2 t} \, dt$, equal to

that of the sine curve in **a**

Challenge

$\dfrac{62}{5}$

Exercise 4H

1. **a** 45π

   b Rotating about the y–axis

   From the work in part **a** $1 + \left(\dfrac{dy}{dx}\right)^2 = 1 + \dfrac{16}{9} = \dfrac{25}{9}$

   As integration is with respect to y the integrand must be in terms of y

   The limits for y are 3 (when $x = 4$) and 6 (when $x = 8$)

   So area of surface $= \int_3^6 2\pi\left(\dfrac{4}{3}y\right)\left(\dfrac{5}{3}\right) dy$

   $= \dfrac{40}{9}\pi \left[\dfrac{y^2}{2}\right]_3^6$

   $= \dfrac{40 \times 27}{9 \times 2}\pi = 60\pi$

2. $\dfrac{\pi}{27}[10\sqrt{10} - 1]$ or 3.56 (3 s.f.)
3. $\dfrac{2\pi}{3}[5\sqrt{5} - 1]$
4. $\dfrac{592}{3}\pi$

5 **a** 8.84 (3 s.f.)

b Using $\int_{x_1}^{x_2} 2\pi x \sqrt{1 + \left(\frac{dy}{dx}\right)^2} \, dx$,

the area of the surface is $\int_0^1 2\pi x \cosh x \, dx$

$= 2\pi \left\{ [x \sinh x]_0^1 - \int_0^1 \sinh x \, dx \right\}$

$= 2\pi \left\{ \sinh 1 - [\cosh x]_0^1 \right\}$

$= 2\pi (\sinh 1 - \cosh 1 + 1)$

$= 2\pi \left\{ \frac{1}{2} \left(e - \frac{1}{e} - e - \frac{1}{e} \right) + 1 \right\}$

$= 2\pi \left(1 - \frac{1}{e} \right)$

$= 2\pi \left(\frac{e-1}{e} \right)$

6 **b** $23\frac{1}{9}\pi$

7 $\frac{384\pi}{5}$ (241 (3 s.f.))

8 **a** $4\pi R^2$

9 $x = at^2, y = 2at$, so $\frac{dx}{dt} = 2at, \frac{dy}{dt} = 2a$

$\left(\frac{dx}{dt}\right)^2 + \left(\frac{dy}{dt}\right)^2 = 4a^2t^2 + 4a^2 = 4a^2(1+t^2)$

$x = 4a$ when $t = \pm 2$

Using $\int_{t_1}^{t} 2\pi y \sqrt{\left(\frac{dx}{dt}\right)^2 + \left(\frac{dy}{dt}\right)^2} \, dt$

Surface area $= 2\pi \int_0^2 4a^2(1+t^2) \, dt$

$= 8\pi a^2 \left[\frac{1}{3}(1+t^2)^{\frac{3}{2}} \right]_0^2$

$= 8\pi a^2 \left[5^{\frac{3}{2}} - 1 \right]$

$= 8\pi a^2 (5\sqrt{5} - 1)$

10 $x = \operatorname{sech} t, y = \tanh t$, so $\frac{dx}{dt} = -\operatorname{sech} t \tanh t, \frac{dy}{dt} = \operatorname{sech}^2 t$

$\left(\frac{dx}{dt}\right)^2 + \left(\frac{dy}{dt}\right)^2 = \operatorname{sech}^2 t \tanh^2 t + \operatorname{sech}^4 t$

$= \operatorname{sech}^2 t (\tanh^2 t + \operatorname{sech}^2 t) = \operatorname{sech}^2 t$

Using $\int_{t_1}^{t} 2\pi y \sqrt{\left(\frac{dx}{dt}\right)^2 + \left(\frac{dy}{dt}\right)^2} \, dt$

Surface area $= 2\pi \int_0^{\ln 2} \tanh t \operatorname{sech} t \, dt$

$= 2\pi [-\operatorname{sech} t]_0^{\ln 2}$

$= 2\pi \left[-\frac{2}{e^t + e^{-t}} \right]_0^{\ln 2}$

$= \frac{2\pi}{5}$

11 **a** $x = 3t^2, y = 2t^3$, so $\frac{dx}{dt} = 6t, \frac{dy}{dt} = 6t^2$

$\left(\frac{dx}{dt}\right)^2 + \left(\frac{dy}{dt}\right)^2 = 36t^2(t^2 + 1)$

Using $\int_{t_1}^{t} 2\pi x \sqrt{\left(\frac{dx}{dt}\right)^2 + \left(\frac{dy}{dt}\right)^2} \, dt$

Surface area $= 2\pi \int_0^2 3t^2 \times 6t\sqrt{1+t^2} \, dt$

$= 36\pi \int_0^2 t^3 \sqrt{1+t^2} \, dt$

b $\frac{24\pi}{5}[25\sqrt{5} + 1]$

12 $\frac{11\pi}{9}$

13 $\frac{93\pi a^2}{80}$

14 $y = e^x, \frac{dy}{dx} = e^x$

Using $\int_{x_1}^{x_2} 2\pi y \sqrt{1 + \left(\frac{dy}{dx}\right)^2} \, dx$,

the area of the surface is $2\pi \int_0^{\ln 2} e^x \sqrt{1 + e^{2x}} \, dx$

Make the substitution $e^x = \sinh u$, so $e^x dx = \cosh u \, du$
Limits: when $x = \ln 2$, $u = \operatorname{arsinh} e^{\ln 2} = \operatorname{arsinh} 2$
when $x = 0$, $u = \operatorname{arsinh} e^0 = \operatorname{arsinh} 1$
Then the area of the surface is $2\pi \int_{\operatorname{arsinh} 1}^{\operatorname{arsinh} 2} \cosh^2 u \, du$

$= \pi \int_{\operatorname{arsinh} 1}^{\operatorname{arsinh} 2} (1 + \cosh 2u) \, du$

$= \pi \left[u + \frac{\sinh 2u}{2} \right]_{\operatorname{arsinh} 2}^{\operatorname{arsinh} 1}$

$= \pi [u + \sinh u \cosh u]_{\operatorname{arsinh} 1}^{\operatorname{arsinh} 2}$

$= \pi [\operatorname{arsinh} 2 + 2\sqrt{5} - (\operatorname{arsinh} 1 + (1)(\sqrt{2}))]_{\operatorname{arsinh} 1}^{\operatorname{arsinh} 2}$

$= \pi (\operatorname{arsinh} 2 - \operatorname{arsinh} 1 + 2\sqrt{5} - \sqrt{2})$

Chapter Review 4

1 Volume $= \pi \int_0^1 y^2 \, dx = \pi \int_0^1 \tanh^2 x \, dx$

$= \pi \int_0^1 (1 - \operatorname{sech}^2 x) \, dx$

$= \pi [x - \tanh x]_1^0$

$= \pi (1 - \tanh 1)$

$= \pi \left(1 - \frac{e^2 - 1}{e^2 + 1} \right)$

$= \frac{2\pi}{1 + e^2}$

2 **a** $a = 2$
  $b = 1$
  $c = 16$

b $\frac{\pi}{32}$

3 **a** $\frac{1}{20} \cosh 10x - \frac{1}{4} \cosh 2x + C$

b $-\frac{1}{2} \ln (1 + 2\operatorname{sech} x) + C$

c $\frac{1}{4} e^{2x} - \frac{1}{2} x + C$

4 960 (2 s.f.)

5 **a** $\frac{1}{2} \arctan 2x + \frac{1}{4} \ln(1 + 4x^2) + C$

b $\frac{1}{8}[\pi + 2\ln 2]$

6 594 (3 s.f.)

7 Let $u = \operatorname{artanh} x$, $\frac{dv}{dx} = 1$

So $\frac{du}{dx} = \frac{1}{1-x^2}$, $v = x$

Then $\int_0^{\frac{1}{2}} \operatorname{artanh} x \, dx = [x \operatorname{artanh} x]_0^{\frac{1}{2}} - \int_0^{\frac{1}{2}} \frac{x}{1-x^2} \, dx$

$= [x \operatorname{artanh} x]_0^{\frac{1}{2}} + \frac{1}{2} \int_0^{\frac{1}{2}} \frac{-2x}{1-x^2} \, dx$

$= \left[x \operatorname{artanh} x + \frac{1}{2} \ln(1-x^2) \right]_0^{\frac{1}{2}}$

$= \frac{1}{2} \operatorname{artanh} \left(\frac{1}{2} \right) + \frac{1}{2} \ln \left(\frac{3}{4} \right)$

$= \frac{1}{2}\left(\frac{1}{2}\ln\left(\frac{\frac{3}{2}}{\frac{1}{2}}\right)\right) + \frac{1}{2}\ln\left(\frac{3}{4}\right)$

$= \frac{1}{4}\ln 3 + \frac{1}{2}\ln\left(\frac{3}{4}\right)$

$= \frac{1}{4}\left(\ln 3 + 2\ln\left(\frac{3}{4}\right)\right)$

$= \frac{1}{4}\left(\ln 3 + \ln\left(\frac{9}{16}\right)\right)$

$= \frac{1}{4}\ln\left(\frac{27}{16}\right)$ so $a = 27$ and $b = 16$

8 a i 1   **ii** $\frac{\pi}{2} - 1$

b Integrating by parts with $u = x^n$ and $\frac{dv}{dx} = \cos x$

$\frac{du}{dx} = nx^{n-1}$, $v = \sin x$

So $I_n = \int_0^{\frac{\pi}{2}} x^n \cos x\, dx = [x^n \sin x]_0^{\frac{\pi}{2}} - n\int_0^{\frac{\pi}{2}} x^{n-1}\sin x\, dx$

$= \left(\frac{\pi}{2}\right)^n - n\int_0^{\frac{\pi}{2}} x^{n-1}\sin x\, dx$ (1)

Integrating by parts on $\int_0^{\frac{\pi}{2}} x^{n-1}\sin x\, dx$ with

$u = x^{n-1}$ and $\frac{dv}{dx} = \sin x$

$\frac{du}{dv} = (n-1)x^{n-2}$, $v = \cos x$

gives $\int_0^{\frac{\pi}{2}} x^{n-1}\sin x\, dx = [-x^{n-1}\cos x]_0^{\frac{\pi}{2}}$
$+ (n-1)\int_0^{\frac{\pi}{2}} x^{n-2}\cos x\, dx$

$= (n-1)I_{n-2}$ as $[-x^{n-1}\cos x]_0^{\frac{\pi}{2}} = 0$

Substituting in (1)

$I_n = \left(\frac{\pi}{2}\right)^n - n(n-1)I_{n-2}$

c $\int_0^{\frac{\pi}{2}} x^3 \cos x\, dx = I_3 = \left(\frac{\pi}{2}\right)^3 - 3(2)I_1$

$= \left(\frac{\pi}{2}\right)^3 - 6\left(\frac{\pi}{2} - 1\right)$

$= \frac{\pi^3}{8} - 3\pi + 6$

$= \frac{1}{8}(\pi^3 - 24\pi + 48)$

d $\frac{\pi^4}{16} - 3\pi^2 + 24$

9 a $\text{arsinh}\left(\frac{x-1}{3}\right) + C$

b $\frac{1}{3}\arctan\left(\frac{x-1}{3}\right) + C$

c Using the substitution $x = \sin\theta$, so $dx = \cos\theta\, d\theta$

$\int_0^{\frac{1}{2}} \frac{x^4 dx}{\sqrt{(1-x^2)}} = \int_0^{\frac{\pi}{6}} \frac{\sin^4\theta\cos\theta\, d\theta}{\cos\theta}$

$= \int_0^{\frac{\pi}{6}} \sin^4\theta\, d\theta$

$= \frac{1}{4}\int_0^{\frac{\pi}{6}} (1 - 2\cos 2\theta + \cos^2 2\theta)\, d\theta$

$= \frac{1}{4}\int_0^{\frac{\pi}{6}} \left(1 - 2\cos 2\theta + \frac{1 + \cos 4\theta}{2}\right) d\theta$

$= \frac{1}{4}\left[\frac{3\theta}{2} - \sin 2\theta + \frac{\sin 4\theta}{8}\right]_0^{\frac{\pi}{6}}$

$= \frac{1}{4}\left(\frac{\pi}{4} - \frac{\sqrt{3}}{2} + \frac{\sqrt{3}}{16}\right)$

$= \frac{(4\pi - 7\sqrt{3})}{64}$

10 a Using integration by parts on I_n, with $u = x^n$
and $\frac{dv}{dx} = (1-x)^{\frac{1}{3}}$ so $\frac{du}{dx} = nx^{n-1}$ and $v = -\frac{3}{4}(1-x)^{\frac{4}{3}}$

$I_n = -\frac{3}{4}\left[x^n(1-x)^{\frac{4}{3}}\right]_0^8 + \frac{3n}{4}\int_0^8 x^{n-1}(1-x)^{\frac{4}{3}}\, dx$

$= \frac{3n}{4}\int_0^8 x^{n-1}(1-x)^{\frac{4}{3}}\, dx$

$= 6n\int_0^8 x^{n-1}(1-x)^{\frac{1}{3}}\, dx - \frac{3n}{4}\int_0^8 x^n(1-x)^{\frac{1}{3}}\, dx$

$\Rightarrow 4I_n = 6nI^{n-4} - \frac{3n}{4}I_n \Rightarrow I_n = \frac{24n}{3n+4}I^{n-1}$

b $\frac{39}{70}$

11 a $x = t - \ln t$, so $\frac{dx}{dt} = 1 - \frac{1}{t}$

$y = 4\sqrt{t}$, so $\frac{dy}{dt} = \frac{2}{\sqrt{t}}$

$\left(\frac{dx}{dt}\right)^2 + \left(\frac{dy}{dt}\right)^2 = 1 - \frac{2}{t} + \frac{1}{t^2} + \frac{4}{t} = 1 + \frac{2}{t} + \frac{1}{t^2} = \left(1 + \frac{1}{t}\right)^2$

Arc length $= \int_1^4 \sqrt{\left(1 + \frac{1}{t}\right)^2}\, dt = \int_1^4 \left(1 + \frac{1}{t}\right)\, dt$

$= [t + \ln t]_1^4 = (4 + \ln 4) - 1 = 3 + \ln 4$

b $\frac{160\pi}{3}$

12 Area $= \int_0^3 y\, dx = \int_0^3 x^2 \text{arsinh}\, x\, dx$

Using the integration by parts on I_n, with $u = \text{arsinh}\, x$
and $\frac{dv}{dx} = x^2$ so $\frac{du}{dx} = \frac{1}{\sqrt{1+x^2}}$ and $v = \frac{x^3}{3}$

$\int x^2 \text{arsinh}\, x\, dx = \frac{1}{3}x^3 \text{arsinh}\, x - \frac{1}{3}\int \frac{x^3}{\sqrt{1+x^2}}\, dx$

Let $x = \sinh u$ so $dx = \cosh u\, du$

$\int_0^3 x^2 \text{arsinh}\, x\, dx = 9\text{arsinh}\, 3 - \frac{1}{3}\int_0^{\text{arsinh}\, 3} \frac{\sinh^3 u}{\cosh u}\cosh u\, du$

$= 9\text{arsinh}\, 3 - \frac{1}{3}\int_0^{\text{arsinh}\, 3} \sinh^2 u\, du$

$= 9\text{arsinh}\, 3 - \frac{1}{3}\int_0^{\text{arsinh}\, 3} \sinh u(\cosh^2 u - 1)$

$= 9\text{arsinh}\, 3 - \frac{1}{3}\left[\frac{1}{3}\cosh^3 u - \cosh u\right]_0^{\text{arsinh}\, 3}$

$= 9\text{arsinh}\, 3 - \frac{1}{3}\left[\frac{1}{3}\cosh 3u - \cosh u\right]_0^{\text{arsinh}\, 3}$

When $x = 3$, $\sinh u = 3$ so $\cosh u = \sqrt{1 + \sinh^2 u} = \sqrt{10}$

So $\int_0^3 x^2 \text{arsinh}\, x\, dx = 9\ln(3 + \sqrt{10})$

$-\frac{1}{3}\left[\frac{10}{3}\sqrt{10} - \sqrt{10} - \left(\frac{1}{3} - 1\right)\right]$

$= 9\ln(3 + \sqrt{10}) - \frac{1}{9}[7\sqrt{10} + 2]$

13 a $\frac{\pi}{8}$

b i $\arcsin\left(\frac{x-2}{2}\right) + C$

ii $2(4x - x^2)^{\frac{1}{2}} + C$

iii $\frac{\pi}{3} - 2\sqrt{3}$

14 a $y = 2\sqrt{x}$ represents the section of curve for $x \geq 0$, $y \geq 0$,
so $\frac{dy}{dx} = \frac{1}{\sqrt{x}}$

b Using $2\pi \int_x^{x_2} y\sqrt{1+\left(\dfrac{dy}{dx}\right)^2}\,dx$

area of surface $= 2\pi \int_0^1 2\sqrt{x}\sqrt{1+\dfrac{1}{x}}\,dx$

$= 4\pi \int_0^1 \sqrt{x}\sqrt{\dfrac{x+1}{x}}\,dx$

$= 4\pi \int_0^1 \sqrt{1+x}\,dx$

b $\dfrac{8\pi}{3}(2\sqrt{2}-1)$

c Using the symmetry of the parabola, arc length is $2 \times$ the length of arc from origin to (1, 2)

so arc length $= 2\int_{x_1}^{x_2}\sqrt{1+\left(\dfrac{dy}{dx}\right)^2}\,dx$

$= 2\int_0^1 \sqrt{\left(\dfrac{x+1}{1}\right)}\,dx$

15 a Using integration by parts with $u = \operatorname{arcosh} x$ and $\dfrac{dv}{dx} = x$, $\dfrac{du}{dx} = \dfrac{1}{\sqrt{x^2-1}}$ and $v = \dfrac{x^2}{2}$

So $\int x\operatorname{arcosh} x\,dx = \dfrac{x^2}{2}\operatorname{arcosh} x - \int \dfrac{x^2}{2\sqrt{x^2-1}}\,dx$ *

Substituting $x = \cosh u$ in $\int \dfrac{x^2}{\sqrt{x^2-1}}\,dx$ gives

$\int \dfrac{x^2}{\sqrt{x^2-1}}\,dx = \int \dfrac{\cosh^2 u}{\sinh u}\sinh u\,du$

$= \int \cosh^2 u\,du$

$= \dfrac{1}{2}\int (1+\cosh 2u)\,du$

$= \dfrac{1}{2}[u + \sinh u \cosh u] + C$

$= \dfrac{1}{2}[\operatorname{arcosh} x + x\sqrt{x^2-1}] + C$

So $\int x\operatorname{arcosh} x\,dx = \dfrac{x^2}{2}\operatorname{arcosh} x$
$- \dfrac{1}{4}[\operatorname{arcosh} x + x\sqrt{x^2-1}] + C$
$= \dfrac{1}{4}(2x^2-1)\operatorname{arcosh} x - \dfrac{1}{4}x\sqrt{x^2-1} + C$

b $\dfrac{1}{2}(2x-1)\operatorname{arcosh}\sqrt{x} - \dfrac{1}{2}\sqrt{x}\sqrt{x-1} + C$

16 a $I_n - I^{n-4} = \int \dfrac{[\sin(2n+1)x - \sin(2n-1)x]}{\sin x}\,dx$

$= \int \dfrac{2\cos 2nx \sin x}{\sin x}\,dx$

$= \int 2\cos 2nx\,dx$

$= \dfrac{\sin 2nx}{n}$

b $\dfrac{\sin 10x}{5} + \dfrac{\sin 8x}{4} + \dfrac{\sin 6x}{3} + \dfrac{\sin 4x}{2} + \sin 2x + x + C$

c $\int_0^{\frac{\pi}{2}} \dfrac{\sin(2n+1)x}{\sin x}\,dx - \int_0^{\frac{\pi}{2}} \dfrac{\sin(2n-1)x}{\sin x}\,dx = \left[\dfrac{\sin 2nx}{n}\right]_0^{\frac{\pi}{2}}$

$= 0$ if n is any a positive integer

$\Rightarrow \int_0^{\frac{\pi}{2}} \dfrac{\sin(2n+1)x}{\sin x}\,dx = \int_0^{\frac{\pi}{2}} \dfrac{\sin(2n-1)x}{\sin x}\,dx$

$= \ldots \int_0^{\frac{\pi}{2}} \dfrac{\sin x}{\sin x}\,dx = \dfrac{\pi}{2}$

17 a The point A on the curve has coordinates (1, 0)
Using symmetry, the length of the loop is

$2\int_0^1 \sqrt{1+\left(\dfrac{dy}{dx}\right)^2}\,dx.$

As $y^2 = \dfrac{1}{3}x(x-1)^2 = \dfrac{1}{3}(x^3 - 2x^2 + x)$

$2y\dfrac{dy}{dx} = \dfrac{1}{3}(3x^2 - 4x + 1) = \dfrac{1}{3}(3x-1)(x-1)$

So $\dfrac{dy}{dx} = \dfrac{\frac{1}{3}(3x-1)(x-1)}{\pm 2\sqrt{\frac{x}{3}}(x-1)} = \pm\dfrac{1}{2\sqrt{3}}\dfrac{(3x-1)}{\sqrt{x}}$

and $1+\left(\dfrac{dy}{dx}\right)^2 = 1 + \dfrac{9x^2 - 6x + 1}{12x} = \dfrac{9x^2 + 6x + 1}{12x}$

$= \dfrac{(3x+1)^2}{12x}$

Therefore, arc length $= 2\int_0^1 \dfrac{3x+1}{2\sqrt{3}\sqrt{x}}\,dx$

$= \dfrac{1}{\sqrt{3}}\int_0^1 \left(3\sqrt{x} + \dfrac{1}{\sqrt{x}}\right)dx$

$= \dfrac{1}{\sqrt{3}}\left[2x^{\frac{3}{2}} + 2\sqrt{x}\right]_0^1$

$= \dfrac{4}{\sqrt{3}} = \dfrac{4\sqrt{3}}{3}$

b $\dfrac{\pi}{3}$

18 a $\dfrac{2}{\sqrt{3}}\arctan(\sqrt{3}\,e^x) + C$

b $x^2 - 2x + 10 = (x-1)^2 + 9$
So let $x - 1 = 3\sinh u$, then $dx = 3\cosh u\,du$

and $\int \dfrac{3x-1}{\sqrt{x^2-2x+10}}\,dx = \int \dfrac{9\sinh u + 2}{\sqrt{9\sinh^2 u + 9}}3\cosh u\,du$

$= \int \dfrac{9\sinh u + 2}{3\cosh u}3\cosh u\,du$

$= 9\cosh u + 2u + C$

$= 9\sqrt{1+\left(\dfrac{x-1}{3}\right)^2} + 2\operatorname{arsinh}\left(\dfrac{x-1}{3}\right) + C$

So $\int_1^4 \dfrac{3x-1}{\sqrt{x^2-2x+10}} = [9\sqrt{2} + 2\operatorname{arsinh} 1] - [9]$

$= 9(\sqrt{2}-1) + 2\operatorname{arsinh} 1$

19 a $\int \sec^n x\,dx = \int \sec^{n-2} x \sec^2 x\,dx$

Let $u = \sec^{n-2} x$ and $\dfrac{dv}{dx} = \sec^2 x$

$\dfrac{du}{dx} = (n-2)\sec^{n-3} x (\sec x \tan x) = (n-2)\sec^{n-2} x \tan x$

and $v = \tan x$

Integrating by parts

$\int \sec^n x\,dx =$

$I_n = \sec^{n-2} x \tan x - (n-2)\int \sec^{n-2} x \tan^2 x\,dx$

$= \sec^{n-2} x \tan x - (n-2)\int \sec^{n-2} x (\sec^2 x - 1)\,dx$

$= \sec^{n-2} x \tan x - (n-2)\int \sec^n x\,dx + (n-2)\int \sec^{n-2} x\,dx$

$I_n = \sec^{n-2} x \tan x - (n-2)I_n + (n-2)I_{n-2}$

So $(n-1)I_n = \sec^{n-2} x \tan x + (n-2)I_{n-2}, n \geq 2$

b $\dfrac{1}{4}\sec^3 x \tan x + \dfrac{3}{8}\sec x \tan x + \dfrac{3}{8}\ln|\sec x + \tan x| + C$

c $\int_0^{\frac{\pi}{4}} \sec^5 x\,dx = \dfrac{1}{4}(\sqrt{2})^3 + \dfrac{3}{8}(\sqrt{2}) + \dfrac{3}{8}\ln(\sqrt{2}+1)$

$= \dfrac{1}{8}(7\sqrt{2} + 3\ln(\sqrt{2}+1))$

Online Worked solutions are available in SolutionBank.

20 a Let $x = a\sin\theta$, then $\dfrac{dx}{d\theta} = a\cos\theta$

So $\int \sqrt{a^2 - x^2}\,dx = \int a^2 \cos^2\theta\,d\theta$

$= \dfrac{a^2}{2}\int (1 + \cos 2\theta)\,d\theta$

$= \dfrac{a^2}{2}\left(\theta + \dfrac{\sin 2\theta}{2}\right) + C$

$= \dfrac{a^2}{2}(\theta + \sin\theta\cos\theta) + C$

$= \dfrac{a^2}{2}\left(\arcsin\left(\dfrac{x}{a}\right) + \dfrac{x}{a}\sqrt{1 - \left(\dfrac{x}{a}\right)^2}\right) + C$

$= \dfrac{a^2}{2}\arcsin\left(\dfrac{x}{a}\right) + \dfrac{x}{2}\sqrt{a^2 - x^2} + C$

b Area enclosed by the ellipse = $4 \times$ area enclosed by arc in first quadrant and the positive coordinate axes (symmetry)

$= 4\int_0^a y\,dx$

$\dfrac{x^2}{a^2} + \dfrac{y^2}{b^2} = 1 \Rightarrow y = \pm\dfrac{b}{a}\sqrt{a^2 - x^2}$

So area $= 4\dfrac{b}{a}\left[\dfrac{a^2}{2}\arcsin\left(\dfrac{x}{a}\right) + \dfrac{x}{2}\sqrt{a^2 - x^2}\right]_0^a$

$= 2ab\arcsin 1$

$= \pi ab$

Challenge

Length $= a$

Review exercise 1

1 $\dfrac{1}{2}\ln 3$

2 $\ln\left(\dfrac{1}{3}\right), \ln 3$

3 $p = 3, q = \dfrac{20}{3}$

4 $\ln\dfrac{1}{3}, \ln 7$

5 $x = \dfrac{1}{2}\ln\dfrac{1}{3}, \dfrac{1}{2}\ln\dfrac{3}{5}$

6 a $k \geq \sqrt{3}$ **b** $0, -\ln 3$

7 a $\cosh^2 x - \sinh^2 x = \left(\dfrac{e^x + e^{-x}}{2}\right)^2 - \left(\dfrac{e^x - e^{-x}}{2}\right)^2$

$= \dfrac{e^{2x} + 2 + e^{-2x} - (e^{2x} - 2 + e^{-2x})}{4} = \dfrac{4}{4} = 1$

b $k = -1, a = 2$

8 a $2\cosh^2 x - 1 = 2\left(\dfrac{e^x + e^{-x}}{2}\right)^2 - 1 = \dfrac{e^{2x} + 2 + e^{-2x}}{2} - 1$

$= \dfrac{e^{2x} + e^{-2x}}{2} = \cosh 2x$

b $\pm\ln(3 + \sqrt{8})$

9 a $4\cosh^3 x - 3\cosh x = 4\left(\dfrac{e^x + e^{-x}}{2}\right)^3 - 3\left(\dfrac{e^x + e^{-x}}{2}\right)$

$= \dfrac{e^{3x} + 3e^x + 3e^{-x} + e^{-3x}}{2} - \dfrac{3e^x + 3e^{-x}}{2} = \dfrac{e^{3x} + e^{-3x}}{2}$

$= \cosh 3x$

b $\ln(\sqrt{2} \pm 1)$

10 a $\cosh A \cosh B - \sinh A \sinh B$

$= \left(\dfrac{e^A + e^{-A}}{2}\right)\left(\dfrac{e^B + e^{-B}}{2}\right) - \left(\dfrac{e^A - e^{-A}}{2}\right)\left(\dfrac{e^B - e^{-B}}{2}\right)$

$= \dfrac{1}{4}(e^{A+B} + e^{-A+B} + e^{A-B} + e^{-A-B} - e^{A+B} + e^{-A+B}$

$+ e^{A-B} - e^{-A-B})$

$= \dfrac{1}{4}(2e^{-A+B} + 2e^{A-B}) = \dfrac{e^{A-B} + e^{-(A-B)}}{2} = \cosh(A - B)$

b $\cosh x \cosh 1 - \sinh x \sinh 1 = \sinh x$

$\cosh 2\cosh 1 = \sinh x(1 + \sinh 1)$

$\tanh x = \dfrac{\cosh 1}{1 + \sinh 1} = \dfrac{\dfrac{e + e^{-1}}{2}}{1 + \dfrac{e - e^{-1}}{2}} = \dfrac{e + e^{-1}}{2 + e - e^{-1}}$

$= \dfrac{e^2 + 1}{e^2 + 2e - 1}$

11 a Let $\text{arsinh}\,x \Rightarrow x = \sinh y = \dfrac{e^y - e^{-y}}{2}$

$\Rightarrow 2x = e^y - e^{-y} \Rightarrow e^{2y} - 2xe^y - 1 = 0$

$\Rightarrow e^y = \dfrac{2x + \sqrt{4x^2 + 4}}{2} = x + \sqrt{x^2 + 1}$

$\Rightarrow y = \ln(x + \sqrt{x^2 + 1})$

b $\text{arsinh}(\cot\theta) = \ln[\cot\theta + \sqrt{1 + \cot^2\theta}]$

$= \ln(\cot\theta + \text{cosec}\,\theta)$

$= \ln\left(\dfrac{\cos\theta + 1}{\sin\theta}\right) = \ln\left(\dfrac{2\cos^2\dfrac{\theta}{2}}{2\sin\dfrac{\theta}{2}\cos\dfrac{\theta}{2}}\right) = \ln\left(\cot\dfrac{\theta}{2}\right)$

12 a Let $y = \text{artanh}\,x$

$x = \tanh y = \dfrac{e^{2y} - 1}{e^{2y} + 1}$

$e^{2y} = \dfrac{1 + x}{1 - x} \Rightarrow 2y = \ln\left(\dfrac{1 + x}{1 - x}\right)$

$\Rightarrow y = \dfrac{1}{2}\ln\left(\dfrac{1 + x}{1 - x}\right)$ for $|x| < 1$

b

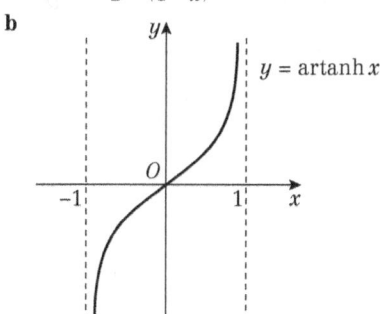

c $\dfrac{1}{2}, \dfrac{1}{3}$

13 a $\ln\left(\dfrac{1 - \sqrt{1 - x^2}}{x}\right) + \ln\left(\dfrac{1 + \sqrt{1 - x^2}}{x}\right)$

$= \ln\left[\left(\dfrac{1 - \sqrt{1 - x^2}}{x}\right)\left(\dfrac{1 + \sqrt{1 - x^2}}{x}\right)\right]$

$= \ln\left(\dfrac{1 - (1 - x^2)}{x^2}\right) = \ln\dfrac{x^2}{x^2} = \ln 1 = 0$

b $y = \text{arcosh}\left(\dfrac{1}{x}\right) \Rightarrow \cosh y = \dfrac{1}{x} \Rightarrow \dfrac{e^y + e^{-y}}{2} = \dfrac{1}{x}$

$\Rightarrow xe^y + xe^{-y} - 2 = 0 \Rightarrow xe^{2y} - 2e^y + x = 0$

$\Rightarrow e^y = \dfrac{2 \pm \sqrt{4 - 4x^2}}{2x} = \dfrac{1 \pm \sqrt{1 - x^2}}{x}$

$\Rightarrow y = \ln\left(\dfrac{1 + \sqrt{1 - x^2}}{x}\right)$

c $\pm\ln\left(\dfrac{3 + \sqrt{5}}{2}\right)$

14 a $\cosh 3\theta = 4\cosh^3\theta - 3\cosh\theta$

$\cosh 5\theta = 16\cosh^5\theta - 20\cosh^3\theta + 5\cosh\theta$

b ± 0.96

15 a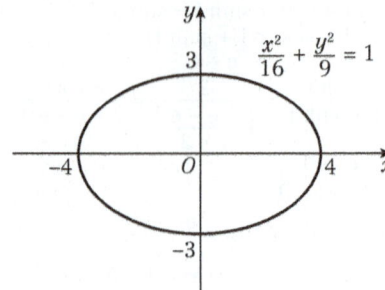

b $\frac{\sqrt{7}}{4}$ **c** $(\pm\sqrt{7}, 0)$

16 a $\frac{\sqrt{5}}{2}$ **b** $4\sqrt{5}$

c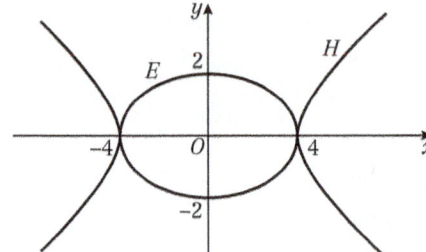

17 a $(\pm\sqrt{5}, 0)$

b The directrices are $x = \pm\frac{9}{\sqrt{5}}$.
Let the line through P parallel to the x–axis intersect the directrices at N and N'.
Then $NN' = 2 \times \frac{9}{\sqrt{5}} = \frac{18}{\sqrt{5}}$
$SP = ePN$ and $S'P = ePN'$, so
$SP + S'P = ePN + ePN' = e(PN + PN') = eN'N$
$= \frac{\sqrt{5}}{3} \times \frac{18}{\sqrt{5}} = 6$

18 a $\frac{1}{2}$ **b** $y = -\frac{1}{2}x + 2$ **c** 2

19 a $x^2 + 4y^2 = a^2$ **b** $x = \pm\frac{2a}{\sqrt{3}}$ **c** $b = \frac{a}{2}$

d P is $\left(\frac{a}{\sqrt{2}}, \frac{a}{2\sqrt{2}}\right)$ and Q is $\left(0, \frac{a}{2}\right)$.
Gradient of PQ is $\frac{\frac{a}{2\sqrt{2}} - \frac{a}{2}}{\frac{a}{\sqrt{2}}} = \frac{1-\sqrt{2}}{2}$
So equation of line containing chord PQ is
$y = \frac{1-\sqrt{2}}{2}x + \frac{a}{2} \Rightarrow (\sqrt{2} - 1)x + 2y - a = 0$

20 a $\frac{\sqrt{5}}{3}$ **b** $(\pm\sqrt{5}, 0)$, $x = \pm\frac{9}{\sqrt{5}}$

c Gradient of ellipse at P is $\frac{2\cos\theta}{-3\sin\theta}$, so equation of tangent is $y - 2\sin\theta = -\frac{2\cos\theta}{3\sin\theta}(x - 3\cos\theta)$,
which can be rearranged to $\frac{x\cos\theta}{3} + \frac{y\sin\theta}{2} = 1$.

d Equation of perpendicular line is $y = \frac{3\sin\theta}{2\cos\theta}x$.
So foot of perpendicular, (x, y), satisfies
$2x\cos\theta + 3y\sin\theta = 6$
$2y\cos\theta - 3x\sin\theta = 0$
Solve these simultaneously to find
$\cos\theta = \frac{3x}{x^2+y^2}$ and $\sin\theta = \frac{2y}{x^2+y^2}$

Therefore $\left(\frac{3x}{x^2+y^2}\right)^2 + \left(\frac{2y}{x^2+y^2}\right)^2 = 1$
Rearranging, this gives that the locus of the foot of the perpendicular as $(x^2 + y^2)^2 = 9x^2 + 4y^2$.

21 a $\frac{dx}{d\theta} = -a\sin\theta$, $\frac{dy}{d\theta} = b\cos\theta \Rightarrow \frac{dy}{dx} = -\frac{b\cos\theta}{a\sin\theta}$
The gradient of the normal is $\frac{a\sin\theta}{b\cos\theta}$
So the equation of the normal is
$y - b\sin\theta = \frac{a\sin\theta}{b\cos\theta}(x - a\cos\theta)$
$\Rightarrow ax\sec\theta - by\csc\theta = a^2 - b^2$

b $y = 0 \Rightarrow ax\sec\theta = a^2 - b^2 \Rightarrow x = \frac{a^2-b^2}{a}\cos\theta$
So G has coordinates $\left(\frac{a^2-b^2}{a}\cos\theta, 0\right)$
Midpoint has coordinates
$\left(\frac{a\cos\theta + \frac{a^2-b^2}{a}\cos\theta}{2}, \frac{b\sin\theta + 0}{2}\right)$
$= \left(\frac{2a^2-b^2}{2a}\cos\theta, \frac{b}{2}\sin\theta\right)$

c $x = \frac{2a^2-b^2}{2a}\cos\theta \Rightarrow \cos\theta = \frac{2ax}{2a^2-b^2}$
$y = \frac{b}{2}\sin\theta \Rightarrow \sin\theta = \frac{2y}{b}$
So using $\cos^2\theta + \sin^2\theta \equiv 1$, M has locus
$\frac{4a^2x^2}{(2a^2-b^2)^2} + \frac{4y^2}{b^2} = 1$, which is an ellipse.

d H has coordinates $\left(0, -\frac{a^2-b^2}{b}\csc\theta\right)$.
$A_1 = $ Area of $\triangle OMG = \frac{1}{2} \times \frac{a^2-b^2}{a\sec\theta} \times \frac{b}{2}\sin\theta$
$= \frac{b}{4a}(a^2-b^2)\sin\theta\cos\theta$
$A_2 = $ Area of $\triangle OGH = \frac{1}{2} \times \frac{a^2-b^2}{b\csc\theta} \times \frac{a^2-b^2}{a\sec\theta}$
$= \frac{(a^2-b^2)^2}{2ab}\sin\theta\cos\theta$
So $A_1 : A_2 = b^2 : 2(a^2 - b^2)$

22 Area of triangle to left of P is
$\frac{1}{2} \times 4 \times 2\sqrt{3} = 4\sqrt{3}$
Area to right of P is
$\int_4^8 \sqrt{16 - \left(\frac{x}{2}\right)^2}\,dx = \frac{16\pi}{3} - 4\sqrt{3}$
So total area is $4\sqrt{3} + \frac{16\pi}{3} - 4\sqrt{3} = \frac{16\pi}{3}$ and $a = \frac{16}{3}$

23 a Substitute $y = mx + c$ into $\frac{x^2}{a^2} + \frac{y^2}{b^2} = 1$:
$\frac{x^2}{a^2} + \frac{(mx+c)^2}{b^2} = 1$
$\Rightarrow (a^2m^2 + b^2)x^2 + 2a^2mcx + a^2(c^2 - b^2) = 0$
As the line is a tangent, need to have "$b^2 - 4ac = 0$".
$4a^4m^2c^2 - 4a^2(a^2m^2 + b^2)(c^2 - b^2) = 0$
$\Rightarrow 4(a^2m^2b^2 - b^2c^2 + b^4) = 0 \Rightarrow c^2 = a^2m^2 + b^2$

b $y = -3x + 13$ and $y = -\frac{3}{7}x + \frac{37}{7}$

24 a Substitute $y = mx + c$ into $\frac{x^2}{a^2} + \frac{y^2}{b^2} = 1$:
$\frac{x^2}{a^2} + \frac{(mx+c)^2}{b^2} = 1$
$\Rightarrow (a^2m^2 + b^2)x^2 + 2a^2mcx + a^2(c^2 - b^2) = 0$

b As the line is a tangent, need to have "$b^2 - 4ac = 0$".
$4a^4m^2c^2 - 4a^2(a^2m^2 + b^2)(c^2 - b^2) = 0$
$\Rightarrow 4(a^2m^2b^2 - b^2c^2 + b^4) = 0 \Rightarrow c^2 = a^2m^2 + b^2$

c $\dfrac{b^2 + a^2m^2}{2m}$

d $T = \dfrac{b^2 + a^2m^2}{2m} = \dfrac{1}{2}b^2m^{-1} + \dfrac{1}{2}a^2m$

For a minimum, $\dfrac{dT}{dm} = -\dfrac{1}{2}b^2m^{-2} + \dfrac{1}{2}a^2 = 0$

$\dfrac{b^2}{m^2} = a^2 \Rightarrow m^2 = \dfrac{b^2}{a^2}$

As L has a positive gradient, $m = \dfrac{b}{a}$

At $m = \dfrac{b}{a}$, $\dfrac{d^2T}{dm^2} = \dfrac{b^2}{m^3} = \dfrac{a^3}{b} > 0$ and so this gives a minimum value of

$T = \dfrac{b^2 + a^2\left(\dfrac{b}{a}\right)^2}{2\left(\dfrac{b}{a}\right)} = \dfrac{2b^2}{2\left(\dfrac{b}{a}\right)} = ab$

e $-\dfrac{a}{\sqrt{2}}$

25 a Tangent at P: $x\cosh t - y\sinh t = 1$
Normal at P: $x\sinh t + y\cosh t = 2\sinh t\cosh t$

b Substitute $y = 0$ into the equation of the normal:
$x\sinh t = 2\sinh t\cosh t \Rightarrow x = 2\cosh t$, so G is $(2\cosh t, 0)$.
Q has $x = \cosh t$, and the asymptote in the first quadrant is $y = x$, so Q is $(\cosh t, \cosh t)$.
Gradient of GQ is $\dfrac{0 - \cosh t}{2\cosh t - \cosh t} = -1$
So GQ is perpendicular to the asymptote $y = x$.

c Substitute $y = 0$ into the equation of the tangent:
$x\cosh t = 1 \Rightarrow x = \dfrac{1}{\cosh t}$, so $T = \left(\dfrac{1}{\cosh t}, 0\right)$
Substitute $x = 0$ into the equation of the normal:
$y\cosh t = 2\sinh t\cosh t \Rightarrow y = 2\sinh t$, so R is $(0, 2\sinh t)$.
$TG = 2\cosh t - \dfrac{1}{\cosh t}$
$TR^2 = OR^2 + OT^2 = (2\sinh t)^2 + \left(\dfrac{1}{\cosh t}\right)^2$
$= 4(\cosh^2 t - 1) + \dfrac{1}{\cosh^2 t} = \left(2\cosh t - \dfrac{1}{\cosh t}\right)^2$
$= TG^2$
So $TR = TG$ and R lies on the circle with centre T and radius TG.

26 Let the point P have coordinates $(a\cosh t, b\sinh t)$
$\dfrac{dx}{dt} = a\sinh t$, $\dfrac{dy}{dt} = b\cosh t \Rightarrow \dfrac{dy}{dx} = \dfrac{b\cosh t}{a\sinh t}$
Equation of tangent is $y - b\sinh t = \dfrac{b\cosh t}{a\sinh t}(x - a\cosh t)$
$\Rightarrow ay\sinh t = bx\cosh t - ab(\cosh^2 t - \sinh^2 t)$
$= bx\cosh t - ab$
For T, $y = 0$, so $bx\cosh t = ab \Rightarrow x = \dfrac{a}{\cosh t}$
The coordinates of N are $(a\cosh t, 0)$
$OT \times ON = \dfrac{a}{\cosh t} \times a\cosh t = a^2$

27 a $\dfrac{dx}{dt} = a\sec t\tan t$, $\dfrac{dy}{dt} = b\sec^2 t$
$\dfrac{dy}{dx} = \dfrac{b\sec^2 t}{a\sec t\tan t} = \dfrac{b}{a\sin t}$
The gradient of the normal is $-\dfrac{a\sin t}{b}$
The equation of the normal is
$y - b\tan t = -\dfrac{a\sin t}{b}(x - a\sec t)$
$\Rightarrow ax\sin t + by = (a^2 + b^2)\tan t$

b $\dfrac{\pi}{3}, \dfrac{2\pi}{3}, \dfrac{4\pi}{3}, \dfrac{5\pi}{3}$

28 a $\dfrac{x^2}{a^2} - \dfrac{y^2}{a^2} = 1$, $b^2 = a^2(e^2 - 1)$
$b^2 = a^2 \Rightarrow a^2 = a^2(e^2 - 1)$
$\Rightarrow 1 = e^2 - 1 \Rightarrow e^2 = 2 \Rightarrow e = \sqrt{2}$

b $(a\sqrt{2}, 0)$, $x = \dfrac{a\sqrt{2}}{2}$

c P is on the line with gradient -1 through $(a\sqrt{2}, 0)$,
$y = -x + a\sqrt{2}$, which intersects $y = x$ at $P\left(\dfrac{a\sqrt{2}}{2}, \dfrac{a\sqrt{2}}{2}\right)$.
Q is on the line with gradient 1 through $(a\sqrt{2}, 0)$,
$y = x - a\sqrt{2}$, which intersects $y = -x$ at $Q\left(\dfrac{a\sqrt{2}}{2}, -\dfrac{a\sqrt{2}}{2}\right)$.
P and Q both have $x = \dfrac{a\sqrt{2}}{2}$, so lie on directrix L.

d SP has equation $x + y = a\sqrt{2}$
So R is where $x^2 - (a\sqrt{2} - x)^2 = a^2$
$\Rightarrow x = \dfrac{3\sqrt{2}}{4}a$, $y = \dfrac{\sqrt{2}}{4}a$
$x^2 - y^2 = a^2 \Rightarrow \dfrac{dy}{dx} = \dfrac{x}{y}$, so at R, $\dfrac{dy}{dx} = 3$.
Therefore the tangent is
$y - \dfrac{\sqrt{2}}{4}a = 3\left(x - \dfrac{3\sqrt{2}}{4}a\right)$
$\Rightarrow y = 3x - 2a\sqrt{2}$
$x = \dfrac{a\sqrt{2}}{2} \Rightarrow y = -\dfrac{a\sqrt{2}}{2}$, which is the y-coordinate of Q, so the tangent passes through Q.

29 Let the equation of the tangent be $y = mx + c$.
$x^2 - 4(mx + c)^2 = 4 \Rightarrow (4m^2 - 1)x^2 + 8mcx + 4(c^2 + 1) = 0$
As the line is a tangent, this equation will have repeated roots, so $b^2 - 4ac = 0$:
$64m^2c^2 - 16(4m^2 - 1)(c^2 + 1) = 0 \Rightarrow 16c^2 - 64m^2 + 16 = 0$
$\Rightarrow c = \pm\sqrt{4m^2 - 1}$, so the equations of the tangents are
$y = mx \pm \sqrt{4m^2 - 1}$, where $|m| > \dfrac{1}{2}$

30 a $ay\sin t + bx\cos t = ab$
b $ax\sin t - by\cos t = (a^2 - b^2)\sin t\cos t$
c $\left(\dfrac{a^2 - b^2}{2a}\cos t, \dfrac{b}{2\sin t}\right)$
d $x = \dfrac{a^2 - b^2}{2a}\cos t \Rightarrow \cos t = \dfrac{2ax}{a^2 - b^2}$ and
$y = \dfrac{b}{2\sin t} \Rightarrow \sin t = \dfrac{b}{2y}$
So using $\cos^2 t + \sin^2 t \equiv 1$, M has locus
$\left(\dfrac{2ax}{a^2 - b^2}\right)^2 + \left(\dfrac{b}{2y}\right)^2 = 1$

31 a Tangent: $bx - ay\sin\theta = ab\cos\theta$
Normal: $ax\sin\theta + by = (a^2 + b^2)\tan\theta$

b Find the coordinates of P and Q by substituting $x = 0$ into the equations of the two lines.
$-ay\sin\theta = ab\cos\theta \Rightarrow y = -b\cot\theta$
$by = (a^2 + b^2)\tan\theta \Rightarrow y = \dfrac{a^2 + b^2}{b}\tan\theta$
So P is $(0, -b\cot\theta)$ and Q is $\left(0, \dfrac{a^2 + b^2}{b}\tan\theta\right)$.
The focus S with $x > 0$ is $(ae, 0)$.
PS has gradient $m = \dfrac{-b\cot\theta - 0}{0 - ae} = \dfrac{b}{ae}\cot\theta$

QS has gradient $m' = \dfrac{\dfrac{a^2 + b^2}{b}\tan\theta - 0}{0 - ae} = -\dfrac{a^2 + b^2}{abe}\tan\theta$

$mm' = -\dfrac{a^2 + b^2}{a^2 e^2} = -1$, since $b^2 = a^2(e^2 - 1)$, so PS and QS are perpendicular. Thus PSQ is a right-angled triangle, and PQ is the diameter of a circle, C, through S. By symmetry, C also passes through the other focus, $(-ae, 0)$.

32 a $\cosh 2x = \dfrac{e^{2x} + e^{-2x}}{2} = \dfrac{e^{2\ln k} + e^{-2\ln k}}{2} = \dfrac{e^{\ln k^2} + e^{\ln \frac{1}{k^2}}}{2}$

$= \dfrac{1}{2}\left(k^2 + \dfrac{1}{k^2}\right) = \dfrac{k^4 + 1}{2k^2}$

b $\dfrac{128}{289}$

33 a $\dfrac{1}{4}\ln(2 + \sqrt{3})$

b $\tanh^2 4x = 1 - \text{sech}^2 4x = 1 - \dfrac{1}{4} = \dfrac{3}{4}$

As $x \geq 0$, $\tanh 4x = \dfrac{\sqrt{3}}{2}$

At $x = \dfrac{1}{4}\ln(2 + \sqrt{3})$,

$y = -x + \tanh 4x = -\dfrac{1}{4}\ln(2 + \sqrt{3}) + \dfrac{\sqrt{3}}{2}$

$= \dfrac{1}{4}(2\sqrt{3} - \ln(2 + \sqrt{3}))$

34 $x = \dfrac{a}{\sinh \theta} \Rightarrow \dfrac{dx}{d\theta} = -\dfrac{a\cosh\theta}{\sinh^2\theta}$

$\int \dfrac{1}{x\sqrt{x^2 + a^2}} dx = \int \dfrac{-\dfrac{a\cosh\theta}{\sinh^2\theta}}{a^2\dfrac{\sqrt{1 + \sinh^2\theta}}{\sinh^2\theta}} d\theta = -\dfrac{1}{a}\int 1\, d\theta$

$= -\dfrac{1}{a}\theta + c = -\dfrac{1}{a}\text{arsinh}\left(\dfrac{a}{x}\right) + c$

35 a $y = \text{artanh}\, x \Rightarrow \tanh y = x$

Differentiate implicitly with respect to x

$\text{sech}^2 y \dfrac{dy}{dx} = 1 \Rightarrow \dfrac{dy}{dx} = \dfrac{1}{\text{sech}^2 y} = \dfrac{1}{1 - \tanh^2 y} = \dfrac{1}{1 - x^2}$

b $x\,\text{artanh}\,x + \dfrac{1}{2}\ln(1 - x^2) + A$

36 a Let $y = \text{arsinh}\, x \Rightarrow x = \sinh y = \dfrac{e^y - e^{-y}}{2}$

$\Rightarrow 2x = e^y - e^{-y} \Rightarrow e^{2y} - 2xe^y - 1 = 0$

$\Rightarrow e^y = \dfrac{2x + \sqrt{4x^2 + 4}}{2} = x + \sqrt{x^2 + 1}$

$\Rightarrow y = \ln(x + \sqrt{x^2 + 1})$

b $y = \text{arsinh}\, x \Rightarrow \sinh y = x$

Differentiating implicitly with respect to x

$\cosh y \dfrac{dy}{dx} = 1 \Rightarrow \dfrac{dy}{dx} = \dfrac{1}{\cosh y} = \dfrac{1}{\sqrt{1 + \sinh^2 y}} = \dfrac{1}{\sqrt{1 + x^2}}$

$\Rightarrow \dfrac{d}{dx}(\text{arsinh}\, x) = (1 + x^2)^{-\frac{1}{2}}$

c $y = (\text{arsinh}\, x)^2$, $\dfrac{dy}{dx} = 2\,\text{arsinh}\, x\,(1 + x^2)^{-\frac{1}{2}}$

$\dfrac{d^2y}{dx^2} = 2(1 + x^2)^{-1} - 2x\,\text{arsinh}\, x\,(1 + x^2)^{-\frac{3}{2}}$

$\Rightarrow (1 + x^2)\dfrac{d^2y}{dx^2} + x\dfrac{dy}{dx} - 2$

$= 2 - 2x\,\text{arsinh}\, x\,(1 + x^2)^{-\frac{1}{2}} + 2x\,\text{arsinh}\, x\,(1 + x^2)^{-\frac{1}{2}} - 2$

$= 0$

d $\ln(1 + \sqrt{2}) - \sqrt{2} + 1$

37 a $p = 2$, $q = 1$, $r = 4$

b $\dfrac{1}{4}\arctan\left(\dfrac{2x + 1}{2}\right) + c$

c $\int \dfrac{2}{\sqrt{4x^2 + 4x + 5}}\, dx = \int \dfrac{2}{\sqrt{(2x + 1)^2 + 4}}\, dx$

Let $2x + 1 = 2\sinh\theta \Rightarrow \dfrac{dx}{d\theta} = \cosh\theta$

$\int \dfrac{2}{\sqrt{(2x+1)^2 + 4}} dx = \int \dfrac{2}{\sqrt{4\sinh^2\theta + 4}}\cosh\theta\, d\theta = \int 1\, d\theta$

$= \theta + c = \text{arsinh}\left(\dfrac{2x+1}{2}\right) + c$

Using $\text{arsinh}\,x = \ln(x + \sqrt{x^2 + 1})$

$\int \dfrac{2}{\sqrt{4x^2 + 4x + 5}}\, dx = \ln\left(\dfrac{2x+1}{2} + \sqrt{\left(\dfrac{2x+1}{2}\right)^2 + 1}\right) + c$

$= \ln\left(\dfrac{2x+1}{2} + \dfrac{1}{2}\sqrt{4x^2 + 4x + 5}\right) + c$

$= \ln(2x + 1 + \sqrt{4x^2 + 4x + 5}) - \ln 2 + c$

$= \ln(2x + 1 + \sqrt{4x^2 + 4x + 5}) + k$

38 $\dfrac{\sqrt{4x^2 + 9}}{4} + \text{arsinh}\left(\dfrac{2x}{3}\right) + c$

39 $\int_2^5 \dfrac{1}{\sqrt{x^2 - 4x + 8}}\, dx = \int_2^5 \dfrac{1}{\sqrt{(x-2)^2 + 2^2}}\, dx$

$= \left[\text{arsinh}\left(\dfrac{x-2}{2}\right)\right]_2^5 = \text{arsinh}\left(\dfrac{3}{2}\right) - \text{arsinh}\, 0$

$= \text{arsinh}\left(\dfrac{3}{2}\right)$, so $k = \dfrac{3}{2}$

40 $\int x\,\text{arcosh}\, x\, dx = \dfrac{x^2}{2}\text{arcosh}\, x - \int \dfrac{x^2}{2\sqrt{x^2 - 1}}\, dx$

Use integration by substitution to evaluate $\int \dfrac{x^2}{2\sqrt{x^2 - 1}}\, dx$

Let $x = \cosh\theta \Rightarrow \dfrac{dx}{d\theta} = \sinh\theta$

$\int \dfrac{x^2}{2\sqrt{x^2 - 1}}\, dx = \int \dfrac{(\cosh\theta)^2}{2\sqrt{(\cosh\theta)^2 - 1}}\sinh\theta\, d\theta$

$= \dfrac{1}{2}\int \cosh^2\theta\, d\theta = \dfrac{1}{4}\int(\cosh 2\theta + 1)\, d\theta$

$= \dfrac{\sinh 2\theta}{8} + \dfrac{\theta}{4} = \dfrac{\sinh\theta\cosh\theta}{4} + \dfrac{\theta}{4}$

$= \dfrac{x\sqrt{x^2 - 1}}{4} + \dfrac{1}{4}\text{arcosh}\, x$

Area of $R = \left[\dfrac{x^2}{2}\text{arcosh}\, x - \dfrac{x\sqrt{x^2-1}}{4} - \dfrac{1}{4}\text{arcosh}\, x\right]_1^2$

$= \left(\dfrac{7}{4}\text{arcosh}\, 2 - \dfrac{\sqrt{3}}{2}\right) - 0$

$= \dfrac{7}{4}\ln(2 + \sqrt{3}) - \dfrac{\sqrt{3}}{2}$

41 a Use integration by parts with $u = \sec^{n-2}x$ and $\dfrac{dv}{dx} = \sec^2 x$

$I_n = \sec^{n-2}x\tan x - (n-2)\int \sec^{n-2}x(\sec^2 - 1)\, dx$

$\Rightarrow (n-1)I_n = \sec^{n-2}x\tan x + (n-2)I_{n-2}$

b $\dfrac{1}{3}\sec^2 x\tan x + \dfrac{2}{3}\tan x + c$

42 a $I_n = [x^n \sin x]_0^{\frac{\pi}{4}} - n\int_0^{\frac{\pi}{4}} x^{n-1}\sin x\, dx$

$= [x^n \sin x]_0^{\frac{\pi}{4}} - n[-x^{n-1}\cos x]_0^{\frac{\pi}{4}} - n(n-1)\int_0^{\frac{\pi}{4}} x^{n-2}\cos x\, dx$

$= \dfrac{1}{\sqrt{2}}\left(\dfrac{\pi}{4}\right)^{n-1}\left(\dfrac{\pi}{4} + n\right) - n(n-1)I_{n-2}$

b 0.0471

43 a $I_n = -\frac{3}{4}[x^n(a-x)^{\frac{4}{3}}]_0^a + \frac{3}{4}n\int_0^a x^{n-1}(a-x)^{\frac{4}{3}}dx$
$= \frac{3an}{4}I_{n-1} - \frac{3n}{4}I_n \Rightarrow I_n = \frac{3an}{3n+4}I_{n-1}$
b $\frac{2}{7}\sqrt{5}$

44 a $\frac{8}{27a}((1+9a)^{\frac{3}{2}}-1)$ **b** 3.6967

45 a $x = \frac{1}{2}y^2 - 8 \Rightarrow \frac{dx}{dy} = y$
$\Rightarrow L = \int_0^3 \sqrt{1+\left(\frac{dx}{dy}\right)^2}\,dy = \int_0^3 \sqrt{1+y^2}\,dy$
b $\frac{1}{2}\ln(3+\sqrt{10}) + \frac{3}{2}\sqrt{10}$

46 $s = \int_0^2 \sqrt{(2t)^2 + (t^2)^2}\,dt = \int_0^2 t\sqrt{4+t^2}\,dt$
$= \frac{1}{2}\int_4^8 \sqrt{u}\,du = \frac{1}{3}[u^{\frac{3}{2}}]_4^8 = \frac{8^{\frac{3}{2}}-8}{3}$

47 a Use substitution $\theta = \tan x$ with $\int_0^{4\pi}\sqrt{\theta^2+1}\,d\theta$ to get
$W = \int_0^{\arctan(4\pi)} \sec^2 x\sqrt{1+\tan^2 x}\,dx = \int_0^{\arctan(4\pi)} \sec^3 x\,dx$
b 80.82 (2 d.p.)

48 $\frac{8\pi(5\sqrt{5}-2\sqrt{2})}{3}$

49 a $a = 36$
b
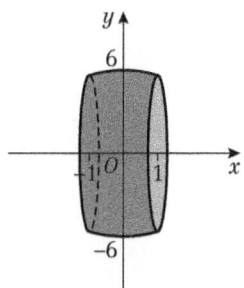

50 Area $= 2\pi\left(1 - \frac{1}{\sqrt{2}}\right)$

51 22.943 (3 d.p.)

Challenge

1 a
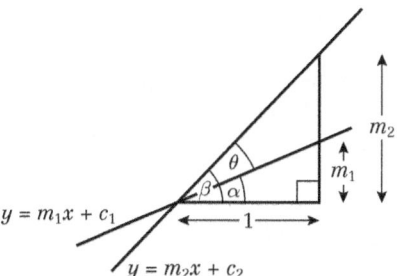

Using the identity for $\tan(A \pm B)$:
$\tan\theta = \tan(\beta - \alpha) = \frac{\tan\beta - \tan\alpha}{1 + \tan\beta\tan\alpha} = \frac{m_2 - m_1}{1 + m_1 m_2}$
as required.

b
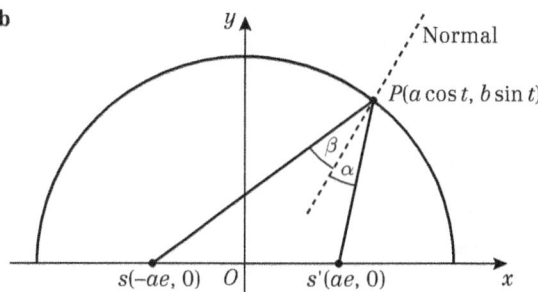

At P, $\frac{dy}{dx} = \frac{-b\cos t}{a\sin t}$, so gradient of normal is $\frac{a\sin t}{b\cos t}$

Gradient of PS' is $\frac{b\sin t}{a\cos t - ae}$ and gradient of

PS is $\frac{b\sin t}{a\cos t + ae}$

So using the result from part **a**,

$\tan\alpha = \dfrac{\dfrac{a\sin t}{b\cos t} - \dfrac{b\sin t}{a\cos t - ae}}{1 + \left(\dfrac{a\sin t}{b\cos t}\right)\left(\dfrac{b\sin t}{a\cos t - ae}\right)}$

$= \dfrac{a\sin t(a\cos t - ae) - b^2\sin t\cos t}{b\cos t(a\cos t - ae) + ab\sin^2 t}$

$= \dfrac{(a^2-b^2)\sin t\cos t - a^2 e\sin t}{ab(\cos^2 t + \sin^2 t) - abe\cos t}$

$= \dfrac{a^2 e^2\sin t\cos t - a^2 e\sin t}{ab - abe\cos t}$

$= \dfrac{a^2 e\sin t(e\cos t - 1)}{ab(1 - e\cos t)} = \dfrac{-ae\sin t}{b}$

Similarly, $\tan\beta = \dfrac{-ae\sin t}{b}$

So $\tan\alpha = \tan\beta$, and hence $\alpha = \beta$ as required.

CHAPTER 5
Prior knowledge check
1 -1
2 $\frac{x-1}{2} = \frac{y-4}{3} = \frac{x+2}{5}$ $(= \lambda)$
3 $\angle QPR = 27.2°$

Exercise 5A
1 a $5\mathbf{i}$ **b** $-3\mathbf{j}$
c $3\mathbf{j}$ **d** $-3\mathbf{j} - 3\mathbf{k}$
e $-2\mathbf{i} - 6\mathbf{k}$ **f** $2\mathbf{i} + 6\mathbf{k}$
g $\begin{pmatrix} 5 \\ -16 \\ -7 \end{pmatrix}$ **h** $\begin{pmatrix} 9 \\ 0 \\ -3 \end{pmatrix}$
i $\begin{pmatrix} -9 \\ -7 \\ -11 \end{pmatrix}$ **j** $\begin{pmatrix} 2 \\ -4 \\ -3 \end{pmatrix}$

2 a $-6\mathbf{i} + (3\lambda + 1)\mathbf{j} - 2\mathbf{k}$ **b** $(7\lambda - 3)\mathbf{i} + \mathbf{j} + (1 - 2\lambda)\mathbf{k}$

3 $-\frac{1}{3}\mathbf{i} - \frac{2}{3}\mathbf{j} + \frac{2}{3}\mathbf{k}$ or $\frac{1}{3}\mathbf{i} + \frac{2}{3}\mathbf{j} - \frac{2}{3}\mathbf{k}$

4 $\frac{1}{7}(-\mathbf{i} + 4\sqrt{2}\mathbf{j} + 4\mathbf{k})$

5 $\frac{1}{11}(6\mathbf{i} + 6\mathbf{j} + 7\mathbf{k})$

6 $\begin{pmatrix} \frac{4}{9} \\ \frac{4}{9} \\ -\frac{7}{9} \end{pmatrix}$

ANSWERS

7 $-\mathbf{i} - 2\sqrt{2}\mathbf{j} + 4\mathbf{k}$
8 $\sqrt{8}$ or $2\sqrt{2}$ or 2.83 (to 3 s.f.)
9 a -14 b $-8\mathbf{i} - 24\mathbf{j} - 8\mathbf{k}$ c $\frac{1}{\sqrt{11}}(-\mathbf{i} - 3\mathbf{j} - \mathbf{k})$
10 a $\frac{\sqrt{221}}{15}$ b 1 c $\frac{\sqrt{21}}{11}$
11 Any multiple of $(\mathbf{i} + \mathbf{j} - \mathbf{k})$
12 $u = -1, v = 4$ and $w = 11$
13 a $a = 1$ and $b = -1$ b $-\frac{5}{6}$
14 $\lambda = \frac{3}{2}$ and $\mu = -\frac{3}{2}$
15 Given that $\mathbf{a} + \mathbf{b} + \mathbf{c} = \mathbf{0}$ (1)
   Take the vector product of this with $\mathbf{a}$.
   $\mathbf{a} \times (\mathbf{a} + \mathbf{b} + \mathbf{c}) = \mathbf{a} \times \mathbf{0}$
   $\mathbf{a} \times \mathbf{a} + \mathbf{a} \times \mathbf{b} + \mathbf{a} \times \mathbf{c} = \mathbf{0}$
   But $\mathbf{a} \times \mathbf{a} = \mathbf{0}$ and $\mathbf{a} \times \mathbf{c} = -\mathbf{c} \times \mathbf{a}$
   Therefore $\mathbf{a} \times \mathbf{b} - \mathbf{c} \times \mathbf{a} = \mathbf{0}$
   So $\mathbf{a} \times \mathbf{b} = \mathbf{c} \times \mathbf{a}$
   Take the vector product of (1) with $\mathbf{b}$.
   $\mathbf{b} \times (\mathbf{a} + \mathbf{b} + \mathbf{c}) = \mathbf{b} \times \mathbf{0}$
   $\mathbf{b} \times \mathbf{a} + \mathbf{b} \times \mathbf{b} + \mathbf{b} \times \mathbf{c} = \mathbf{0}$
   But $\mathbf{b} \times \mathbf{b} = \mathbf{0}$ and $\mathbf{b} \times \mathbf{a} = -\mathbf{a} \times \mathbf{b}$
   Therefore $-\mathbf{a} \times \mathbf{b} + \mathbf{b} \times \mathbf{c} = \mathbf{0}$
   So $\mathbf{b} \times \mathbf{c} = \mathbf{a} \times \mathbf{b}$
   Therefore $\mathbf{a} \times \mathbf{b} = \mathbf{b} \times \mathbf{c} = \mathbf{c} \times \mathbf{a}$

Challenge
$\mathbf{a} \times \mathbf{b} = \mathbf{c} \times \mathbf{a}$
$\mathbf{a} \times \mathbf{b} - \mathbf{c} \times \mathbf{a} = 0$
$\mathbf{a} \times \mathbf{b} + \mathbf{a} \times \mathbf{c} = 0$
$\mathbf{a} \times (\mathbf{b} + \mathbf{c}) = 0$
As $\mathbf{a} \neq \mathbf{0}$ and $\mathbf{b}$ and $\mathbf{c}$ are non-parallel
$\mathbf{a}$ is parallel to $\mathbf{b} + \mathbf{c}$

Exercise 5B
1 a 4.5 b $\frac{5\sqrt{2}}{2}$ c 16.5
2 a $2\sqrt{13}$ b 8.5
3 $\frac{3}{2}\sqrt{2}$
4 $\frac{5}{2}\sqrt{3}$
5 $5\sqrt{2}$
6 $10\sqrt{2}$
7 $3\sqrt{2}$
8 $\frac{\sqrt{3}}{2}a^2$
9 a Area of parallelogram $ABCD = 2 \times$ area of triangle ABC
   $= 2 \times \frac{1}{2}|\overrightarrow{AB} \times \overrightarrow{AC}|$
   $= |\overrightarrow{AB} \times \overrightarrow{AC}|$
   As $\overrightarrow{AB} = (\mathbf{b} - \mathbf{a})$ and $\overrightarrow{AC} = (\mathbf{c} - \mathbf{a})$
   Area $= |(\mathbf{b} - \mathbf{a}) \times (\mathbf{c} - \mathbf{a})|$
 b $(\mathbf{b} - \mathbf{a}) \times (\mathbf{c} - \mathbf{a}) - (\mathbf{b} - \mathbf{a}) \times (\mathbf{d} - \mathbf{a}) = 0$
   $\Rightarrow (\mathbf{b} - \mathbf{a}) \times \mathbf{c} + (\mathbf{b} - \mathbf{a}) \times (\mathbf{a} - \mathbf{a}) - (\mathbf{b} - \mathbf{a}) \times \mathbf{d} = 0$
   $\Rightarrow (\mathbf{b} - \mathbf{a}) \times (\mathbf{c} - \mathbf{d}) = 0$
   $\overrightarrow{AB}$ is parallel to $\overrightarrow{DC}$
10 a $5\mathbf{i} - 5\mathbf{j} + 15\mathbf{k}$ b $\frac{5\sqrt{11}}{2}$

11 a $-15\mathbf{i} + 17\mathbf{j} + 20\mathbf{k}$
   b $21.54\,\mathrm{m}^2$
   c The area of fabric needed will be larger as there will need to be excess fabric to attach to the masts and some slack in the sail to fill with air.
12 a $(2, -5, 1)$ b £4481

Challenge
$|\mathbf{p} \times (\mathbf{q} + \mathbf{r})| = ABFE$ as $BF = \mathbf{q} + \mathbf{r}$
$|\mathbf{p} \times \mathbf{q}| = ABCD$
$|\mathbf{p} \times \mathbf{r}| = CDEF$
$|\mathbf{p} \times \mathbf{q}| + |\mathbf{p} \times \mathbf{r}| = ABFE + BCF - ADE$
By definition, $AD = \mathbf{q}$ and $DE = \mathbf{r}$
$|\mathbf{p} \times \mathbf{q}| + |\mathbf{p} \times \mathbf{r}| = ABFE + \frac{1}{2}|\mathbf{q} \times \mathbf{r}| - \frac{1}{2}|\mathbf{q} \times \mathbf{r}| = |\mathbf{p} \times (\mathbf{q} + \mathbf{r})|$

Exercise 5C
1 a 21 b 21 c 21
2 0; $\mathbf{a}$ is parallel to the plane containing $\mathbf{b}$ and $\mathbf{c}$
3 17
4 18
5 $\frac{3}{2}$
6 a 3 b $\pm\frac{1}{3}(\mathbf{i} + 2\mathbf{j} - 2\mathbf{k})$ c $\frac{7}{3}$
7 a The distance between any two vertices is 2.
  b $\frac{2}{3}\sqrt{2}$
8 a $\overrightarrow{AB} = -2\mathbf{i} - \mathbf{j} + 3\mathbf{k}$
   $\overrightarrow{AC} = \mathbf{i} - 3\mathbf{j} + 2\mathbf{k}$
   $\overrightarrow{AB} \times \overrightarrow{AC} = 7\mathbf{i} + 7\mathbf{j} + 7\mathbf{k}$
  b $\frac{7\sqrt{3}}{2}$
  c $\frac{7}{3}$
9 a $\overrightarrow{AB} \times \overrightarrow{BC} = 5\mathbf{i} - \mathbf{j} - 7\mathbf{k}$
   $\overrightarrow{BD} \times \overrightarrow{DC} = 2\mathbf{i} - 8\mathbf{j} + \mathbf{k}$
  b i $\frac{5}{2}\sqrt{3}$ ii $\frac{19}{6}$
10 a $\mathbf{i} + 2\mathbf{j}$
   b $\overrightarrow{OP} = 2\sqrt{5}$
   Area of $OQR = \frac{\sqrt{5}}{2}$
   Volume of tetrahedron $= \frac{5}{3}$
   c $\mathbf{a}.(\mathbf{b} \times \mathbf{c}) = 10$
   This is $6 \times$ volume of tetrahedron so verified.
11 a 12:1 b Ratio will be unchanged as N moves.
12 $\frac{14}{3}$ units^3

Challenge
a Let: $\mathbf{a} = a_1\mathbf{i} + a_2\mathbf{j} + a_3\mathbf{k}$
        $\mathbf{b} = b_1\mathbf{i} + b_2\mathbf{j} + b_3\mathbf{k}$
        $\mathbf{c} = c_1\mathbf{i} + c_2\mathbf{j} + c_3\mathbf{k}$
   $\mathbf{a}.(\mathbf{b} \times \mathbf{c}) = a_1(b_2c_3 - b_3c_2) + a_2(b_3c_1 - b_1c_3) + a_3(b_1c_2 - b_2c_1)$
   $\mathbf{a} \times \mathbf{b} = (a_2b_3 - a_3b_2)\mathbf{i} + (a_3b_1 - a_1b_3)\mathbf{j} + (a_1b_2 - a_2b_1)\mathbf{k}$
   $(\mathbf{a} \times \mathbf{b}).\mathbf{c} = (a_2b_3 - a_3b_2)c_1 + (a_3b_1 - a_1b_3)c_2 + (a_1b_2 - a_2b_1)c_3$
   $= a_2b_3c_1 - a_3b_2c_1 + a_3b_1c_2 - a_1b_3c_2 + a_1b_2c_3 - a_2b_1c_3$
   $= a_1(b_2c_3 - b_3c_2) + a_2(b_3c_1 - b_1c_3) + a_3(b_1c_2 - b_2c_1)$
   Therefore, $\mathbf{a}.(\mathbf{b} \times \mathbf{c}) = (\mathbf{a} \times \mathbf{b}).\mathbf{c}$
b $\mathbf{d}.(\mathbf{a} \times \mathbf{b} + \mathbf{a} \times \mathbf{c}) = \mathbf{d}.(\mathbf{a} \times \mathbf{b}) + \mathbf{d}.(\mathbf{a} \times \mathbf{c})$
   $= (\mathbf{d} \times \mathbf{a}).\mathbf{b} + (\mathbf{d} \times \mathbf{a}).\mathbf{c}$
   $= (\mathbf{d} \times \mathbf{a}).(\mathbf{b} + \mathbf{c})$
   $= \mathbf{d}.(\mathbf{a} \times (\mathbf{b} + \mathbf{c}))$

Online Worked solutions are available in SolutionBank.

c As **d** can be any vector, if $\mathbf{d}.(\mathbf{a} \times \mathbf{b} + \mathbf{a} \times \mathbf{c}) = \mathbf{d}.(\mathbf{a} \times (\mathbf{b} + \mathbf{c}))$, then it follows that $\mathbf{a} \times \mathbf{b} + \mathbf{a} \times \mathbf{c} = \mathbf{a} \times (\mathbf{b} + \mathbf{c})$

Exercise 5D

1 a $\mathbf{r} \times (3\mathbf{i} + \mathbf{j} - 2\mathbf{k}) = -4\mathbf{i} + 10\mathbf{j} - \mathbf{k}$
b $\mathbf{r} \times (\mathbf{i} + \mathbf{j} + 5\mathbf{k}) = 3\mathbf{i} - 13\mathbf{j} + 2\mathbf{k}$
c $\mathbf{r} \times (-\mathbf{i} - 2\mathbf{j} + 3\mathbf{k}) = -4\mathbf{i} - 13\mathbf{j} - 10\mathbf{k}$

2 a $\dfrac{x-2}{3} = \dfrac{y-1}{1} = \dfrac{z-2}{-2} = \lambda$
b $\dfrac{x-2}{1} = \dfrac{y}{1} = \dfrac{z+3}{5} = \lambda$
c $\dfrac{x-4}{-1} = \dfrac{y+2}{-2} = \dfrac{z-1}{3} = \lambda$

3 a $\left(\mathbf{r} - \begin{pmatrix}1\\3\\5\end{pmatrix}\right) \times \begin{pmatrix}5\\1\\-3\end{pmatrix} = 0$
b $\left(\mathbf{r} - \begin{pmatrix}3\\4\\12\end{pmatrix}\right) \times \begin{pmatrix}1\\-1\\-7\end{pmatrix} = 0$
c $\left(\mathbf{r} - \begin{pmatrix}-2\\2\\6\end{pmatrix}\right) \times \begin{pmatrix}5\\5\\5\end{pmatrix} = 0$
d $\left(\mathbf{r} - \begin{pmatrix}1\\1\\1\end{pmatrix}\right) \times \begin{pmatrix}-3\\-1\\5\end{pmatrix} = 0$

4 a $\dfrac{x-1}{5} = \dfrac{y-3}{1} = \dfrac{z-5}{-3} = \lambda$
b $\dfrac{x-3}{1} = \dfrac{y-4}{-1} = \dfrac{z-12}{-7} = \lambda$
c $\dfrac{x+2}{5} = \dfrac{y-2}{5} = \dfrac{z-6}{5} = \lambda$ or $x + 2 = y - 2 = z - 6 = \mu$
d $\dfrac{x-4}{-3} = \dfrac{y-2}{-1} = \dfrac{z+4}{5} = \lambda$

5 a $(\mathbf{r} - (\mathbf{i} + \mathbf{j} - 2\mathbf{k})) \times (2\mathbf{i} - \mathbf{k}) = 0$
b $(\mathbf{r} - (\mathbf{i} + 4\mathbf{j})) \times (3\mathbf{i} + \mathbf{j} - 5\mathbf{k}) = 0$
c $(\mathbf{r} - (3\mathbf{i} + 4\mathbf{j} - 4\mathbf{k})) \times (2\mathbf{i} - 2\mathbf{j} - 3\mathbf{k}) = 0$

6 a $\mathbf{r} \times (2\mathbf{i} + 5\mathbf{j} + \dfrac{3}{2}\mathbf{k}) = -9\mathbf{i} - \dfrac{3}{2}\mathbf{j} + 17\mathbf{k}$
b $\mathbf{r} = 3\mathbf{i} - \mathbf{j} + \dfrac{3}{2}\mathbf{k} + t(2\mathbf{i} + 5\mathbf{j} + \dfrac{3}{2}\mathbf{k})$
or $\mathbf{r} = 3\mathbf{i} - \mathbf{j} + \dfrac{3}{2}\mathbf{k} + s(4\mathbf{i} + 10\mathbf{j} + 3\mathbf{k})$

7 $p = 3$ and $q = 3$
8 $\mathbf{r} = -\mathbf{j} + 2\mathbf{k} + t(\mathbf{i} + \mathbf{j} - \mathbf{k})$

Exercise 5E

1 a $\mathbf{r}.(2\mathbf{i} + \mathbf{j} + \mathbf{k}) = 0$
b $\mathbf{r}.(5\mathbf{i} - \mathbf{j} - 3\mathbf{k}) = 0$
c $\mathbf{r}.(\mathbf{i} + 3\mathbf{j} + 4\mathbf{k}) = -10$
d $\mathbf{r}.(4\mathbf{i} + \mathbf{j} - 5\mathbf{k}) = 9$

2 a $2x + y + z = 0$
b $5x - y - 3z = 0$
c $x + 3y + 4z = -10$
d $4x + y - 5z = 9$

3 a $\mathbf{r} = \mathbf{i} + 2\mathbf{j} + \lambda(2\mathbf{i} - \mathbf{j} - \mathbf{k}) + \mu(3\mathbf{i} + \mathbf{j} + 2\mathbf{k})$
b $\mathbf{r} = 3\mathbf{i} + 4\mathbf{j} + \mathbf{k} + \lambda(24\mathbf{i} - 6\mathbf{j} - \mathbf{k}) + \mu(2\mathbf{i} - 3\mathbf{j} + 3\mathbf{k})$
or $\mathbf{r} = 3\mathbf{i} + 4\mathbf{j} + \mathbf{k} + \lambda 9(4\mathbf{i} + 6\mathbf{j} + \mathbf{k}) + \mu(2\mathbf{i} - 3\mathbf{j} + 3\mathbf{k})$
c $\mathbf{r} = 2\mathbf{i} - \mathbf{j} - \mathbf{k} + \lambda(\mathbf{i} + 2\mathbf{j} + 3\mathbf{k}) 1 \mu(2\mathbf{i} + \mathbf{j} + 2\mathbf{k})$
d $\mathbf{r} = 2\mathbf{i} + \mathbf{j} + 3\mathbf{k} + \lambda(\mathbf{j} + 2\mathbf{k}) + \mu(\mathbf{i} + 3\mathbf{j} + \mathbf{k})$

4 a $x + 7y - 5z = 15$
b $21x - 13y - 6z = 5$
c $x + 4y - 3z = 1$
d $25x + 2y - z = 4$

5 a $3x + y - z = 2$
b $7x - 2y + z = 5$
c $x + 2y - z = 3$
d $2x - 6y - z = 2$

6 a $\mathbf{r}.(2\mathbf{i} - 9\mathbf{j} + 4\mathbf{k}) = -15$
b $\mathbf{r}.(2\mathbf{i} - \mathbf{j} + \mathbf{k}) = 2$
c $\mathbf{r}.(8\mathbf{i} - 5\mathbf{j} + \mathbf{k}) = 22$

7 $-10x - 2y + 16z = 4$

Exercise 5F

1 a The two lines do meet at the point $(3, 1, 10)$
b The lines do not meet.
c The two lines do meet at the point $\left(0, 1\dfrac{1}{2}, 4\dfrac{1}{2}\right)$

2 a $\left(2\dfrac{2}{3}, \dfrac{1}{6}, 4\dfrac{1}{3}\right)$
b There are no values of λ for which the line meets the plane.
The line is parallel to the plane.
c $(1, 2, 0)$

3 a $\mathbf{r} = \left(\dfrac{5}{2}\mathbf{i} + \dfrac{5}{2}\mathbf{k}\right) + \lambda\left(\dfrac{3}{2}\mathbf{i} + \mathbf{j} + \dfrac{5}{2}\mathbf{k}\right)$
b $\mathbf{r} = (3\mathbf{i} - \mathbf{j}) + \lambda\left(\dfrac{2}{3}\mathbf{i} + \dfrac{4}{3}\mathbf{j} + \mathbf{k}\right)$
c $\mathbf{r} = \left(23\mathbf{i} - \dfrac{13}{3}\mathbf{j}\right) + \lambda\left(\mathbf{i} + \dfrac{2}{3}\mathbf{j} + \mathbf{k}\right)$

4 $\alpha = 68.3°$ (to 3 s.f.)
5 $\alpha = 40.2°$ (to 3 s.f.)
6 $\alpha = 4.25°$ (to 3 s.f.)
7 $\alpha = 43.1°$ (to 3 s.f.)
8 $\alpha = 21.7°$ (3 s.f.)

9 a 3 **b** $\dfrac{1}{3}$ **c** $\dfrac{2}{3}$ **d** 4
10 a 3 **b** 1
11 $\dfrac{13}{11}$
12 $x = \dfrac{\sqrt{198}}{5}$ or 2.81 (3 s.f.)

13 a The lines do not meet.
Distance $= \dfrac{4\sqrt{11}}{11}$ or 1.21
b Lines do not meet.
$x = 3\sqrt{2}$ or 4.24 (3 s.f.)
c Lines do not meet.
Shortest distance $= 0.196$ (3 s.f.)

14 3.54 (3 s.f.)

15 a The line $\mathbf{r} = 2\mathbf{i} + 3\mathbf{j} + \mathbf{k} + \lambda(-\mathbf{i} + 2\mathbf{j} + \mathbf{k})$ passes through the point $(2, 3, 1)$.

The point $(2, 3, 1)$ also lies on the plane $\mathbf{r}.(-\mathbf{i} + 2\mathbf{j} + \mathbf{k}) = 4$ as $2 \times 1 + 3 \times 1 - 1 = 4$

So the line and the plane have a point in common.

The line is in the direction $-\mathbf{i} + 2\mathbf{j} + \mathbf{k}$

This direction is parallel to the plane as it is perpendicular to the normal $\mathbf{i} + \mathbf{j} + \mathbf{k}$ as
$-1 \times 1 + 2 \times 1 + 1 \times -1 = 0$

As the line also has a common point with the plane, it lies in the plane.

b $\dfrac{7\sqrt{3}}{3} = 4.04$ (3 s.f.)

Chapter Review 5

1 $3\sqrt{2}$ or 4.24
2 $\frac{7\sqrt{6}}{9}$ or 1.91 (3 s.f.)
3 a $\mathbf{i} - \mathbf{j} - \mathbf{k}$
   b $\frac{2}{3}\sqrt{3}$
4 a $\mathbf{r} = 2\mathbf{i} - 3\mathbf{j} + \mathbf{k} + \lambda(24\mathbf{i} + \mathbf{j} - 2\mathbf{k})$
   b $\frac{5}{2}\sqrt{5}$ or 5.59 (3 s.f.)
5 a $5\mathbf{i} + 5\mathbf{j} - 5\mathbf{k}$
   b A is $\left(1\frac{3}{5}, \frac{1}{5}, 1\frac{4}{5}\right)$
      B is $\left(2\frac{4}{15}, \frac{13}{15}, 1\frac{2}{15}\right)$
6 a $\frac{1}{\sqrt{50}}(3\mathbf{i} + 5\mathbf{j} + 4\mathbf{k})$
   b $3x + 5y + 4z = 30$
   c $3\sqrt{2}$
7 b $\frac{\sqrt{2}}{2}$ or 0.707 (to 3 s.f.)
   c $x + z = 1$
8 a $-15\mathbf{i} - 20\mathbf{j} + 10\mathbf{k}$ or a multiple of $(3\mathbf{i} + 4\mathbf{j} - 2\mathbf{k})$
   b $3x + 4y - 2z - 5 = 0$
   c 5
9 a $-6\mathbf{i} - 4\mathbf{j} + 2\mathbf{k}$
   b $\mathbf{r} \cdot (3\mathbf{i} + 2\mathbf{j} - \mathbf{k}) = 0$
   c (21, 1, 21)
10 a $-\mathbf{i} + 7\mathbf{j} + 5\mathbf{k}$
   b $-x + 7y + 5z = 0$
   c (1, 22, 3)
11 a 73° (nearest degree)
   b $\mathbf{r} \times \left(\mathbf{i} + \frac{1}{2}\mathbf{j} + 3\mathbf{k}\right) = \left(-\frac{5}{2}\mathbf{i} - 16\mathbf{j} + \frac{7}{2}\mathbf{k}\right)$
12 a $-\mathbf{i} + \mathbf{j} + 4\mathbf{k}$
   c 27° (nearest degree)
13 b $2\sqrt{6}$ or 4.90
14 c $\mathbf{r} = 2\mathbf{i} - 2\mathbf{j} + 3\mathbf{k} + \lambda(2\mathbf{i} - \mathbf{j} + \mathbf{k})$
   d $\left(-1, -\frac{1}{2}, \frac{3}{2}\right)$
   e 3.67 (3 s.f.)
   f $\mathbf{r} \cdot (2\mathbf{i} - \mathbf{j} + \mathbf{k}) = 9$
15 a $\mathbf{r} = \mathbf{i} + 2\mathbf{j} + \mathbf{k} + \lambda(2\mathbf{i} + \mathbf{j} + 3\mathbf{k})$
   b (3, 3, 4)
   c $5\mathbf{i} - \mathbf{j} - 3\mathbf{k}$
   d $\frac{\sqrt{35}}{\sqrt{34}}$
   e (5, 4, 7)
16 b $\mathbf{r} = 7\mathbf{i} + 2\mathbf{j} - 6\mathbf{k}$
   c $\frac{14}{15}$
   d $\mathbf{r} = \mathbf{i} - \mathbf{j} + \lambda(2\mathbf{i} + \mathbf{j} - 2\mathbf{k}) + \mu(-3\mathbf{i} + 4\mathbf{k})$
17 $2x - 5y + 3z + 10 = 0$
18 b $-10\mathbf{i} + 10\mathbf{j} - 5\mathbf{k}$
   c $+2x - 2y + z = 10$
   d 15
19 a $\mathbf{r} = a(-4\mathbf{i} + 4\mathbf{j} - \mathbf{k}) + \lambda a(9\mathbf{i} - 6\mathbf{j} + 12\mathbf{k})$
   b $\mathbf{r} = a(5\mathbf{i} - \mathbf{j} - 3\mathbf{k}) + \lambda a(5\mathbf{i} - \mathbf{j} - 3\mathbf{k}) + \mu a(-4\mathbf{i} + 4\mathbf{j} - \mathbf{k})$
   c $\frac{12}{\sqrt{35}\sqrt{33}}$ or 0.353 (3 s.f.)
   d $3x - 2y + 4z = 5a$
   e $\frac{x + 4a}{3} = \frac{y - 4a}{22} = \frac{z + a}{4} = \lambda 9$
20 a $\mathbf{r} \cdot (\mathbf{i} - \mathbf{j} - 2\mathbf{k}) = 27$
      $x - y - 2z + 7 = 0$
   b $\frac{3}{\sqrt{2}\sqrt{14}} = 0.567$ (3 s.f.)
   c (0, 5, 7) and (4, 1, 21)

Challenge
Find the equation of the plane passing through
$A(p, 0, 0)$, $B(0, q, 0)$ and $C(0, 0, r)$:
$\overrightarrow{AB} = \overrightarrow{OB} - \overrightarrow{OA} = -p\mathbf{i} + q\mathbf{j}$
$\overrightarrow{AC} = \overrightarrow{OC} - \overrightarrow{OA} = -p\mathbf{i} + r\mathbf{k}$
$\overrightarrow{AB} \times \overrightarrow{AC} = \begin{vmatrix} \mathbf{i} & \mathbf{j} & \mathbf{k} \\ -p & q & 0 \\ -p & 0 & r \end{vmatrix} = qr\mathbf{i} + pr\mathbf{j} + pq\mathbf{k}$

$\mathbf{r} \cdot (qr\mathbf{i} + pr\mathbf{j} + pq\mathbf{k}) = p\mathbf{i} \cdot (qr\mathbf{i} + pr\mathbf{j} + pq\mathbf{k})$
$qrx + pry + pqz = pqr$
Distance between plane and origin:
$d = \frac{|pqr|}{\sqrt{(qr)^2 + (pr)^2 + (pq)^2}}$
$d^2 = \frac{(pqr)^2}{(qr)^2 + (pr)^2 + (pq)^2}$
$\frac{1}{d^2} = \frac{(qr)^2 + (pr)^2 + (pq)^2}{(pqr)^2} = \frac{1}{p^2} + \frac{1}{q^2} + \frac{1}{r^2}$

CHAPTER 6
Prior knowledge check

1 2
2 $k = 6$
3 $\begin{pmatrix} -1 & 3 \\ 2 & -2 \end{pmatrix}\begin{pmatrix} x \\ y \end{pmatrix} = \begin{pmatrix} x \\ y \end{pmatrix} \Rightarrow \begin{matrix} -x + 3y = x \\ 2x - 2y = y \end{matrix} \Rightarrow \begin{matrix} 3y - 2x = 0 \\ 3y - 2x = 0 \end{matrix} \Rightarrow$ as required

Exercise 6A

1 a $\begin{pmatrix} 3 & -1 \\ 1 & 0 \\ 2 & 4 \end{pmatrix}$ dimension 3 × 2
   b $\begin{pmatrix} 0 & -2 \\ 2 & 0 \end{pmatrix}$ dimension 2 × 2
   c $\begin{pmatrix} 0 & -2 & 1 \\ 2 & 0 & -3 \\ -1 & 3 & 0 \end{pmatrix}$ dimension 3 × 3
   d $(1\ 2\ 4)$ dimension 1 × 3
2 a $\begin{pmatrix} 2 & -3 \\ 4 & 6 \end{pmatrix}$
   b $\begin{pmatrix} 20 & 18 \\ 18 & 45 \end{pmatrix}$
   c $\begin{pmatrix} 13 & -10 \\ -10 & 52 \end{pmatrix}$
3 a $\begin{pmatrix} -9 & 8 \\ 8 & -4 \end{pmatrix}$
4 a $\begin{pmatrix} 1 & 4 & 8 \\ -4 & -7 & 4 \\ 8 & -4 & 1 \end{pmatrix}$
5 a $\begin{pmatrix} -12 & 15 & 21 \\ 15 & -3 & 0 \\ 21 & 0 & 7 \end{pmatrix}$
6 a $\begin{pmatrix} -5 & 3 & 15 \\ 3 & 2 & -5 \\ 1 & 2 & -1 \end{pmatrix}$

Online Worked solutions are available in SolutionBank.

Exercise 6B

1 a 6   **b** −56   **c** 1   **d** 0

2 a 20   **b** 17   **c** 0

3 17

4 −8, −2

5 a $\det(\mathbf{A}) = \begin{vmatrix} 2 & 5 & 3 \\ -2 & 0 & 4 \\ 3 & 10 & 8 \end{vmatrix}$

$= 2\begin{vmatrix} 0 & 4 \\ 10 & 8 \end{vmatrix} - 5\begin{vmatrix} -2 & 4 \\ 3 & 8 \end{vmatrix} + 3\begin{vmatrix} -2 & 0 \\ 3 & 10 \end{vmatrix}$

$= 2(0 - 40) - 5(-16 - 12) + 3(-20 - 0)$

$= -80 + 140 - 60 = 0$

Hence **A** is singular

b $\begin{pmatrix} 7 & 6 & 7 \\ -2 & -10 & -4 \\ 13 & 7 & 12 \end{pmatrix}$

c $\det(\mathbf{AB}) = \begin{vmatrix} 7 & 6 & 7 \\ -2 & -10 & -4 \\ 13 & 7 & 12 \end{vmatrix}$

$= 7\begin{vmatrix} -10 & -4 \\ 7 & 12 \end{vmatrix} - 6\begin{vmatrix} -2 & -4 \\ 13 & 12 \end{vmatrix} + 7\begin{vmatrix} -2 & -10 \\ 13 & 7 \end{vmatrix}$

$= 2(-120 + 28) - 6(-24 + 52) + 7(-14 + 130)$

$= 7 \times (-92) - 6 \times 28 + 7 \times 116$

$= -644 - 168 + 812 = 0$

Hence **AB** is also singular

6 a −10

b $\begin{pmatrix} 4 & 2 & 2 \\ 5 & -3 & -4 \\ -2 & 2 & 3 \end{pmatrix}$

7 a $\begin{vmatrix} 0 & a & -b \\ -a & 0 & c \\ b & -c & 0 \end{vmatrix} = 0\begin{vmatrix} 0 & c \\ -c & 0 \end{vmatrix} - a\begin{vmatrix} -a & c \\ b & 0 \end{vmatrix} + (-b)\begin{vmatrix} -a & 0 \\ b & -c \end{vmatrix}$

$= 0 - a(0 - cb) - b(ac - 0)$

$= abc - abc = 0$

Hence the matrix is singular for all a, b and c.

b $\begin{vmatrix} 2 & -2 & 4 \\ 3 & x & -2 \\ -1 & 3 & x \end{vmatrix} = 2\begin{vmatrix} x & -2 \\ 3 & x \end{vmatrix} - (-2)\begin{vmatrix} 3 & -2 \\ -1 & x \end{vmatrix} + 4\begin{vmatrix} 3 & x \\ -1 & 3 \end{vmatrix}$

$= 2(x^2 - 6) + 2(3x - 2) + 4(9 + x)$

$= 2x^2 + 12 + 6x - 4 + 36 + 4x$

$= 2(x^2 + 10x) + 44$

$= 2\left(x^2 + 5x + \left(\frac{5}{2}\right)^2\right) + 44 - 2x\left(\frac{5}{2}\right)^2$

$= 2\left(x + \frac{5}{2}\right)^2 + 31\frac{1}{2} \geq 31\frac{1}{2}$, for all real x.

8 −1, 0, 3

Exercise 6C

1 a $\begin{pmatrix} 1 & 0 & 0 \\ 0 & \frac{2}{3} & -\frac{1}{3} \\ 0 & -\frac{1}{3} & \frac{2}{3} \end{pmatrix}$

b $\begin{pmatrix} 1 & 0 & 0 \\ 0 & \frac{1}{2} & 0 \\ 0 & 0 & \frac{1}{3} \end{pmatrix}$

c $\begin{pmatrix} 1 & 0 & 0 \\ 0 & \frac{3}{5} & \frac{4}{5} \\ 0 & -\frac{4}{5} & \frac{3}{5} \end{pmatrix}$

2 a $\begin{pmatrix} 4 & -6 & -1 \\ -3 & 4 & 1 \\ -6 & 9 & 2 \end{pmatrix}$

b $\begin{pmatrix} -\frac{3}{5} & -\frac{1}{5} & \frac{7}{5} \\ -\frac{1}{5} & -\frac{2}{5} & \frac{4}{5} \\ \frac{7}{5} & \frac{4}{5} & -\frac{13}{5} \end{pmatrix}$

c $\begin{pmatrix} 2 & -5 & -\frac{19}{2} \\ 1 & -3 & -5 \\ 1 & -3 & -\frac{11}{2} \end{pmatrix}$

3 a $\begin{pmatrix} -1 & 0 & 1 \\ 0 & 1 & 0 \\ 2 & 0 & -1 \end{pmatrix}$

b $\begin{pmatrix} \frac{1}{3} & \frac{1}{2} & -\frac{1}{6} \\ 0 & -\frac{1}{2} & \frac{1}{2} \\ -\frac{1}{3} & \frac{1}{2} & \frac{1}{6} \end{pmatrix}$

4 b $\dfrac{1}{3(k+1)}\begin{pmatrix} 3 & 3 & -3 \\ 1 - 4k & 5 & 3k - 2 \\ k - 1 & -2 & 2 \end{pmatrix}$

5 $a = -4$, $b = 8$, $c = 3$

6 a $\mathbf{A}^2 = \begin{pmatrix} 2 & -1 & 1 \\ 4 & -3 & 0 \\ -3 & 3 & 1 \end{pmatrix}\begin{pmatrix} 2 & -1 & 1 \\ 4 & -3 & 0 \\ -3 & 3 & 1 \end{pmatrix}$

$= \begin{pmatrix} 4 - 4 - 3 & -2 + 3 + 3 & 2 + 0 + 1 \\ 8 - 12 + 0 & -4 + 9 + 0 & 4 + 0 + 0 \\ -6 + 12 - 3 & 3 - 9 + 3 & -3 + 0 + 1 \end{pmatrix}$

$= \begin{pmatrix} -3 & 4 & 3 \\ -4 & 5 & 4 \\ 3 & -3 & -2 \end{pmatrix}$

$\mathbf{A}^3 = \mathbf{A}^2\mathbf{A} = \begin{pmatrix} -3 & 4 & 3 \\ -4 & 5 & 4 \\ 3 & -3 & -2 \end{pmatrix}\begin{pmatrix} 2 & -1 & 1 \\ 4 & -3 & 0 \\ -3 & 3 & 1 \end{pmatrix}$

$= \begin{pmatrix} -6 + 16 - 9 & 3 - 12 + 9 & -3 + 0 + 3 \\ -8 + 20 - 12 & 4 - 15 + 12 & -4 + 0 + 4 \\ 6 - 12 + 6 & -3 + 9 - 6 & 3 + 0 - 2 \end{pmatrix}$

$= \begin{pmatrix} 1 & 0 & 0 \\ 0 & 1 & 0 \\ 0 & 0 & 1 \end{pmatrix} = \mathbf{I}$, as required

b $\mathbf{A}^3 = \mathbf{A}\mathbf{A}^2 = \mathbf{I}$

Comparing with the definition of an inverse $\mathbf{AA}^{-1} = \mathbf{I}$

$\mathbf{A}^{-1} = \mathbf{A}^2 = \begin{pmatrix} -3 & 4 & 3 \\ -4 & 5 & 4 \\ 3 & -3 & 2 \end{pmatrix}$

7 a $\mathbf{A}^2 = \begin{pmatrix} 1 & 1 & 0 \\ 3 & -3 & 1 \\ 0 & 3 & 2 \end{pmatrix}\begin{pmatrix} 1 & 1 & 0 \\ 3 & -3 & 1 \\ 0 & 3 & 2 \end{pmatrix}$

$= \begin{pmatrix} 1 + 3 + 0 & 1 - 3 + 0 & 0 + 1 + 0 \\ 3 - 9 + 0 & 3 + 9 + 3 & 0 - 3 + 2 \\ 0 + 9 + 0 & 0 - 9 + 6 & 0 + 3 + 4 \end{pmatrix}$

$= \begin{pmatrix} 1 & -2 & 1 \\ -6 & 15 & -1 \\ 9 & -3 & 7 \end{pmatrix}$

$\mathbf{A}^3 = \mathbf{A}^2\mathbf{A} = \begin{pmatrix} 4 & -2 & 1 \\ -6 & 15 & -1 \\ 9 & -3 & 7 \end{pmatrix}\begin{pmatrix} 1 & 1 & 0 \\ 3 & -3 & 1 \\ 0 & 3 & 2 \end{pmatrix}$

$= \begin{pmatrix} 4 - 6 + 0 & 4 + 6 + 3 & 0 - 2 + 2 \\ -6 + 45 + 0 & -6 - 45 - 3 & 0 + 15 - 2 \\ 9 - 9 + 0 & 9 + 9 + 21 & 0 - 3 + 14 \end{pmatrix}$

$= \begin{pmatrix} -2 & 13 & 0 \\ 39 & -54 & 13 \\ 0 & 39 & 11 \end{pmatrix}$

$$13\mathbf{A} - 15\mathbf{I} = \begin{pmatrix} 13 & 13 & 0 \\ 39 & -39 & 13 \\ 0 & 39 & 26 \end{pmatrix} - \begin{pmatrix} 15 & 0 & 0 \\ 0 & 15 & 0 \\ 0 & 0 & 15 \end{pmatrix}$$

$$= \begin{pmatrix} -2 & 13 & 0 \\ 39 & -54 & 13 \\ 0 & 39 & 11 \end{pmatrix} = \mathbf{A}^3$$

Hence
$\mathbf{A}^3 = 13\mathbf{A} - 15\mathbf{I}$, as required.

b Multiply the result of part **a** throughout by $\mathbf{A}^{-1}$
$\mathbf{A}^3\mathbf{A}^{-1} = 13\mathbf{A}\mathbf{A}^{-1} - 15\mathbf{I}\mathbf{A}^{-1}$
$\mathbf{A}^2 = 13\mathbf{I} - 15\mathbf{A}^{-1}$
Rearranging
$15\mathbf{A}^{-1} = 13\mathbf{I} - \mathbf{A}^2$, as required.

c $\dfrac{1}{15}\begin{pmatrix} 9 & 2 & -1 \\ 6 & -2 & 1 \\ -9 & 3 & 6 \end{pmatrix}$

8 a $\det(\mathbf{A}) = 2\begin{vmatrix} 3 & -2 \\ 3 & -4 \end{vmatrix} - 0\begin{vmatrix} 4 & -2 \\ 0 & -4 \end{vmatrix} + 1\begin{vmatrix} 4 & 3 \\ 0 & 3 \end{vmatrix}$
$= 2(-12 + 6) - 0 + 1(12 - 0)$
$= -12 + 12 = 0$
Hence **A** is singular.

b $\begin{pmatrix} -6 & 16 & 12 \\ 3 & -8 & -6 \\ -3 & 8 & 6 \end{pmatrix}$

c $\mathbf{AC}^T = \begin{pmatrix} 2 & 0 & 1 \\ 4 & 4 & -2 \\ 0 & 3 & -4 \end{pmatrix}\begin{pmatrix} -6 & 3 & -3 \\ 16 & -8 & 8 \\ 12 & -6 & 6 \end{pmatrix}$

$= \begin{pmatrix} -12 + 0 + 12 & 6 + 0 - 6 & -6 + 0 + 6 \\ -24 + 48 - 24 & 12 - 24 + 12 & -12 + 24 - 12 \\ 0 + 48 - 48 & 0 - 24 + 24 & 0 + 24 - 24 \end{pmatrix}$

$= \begin{pmatrix} 0 & 0 & 0 \\ 0 & 0 & 0 \\ 0 & 0 & 0 \end{pmatrix} = \mathbf{0}$, as required

Exercise 6D

1 a $\begin{pmatrix} 1 & -1 & 0 \\ 0 & 1 & 1 \\ 2 & 0 & -3 \end{pmatrix}$

b $\begin{pmatrix} 2 & -3 & -1 \\ 0 & 2 & 3 \\ 0 & 0 & 5 \end{pmatrix}$

c $\begin{pmatrix} 2 & -5 & -4 \\ 0 & 2 & 8 \\ 4 & -6 & -17 \end{pmatrix}$

2 $a = -3, b = 1, c = -4$

3 $\begin{pmatrix} 3 & 13 & 11 \\ 1 & -1 & 0 \\ 2 & -1 & 1 \end{pmatrix}$

4 $\mathbf{r} = \begin{pmatrix} -2 \\ 20 \\ 15 \end{pmatrix} + t\begin{pmatrix} 8 \\ -24 \\ -4 \end{pmatrix}$

5 a The position vector of A' is $\begin{pmatrix} -1 \\ 7 \\ 2 \end{pmatrix}$ and the position vector of B' is $\begin{pmatrix} 5 \\ 23 \\ 26 \end{pmatrix}$

b $\mathbf{r} = \begin{pmatrix} -1 \\ 7 \\ 2 \end{pmatrix} + t\begin{pmatrix} 6 \\ -10 \\ 24 \end{pmatrix}$

6 $6x + 2y + 9z = 0$

7 $\mathbf{r}.\begin{pmatrix} -10 \\ 49 \\ 27 \end{pmatrix} = 120$

8 $\mathbf{r} = t\begin{pmatrix} 1 \\ -1 \\ 1 \end{pmatrix}$

Exercise 6E

1 a $a = -21, b = 24, c = -23$

b $\mathbf{r} = t\begin{pmatrix} 1 \\ -5 \\ 3 \end{pmatrix}$

2 a $\begin{pmatrix} -4 & 6 & -3 \\ 6 & -7 & 4 \\ -3 & 4 & -2 \end{pmatrix}$

b $a = 2, b = 21, c = 3$

3 a $\begin{pmatrix} 4 & -2 & 1 \\ 3 & -1 & 1 \\ -3 & \frac{3}{2} & -1 \end{pmatrix}$

b $\mathbf{r} = \begin{pmatrix} 1 \\ 3 \\ -1 \end{pmatrix} + t\begin{pmatrix} -3 \\ -2 \\ 2 \end{pmatrix}$

4 a $\begin{pmatrix} 0 & \frac{1}{4} & 0 \\ 1 & -\frac{a}{4} & 2 \\ 0 & 0 & -1 \end{pmatrix}$

b $p = \dfrac{3}{4}, q = -\dfrac{3a}{4}, r = 1$

5 a $\mathbf{SS}^T = \begin{pmatrix} 1 & -\sqrt{2} & 1 \\ \sqrt{2} & 0 & -\sqrt{2} \\ 1 & \sqrt{2} & 1 \end{pmatrix}\begin{pmatrix} 1 & \sqrt{2} & 1 \\ -\sqrt{2} & 0 & \sqrt{2} \\ 1 & -\sqrt{2} & 1 \end{pmatrix}$

$= \begin{pmatrix} 1+2+1 & \sqrt{2}+0-\sqrt{2} & 1-2+1 \\ \sqrt{2}+0-\sqrt{2} & 2+0+2 & \sqrt{2}+0-\sqrt{2} \\ 1-2+1 & \sqrt{2}+0-\sqrt{2} & 1+2+1 \end{pmatrix}$

$= \begin{pmatrix} 4 & 0 & 0 \\ 0 & 4 & 0 \\ 0 & 0 & 4 \end{pmatrix} = 4\mathbf{I}$

$k = 4$

b $a = \dfrac{1}{2}, b = 22, c = -\dfrac{1}{2}$

6 a $a = -2, b = 2, c = 19$

b $\mathbf{r} = \begin{pmatrix} 1 \\ 1 \\ -7 \end{pmatrix} + s\begin{pmatrix} -9 \\ -6 \\ 56 \end{pmatrix} + t\begin{pmatrix} -5 \\ -3 \\ 30 \end{pmatrix}$

7 $a = 2, b = 0, c = 21$

8 a $\begin{pmatrix} -2 & -2 & 5 \\ -1 & 0 & 2 \\ 2 & 1 & -4 \end{pmatrix}$

b $\mathbf{T}^2 = \begin{pmatrix} 3 & 3 & 4 \\ -6 & -7 & -6 \\ 4 & 4 & 3 \end{pmatrix}\begin{pmatrix} 3 & 4 & 4 \\ -6 & -7 & -6 \\ 4 & 4 & 3 \end{pmatrix}$

$= \begin{pmatrix} 9-24+16 & 12-28+16 & 12-24+12 \\ -18+42-24 & -24+49-24 & -24+42-18 \\ 12-24+12 & 16-28+12 & 16-24+9 \end{pmatrix}$

$= \begin{pmatrix} 1 & 0 & 0 \\ 0 & 1 & 0 \\ 0 & 0 & 1 \end{pmatrix} = \mathbf{I}$, as required

c $a = -1, b = 2, c = -1$

Exercise 6F

1 a The eigenvalues are 1 and 6.
    An eigenvector corresponding to the eigenvalue 1 is $\begin{pmatrix} -4 \\ 1 \end{pmatrix}$
    An eigenvector corresponding to the eigenvalue 6 is $\begin{pmatrix} 1 \\ 1 \end{pmatrix}$

  b The eigenvalues are 3 and 5.
    An eigenvector corresponding to the eigenvalue 3 is $\begin{pmatrix} 1 \\ 1 \end{pmatrix}$
    An eigenvector corresponding to the eigenvalue 5 is $\begin{pmatrix} 1 \\ -1 \end{pmatrix}$

  c The eigenvalues are 3 and 4.
    An eigenvector corresponding to the eigenvalue 3 is $\begin{pmatrix} 1 \\ 0 \end{pmatrix}$
    An eigenvector corresponding to the eigenvalue 4 is $\begin{pmatrix} -2 \\ 1 \end{pmatrix}$

2 a 5, 7
  b $y = \frac{1}{2}x,\ y = x$

3 a The eigenvalues are 1, 3 and 4.
    An eigenvector corresponding to the eigenvalue 1 is $\begin{pmatrix} 0 \\ -2 \\ 3 \end{pmatrix}$
    An eigenvector corresponding to the eigenvalue 3 is $\begin{pmatrix} 1 \\ 0 \\ -1 \end{pmatrix}$
    An eigenvector corresponding to the eigenvalue 4 is $\begin{pmatrix} 0 \\ 1 \\ 0 \end{pmatrix}$

  b The eigenvalues are -1, 0 and 4.
    An eigenvector corresponding to the eigenvalue -1 is $\begin{pmatrix} -2 \\ 1 \\ -3 \end{pmatrix}$
    An eigenvector corresponding to the eigenvalue 0 is $\begin{pmatrix} 3 \\ -2 \\ 4 \end{pmatrix}$
    An eigenvector corresponding to the eigenvalue 4 is $\begin{pmatrix} 1 \\ 2 \\ -1 \end{pmatrix}$

4 a $\mathbf{A} - \lambda\mathbf{I} = \begin{pmatrix} 2 & 2 & -2 \\ -3 & 2 & 0 \\ 1 & 4 & -3 \end{pmatrix} - \begin{pmatrix} \lambda & 0 & 0 \\ 0 & \lambda & 0 \\ 0 & 0 & \lambda \end{pmatrix}$
    $= \begin{pmatrix} 2-\lambda & 2 & -2 \\ -3 & 2-\lambda & 0 \\ 1 & 4 & -3-\lambda \end{pmatrix}$
    $\begin{vmatrix} 2-\lambda & 2 & -2 \\ -3 & 2-\lambda & 0 \\ 1 & 4 & -3-\lambda \end{vmatrix}$
    $= (2-\lambda)\begin{vmatrix} 2-\lambda & 0 \\ 4 & -3-\lambda \end{vmatrix}$
    $\quad - 2\begin{vmatrix} -3 & 0 \\ 1 & -3-\lambda \end{vmatrix} + (-2)\begin{vmatrix} -3 & 2-\lambda \\ 1 & 4 \end{vmatrix}$
    $= (2-\lambda^2)(-3-\lambda) - 2(9+3\lambda)$
    $\quad - 2(-12 - 2 + \lambda)$

$= (\lambda^2 - 4\lambda + 4)(-3 - \lambda) - 18 - 6\lambda + 28 - 2\lambda$
$= -\lambda^3 + \lambda^2 + 8\lambda - 12 - 8\lambda + 10$
$= -\lambda^3 + \lambda^2 - 2 = -(\lambda^3 - \lambda^2 + 2)$
$\lambda^3 - \lambda^2 + 2 = \lambda^3 + \lambda^2 - 2\lambda^2 - 2\lambda + 2\lambda + 2$
$= \lambda^2(\lambda + 1) - 2\lambda(\lambda + 1) + 2(\lambda + 1)$
$= (\lambda + 1)(\lambda^2 - 2\lambda + 2)$
$= (\lambda + 1)((\lambda - 1)^2 + 1)$

As $(\lambda - 1)^2 + 1 \geq 1$ for all real λ, $(\lambda - 1)^2 + 1 = 0$ has no real solutions.
Hence $\det(\mathbf{A} - \lambda\mathbf{I}) = 0 \Rightarrow -(\lambda + 1)((\lambda - 1)^2 + 1)$
$\qquad = 0 \Rightarrow \lambda = -1$
The only real eigenvalue of $\mathbf{A}$ is -1.

  b $\begin{pmatrix} 1 \\ 1 \\ \frac{5}{2} \end{pmatrix}$

5 a $\mathbf{A} - \lambda\mathbf{I} = \begin{pmatrix} 2 & 1 & 3 \\ 0 & 2 & 4 \\ 0 & 2 & 0 \end{pmatrix} - \begin{pmatrix} \lambda & 0 & 0 \\ 0 & \lambda & 0 \\ 0 & 0 & \lambda \end{pmatrix} = \begin{pmatrix} 2-\lambda & -1 & 3 \\ 0 & 2-\lambda & 4 \\ 0 & 2 & -\lambda \end{pmatrix}$

    $\begin{vmatrix} 2-\lambda & -1 & 3 \\ 0 & 2-\lambda & 4 \\ 0 & 2 & -\lambda \end{vmatrix}$
    $= (2-\lambda)\begin{vmatrix} 2-\lambda & 4 \\ 2 & -\lambda \end{vmatrix} - (-1)\begin{vmatrix} 0 & 4 \\ 0 & -\lambda \end{vmatrix} + 3\begin{vmatrix} 0 & 2-\lambda \\ 0 & 2 \end{vmatrix}$
    $= (2-\lambda)(-2\lambda + \lambda^2 - 8) + 0 + 0$
    $= (2-\lambda)(\lambda^2 - 2\lambda - 8)$
    $= (2-\lambda)(\lambda - \lambda)(\lambda + 2)$
    $\det(\mathbf{A} - \lambda\mathbf{I}) = 0 > (2-\lambda)(\lambda - 4)(\lambda + 2) = 0 > \lambda$
    $\qquad = 2, 4, -2$
    The eigenvalues of A are 4, as required, 2 and -2

  b $\begin{pmatrix} \frac{1}{2} \\ 2 \\ 1 \end{pmatrix}$

6 a $-1, 6$.
  b An eigenvector corresponding to the eigenvalue -1 is $\begin{pmatrix} -2 \\ 1 \\ 1 \end{pmatrix}$
    An eigenvector corresponding to the eigenvalue 3 is $\begin{pmatrix} 1 \\ -1 \\ 1 \end{pmatrix}$
    An eigenvector corresponding to the eigenvalue 6 is $\begin{pmatrix} 5 \\ 1 \\ 8 \end{pmatrix}$

7 a $\mathbf{A} - \lambda\mathbf{I} = \begin{pmatrix} 2 & 2 & 1 \\ -2 & 4 & 0 \\ 4 & 2 & 5 \end{pmatrix} - \begin{pmatrix} \lambda & 0 & 0 \\ 0 & \lambda & 0 \\ 0 & 0 & \lambda \end{pmatrix}$
    $= \begin{pmatrix} 2-\lambda & 2 & 1 \\ -2 & 4-\lambda & 0 \\ 4 & 2 & 5-\lambda \end{pmatrix}$
    When $\lambda = 2$
    $\mathbf{A} - 2\mathbf{I} = \begin{pmatrix} 0 & 2 & 1 \\ -2 & 2 & 0 \\ 4 & 2 & 3 \end{pmatrix}$
    $\det(A - 2\mathbf{I}) = \begin{vmatrix} 0 & 2 & 1 \\ -2 & 2 & 0 \\ 4 & 2 & 3 \end{vmatrix}$
    $= 0\begin{vmatrix} 2 & 0 \\ 2 & 3 \end{vmatrix} - 2\begin{vmatrix} -2 & 0 \\ 4 & 3 \end{vmatrix} + 1\begin{vmatrix} -2 & 2 \\ 4 & 2 \end{vmatrix}$

$= 0 - 2 \times (-6) + 1 (-4 -8)$
$= 12 - 12 = 0$

Hence 2 is an eigenvector of **A**.

b 4, 5.

c $\begin{pmatrix} \frac{1}{\sqrt{6}} \\ \frac{1}{\sqrt{6}} \\ -\frac{2}{\sqrt{6}} \end{pmatrix}$

8 a $\mathbf{A} - \lambda\mathbf{I} = \begin{pmatrix} 4 & 2 & 1 \\ -2 & 0 & 5 \\ 0 & 3 & 4 \end{pmatrix} - \begin{pmatrix} \lambda & 0 & 0 \\ 0 & \lambda & 0 \\ 0 & 0 & \lambda \end{pmatrix} = \begin{pmatrix} 4-\lambda & 2 & 1 \\ -2 & -\lambda & 5 \\ 0 & 3 & 4-\lambda \end{pmatrix}$

$\begin{vmatrix} 4-\lambda & 2 & 1 \\ -2 & -\lambda & 5 \\ 0 & 3 & 4-\lambda \end{vmatrix}$

$= (4-\lambda)\begin{vmatrix} -\lambda & 5 \\ 3 & 4-\lambda \end{vmatrix} - 2\begin{vmatrix} -2 & 5 \\ 0 & 4-\lambda \end{vmatrix} + 1\begin{vmatrix} -2 & -\lambda \\ 0 & 3 \end{vmatrix}$

$= (4-\lambda)(-4\lambda + \lambda^2 - 15) - 2(-8 + 2\lambda) - 6$
$= -\lambda^3 + 8\lambda^2 - \lambda - 60 + 16 - 4\lambda - 6$
$= -\lambda^3 + 8\lambda^2 - 5\lambda - 50$

$\lambda^3 - 8\lambda^2 + 5\lambda + 50 = \lambda^3 + 2\lambda^2 - 10\lambda^2 - 20\lambda + 25\lambda + 50$
$= \lambda^2(\lambda + 2) - 10\lambda(\lambda + 2) + 25(\lambda + 2)$
$= (\lambda + 2)(\lambda^2 - 10\lambda + 25)$
$= (\lambda + 2)(\lambda - 5)^2$

$\det(\mathbf{A} - \lambda\mathbf{I}) = 0 \Rightarrow -(\lambda + 2)(\lambda - 5)^2 = 0 \Rightarrow \lambda = -2, 5$ repeated

-2 is one eigenvalue of A and the only other distinct eigenvalue is 5.

b An eigenvector corresponding to the eigenvalue -2 is $\begin{pmatrix} \frac{1}{2} \\ -2 \\ 1 \end{pmatrix}$

An eigenvector corresponding to the eigenvalue 5 is $\begin{pmatrix} 5 \\ 1 \\ 3 \end{pmatrix}$

9 a $\pm\sqrt{2}$

b An eigenvector corresponding to the eigenvalue $\sqrt{2}$ is $\begin{pmatrix} 1 \\ 1-\sqrt{2} \\ \sqrt{2}-1 \end{pmatrix}$

An eigenvector corresponding to the eigenvalue $-\sqrt{2}$ is $\begin{pmatrix} 1 \\ 1+\sqrt{2} \\ -1-\sqrt{2} \end{pmatrix}$

An eigenvector corresponding to the eigenvalue 2 is $\begin{pmatrix} 1 \\ -1 \\ -1 \end{pmatrix}$

10 a 4

b $a = 3$ and $b = 4$

c $\mathbf{A} - \lambda\mathbf{I} = \begin{pmatrix} 4 & 1 & 2 \\ 1 & 3 & 0 \\ -1 & 1 & 4 \end{pmatrix} - \begin{pmatrix} \lambda & 0 & 0 \\ 0 & \lambda & 0 \\ 0 & 0 & \lambda \end{pmatrix}$

$= \begin{pmatrix} 4-\lambda & 1 & 2 \\ 1 & 3-\lambda & 0 \\ -1 & 1 & 4-\lambda \end{pmatrix}$

$\begin{vmatrix} 4-\lambda & 1 & 2 \\ 1 & 3-\lambda & 0 \\ -1 & 1 & 4-\lambda \end{vmatrix}$

$= (4-\lambda)\begin{vmatrix} 3-\lambda & 0 \\ 1 & 4-\lambda \end{vmatrix} - 1\begin{vmatrix} 1 & 0 \\ -1 & 4-\lambda \end{vmatrix} + 2\begin{vmatrix} 1 & 3-\lambda \\ -1 & 1 \end{vmatrix}$

$= (4-\lambda)^2(3-\lambda) - 1(4-\lambda) + 2(1 + 3 - \lambda)$

$= (4-\lambda)^2(3-\lambda) - 1(4-\lambda) = (4-\lambda)((4-\lambda)(3-\lambda) + 1)$
$= (4-\lambda)(\lambda^2 - 7\lambda + 13)$

$\det(\mathbf{A} - \lambda\mathbf{I}) = 0 \Rightarrow (4-\lambda)(\lambda^2 - 7\lambda + 13) = 0 \Rightarrow \lambda = 4$ or $\lambda^2 - 7\lambda + 13 = 0$

The discriminant of $\lambda^2 - 7\lambda + 13 = 0$ is given by
$b^2 - 4ac = 49 - 52 = -3 < 0$

There are no real solutions of $\lambda^2 - 7\lambda + 13 = 0$

4 is the only real eigenvalue of A.

Challenge

The matrix has eigenvalue 1 with eigenvector $\begin{pmatrix} 0 \\ 1 \end{pmatrix}$ so all points on the y-axis are invariant. The other eigenvector is $\begin{pmatrix} 1 \\ 1 \end{pmatrix}$ so all lines parallel to $y = x$ stay parallel to $y = x$ under T. Since every line will cross the y-axis at one point, and this point is invariant under T, every line of the form $y = x + k$ is an invariant line of T, and there are infinitely many of these.

Exercise 6G

1 a $\begin{pmatrix} -2 & 0 \\ 0 & 4 \end{pmatrix}$ **b** $\begin{pmatrix} 5 & 0 \\ 0 & 0 \end{pmatrix}$

2 a 2, 5

b A normalised eigenvector corresponding to the eigenvalue 2 is $\begin{pmatrix} -\frac{\sqrt{2}}{\sqrt{3}} \\ \frac{1}{\sqrt{3}} \end{pmatrix}$

A normalised eigenvector corresponding to the eigenvalue 5 is $\begin{pmatrix} \frac{1}{\sqrt{3}} \\ \frac{\sqrt{2}}{\sqrt{3}} \end{pmatrix}$

c $\mathbf{P} = \begin{pmatrix} -\frac{\sqrt{2}}{\sqrt{3}} & \frac{1}{\sqrt{3}} \\ \frac{1}{\sqrt{3}} & \frac{\sqrt{2}}{\sqrt{3}} \end{pmatrix}$ $\mathbf{D} = \begin{pmatrix} 2 & 0 \\ 0 & 5 \end{pmatrix}$

3 a $\mathbf{PP}^T = \begin{pmatrix} \frac{1}{\sqrt{6}} & -\frac{1}{\sqrt{3}} & \frac{1}{\sqrt{2}} \\ \frac{1}{\sqrt{6}} & -\frac{1}{\sqrt{3}} & -\frac{1}{\sqrt{2}} \\ \frac{2}{\sqrt{6}} & \frac{1}{\sqrt{3}} & 0 \end{pmatrix} \begin{pmatrix} \frac{1}{\sqrt{6}} & \frac{1}{\sqrt{6}} & \frac{2}{\sqrt{6}} \\ -\frac{1}{\sqrt{3}} & -\frac{1}{\sqrt{3}} & \frac{1}{\sqrt{3}} \\ \frac{1}{\sqrt{2}} & -\frac{1}{\sqrt{2}} & 0 \end{pmatrix}$

$= \begin{pmatrix} \frac{1}{6}+\frac{1}{3}+\frac{1}{2} & \frac{1}{6}+\frac{1}{3}-\frac{1}{2} & \frac{2}{6}-\frac{1}{3} \\ \frac{1}{6}+\frac{1}{3}-\frac{1}{2} & \frac{1}{6}+\frac{1}{3}+\frac{1}{2} & \frac{2}{6}-\frac{1}{3} \\ \frac{2}{6}-\frac{1}{3} & \frac{2}{6}-\frac{1}{3} & \frac{4}{6}+\frac{1}{3} \end{pmatrix} = \begin{pmatrix} 1 & 0 & 0 \\ 0 & 1 & 0 \\ 0 & 0 & 1 \end{pmatrix} = \mathbf{I}$

Hence **P** is an orthogonal matrix.

4 $\begin{pmatrix} 0 & 0 & 0 \\ 0 & 2 & 0 \\ 0 & 0 & 4 \end{pmatrix}$

5 a $\begin{pmatrix} 0 \\ \frac{1}{\sqrt{2}} \\ -\frac{1}{\sqrt{2}} \end{pmatrix}$

b $\mathbf{P} = \begin{pmatrix} 0 & -\frac{1}{\sqrt{3}} & \frac{2}{\sqrt{6}} \\ \frac{1}{\sqrt{2}} & \frac{1}{\sqrt{3}} & \frac{1}{\sqrt{6}} \\ -\frac{1}{\sqrt{2}} & \frac{1}{\sqrt{3}} & \frac{1}{\sqrt{6}} \end{pmatrix}$ $\mathbf{D} = \begin{pmatrix} 0 & 0 & 0 \\ 0 & -1 & 0 \\ 0 & 0 & 8 \end{pmatrix}$

6 a 3, 6

b An eigenvector corresponding to the eigenvalue 3 is $\begin{pmatrix} 1 \\ 2 \\ 2 \end{pmatrix}$

Online — Worked solutions are available in SolutionBank.

An eigenvector corresponding to the eigenvalue 6 is $\begin{pmatrix} 2 \\ -2 \\ 1 \end{pmatrix}$

An eigenvector corresponding to the eigenvalue 9 is $\begin{pmatrix} 2 \\ 1 \\ -2 \end{pmatrix}$

c $P = \begin{pmatrix} \frac{1}{3} & \frac{2}{3} & \frac{2}{3} \\ \frac{2}{3} & 2\frac{2}{3} & \frac{1}{3} \\ \frac{2}{3} & \frac{1}{3} & 2\frac{2}{3} \end{pmatrix}$ $D = \begin{pmatrix} 3 & 0 & 0 \\ 0 & 6 & 0 \\ 0 & 0 & 9 \end{pmatrix}$

7 a $\det(A - \lambda I) = \begin{vmatrix} 1-\lambda & 2 & 0 \\ 2 & 1-\lambda & \sqrt{5} \\ 0 & \sqrt{5} & 1-\lambda \end{vmatrix}$

Substituting $\lambda = 4$

$\begin{vmatrix} 1-4 & 2 & 0 \\ 2 & 1-4 & \sqrt{5} \\ 0 & \sqrt{5} & 1-4 \end{vmatrix} = \begin{vmatrix} -3 & 2 & 0 \\ 2 & -3 & \sqrt{5} \\ 0 & \sqrt{5} & -3 \end{vmatrix}$

$= (-3)\begin{vmatrix} -3 & \sqrt{5} \\ \sqrt{5} & -3 \end{vmatrix} + 0 \begin{vmatrix} 2 & 0 \\ -3 & \sqrt{5} \end{vmatrix}$

$= (-3)(9 - 5) - 2(-6 - 0)$
$= -12 + 12 = 0$

Hence, by the factor theorem, 4 is an eigenvalue of A.

$\begin{vmatrix} 1-\lambda & 2 & 0 \\ 2 & 1-\lambda & \sqrt{5} \\ 0 & \sqrt{5} & 1-\lambda \end{vmatrix}$

$= (1-\lambda)\begin{vmatrix} 1-\lambda & \sqrt{5} \\ \sqrt{5} & 1-\lambda \end{vmatrix} - 2\begin{vmatrix} 2 & \sqrt{5} \\ 0 & 1-\lambda \end{vmatrix} + 0\begin{vmatrix} 2 & 1-\lambda \\ 0 & \sqrt{5} \end{vmatrix}$

$= (1-\lambda)((1-\lambda)^2 - 5) - 4 + 4\lambda$
$= (1-\lambda)(\lambda^2 - 2\lambda - 4) - 4 + 4\lambda = -\lambda^3 + 3\lambda^2 + 6\lambda - 8$
$= -\lambda^3 + 4\lambda^2 - \lambda^2 + 4\lambda + 2\lambda - 8 = -\lambda^2(\lambda - 4) - \lambda(\lambda - 4) + 2(\lambda - 4)$
$= -(\lambda - 4)(\lambda^2 + \lambda - 2) = -(\lambda - 4)(\lambda + 2)(\lambda - 1)$

$\det(A - \lambda I) = -(\lambda - 4)(\lambda + 2)(\lambda - 1) = 0 \Rightarrow \lambda \Rightarrow 4, -2, 1$
The other two eigenvalues of A are -2 and 1.

b $\begin{pmatrix} \frac{2}{\sqrt{18}} \\ \frac{3}{\sqrt{18}} \\ \frac{\sqrt{5}}{\sqrt{18}} \end{pmatrix}$

c $P = \begin{pmatrix} \frac{2}{\sqrt{18}} & -\frac{2}{\sqrt{18}} & \frac{\sqrt{5}}{3} \\ \frac{3}{\sqrt{18}} & \frac{3}{\sqrt{18}} & 0 \\ \frac{\sqrt{5}}{\sqrt{18}} & -\frac{\sqrt{5}}{\sqrt{18}} & -\frac{2}{3} \end{pmatrix}$ $D = \begin{pmatrix} 4 & 0 & 0 \\ 0 & -1 & 0 \\ 0 & 0 & 1 \end{pmatrix}$

8 a $A - \lambda I = \begin{pmatrix} 2-\lambda & 2 & -3 \\ 2 & 2-\lambda & 3 \\ -3 & 3 & 3-\lambda \end{pmatrix}$

$\det(A - \lambda I) = \begin{vmatrix} 2-\lambda & 2 & -3 \\ 2 & 2-\lambda & 3 \\ -3 & 3 & 3-\lambda \end{vmatrix}$

$= (2-\lambda)\begin{vmatrix} 2-\lambda & 3 \\ 3 & 3-\lambda \end{vmatrix} - 2\begin{vmatrix} 2 & 3 \\ -3 & 3-\lambda \end{vmatrix} + (-3)\begin{vmatrix} 2 & 2-\lambda \\ -3 & 3 \end{vmatrix}$

$= (2-\lambda)((2-\lambda)(3-\lambda) - 9) - 2(6 - 2\lambda + 9) - 3(6 + 6 - 3\lambda)$
$= (2-\lambda)(\lambda^2 - 5\lambda - 3) - 30 + 4\lambda - 36 + 9\lambda$
$= -\lambda^3 + 7\lambda^2 - 7\lambda - 6 - 66 + 13\lambda = -\lambda^3 + 7\lambda^2 + 6\lambda - 72$
$= -\lambda^3 + 6\lambda^2 + \lambda^2 - 6\lambda + 12\lambda - 72$
$= -\lambda^2(\lambda - 6) + \lambda(\lambda - 6) + 12(\lambda - 6) = -(\lambda - 6)(\lambda^2 - \lambda - 12)$
$= -(\lambda - 6)(\lambda - 4)(\lambda + 3)$

$\det(A - \lambda I) = 0 \Rightarrow -(\lambda - 6)(\lambda - 4)(\lambda + 3) = 0 \Rightarrow \lambda = 6, 4, -3$
As $\lambda_1 > \lambda_2 > \lambda_3$, $\lambda_1 = 6$, as required, $\lambda_2 = 4$ and $\lambda_3 = -3$.

b $\det(A) = \begin{vmatrix} 2 & 2 & -3 \\ 2 & 2 & 3 \\ -3 & 3 & 3 \end{vmatrix} = 2\begin{vmatrix} 2 & 3 \\ 3 & 3 \end{vmatrix} - 2\begin{vmatrix} 2 & 3 \\ -3 & 3 \end{vmatrix} + (-3)\begin{vmatrix} 2 & 2 \\ -3 & 3 \end{vmatrix}$

$= 2(6 - 9) - 2(6 + 9) - 3(6 + 6) = -6 - 30 - 36$
$= -72 = 6 \times 4 \times (-3) = \lambda_1 \lambda_2 \lambda_3$, as required

c $\begin{pmatrix} 1 \\ -1 \\ -2 \end{pmatrix}$

d $\begin{pmatrix} \frac{1}{\sqrt{6}} & \frac{1}{\sqrt{2}} & \frac{1}{\sqrt{3}} \\ -\frac{1}{\sqrt{6}} & \frac{1}{\sqrt{2}} & -\frac{1}{\sqrt{3}} \\ -\frac{2}{\sqrt{6}} & 0 & \frac{1}{\sqrt{3}} \end{pmatrix}$

Challenge

Eigen value -1 has eigenvector $\begin{pmatrix} 2 \\ -2 \\ 1 \end{pmatrix}$ and repeated eigenvalue 1 has vectors $\begin{pmatrix} -1 \\ 0 \\ 2 \end{pmatrix}$ and $\begin{pmatrix} 1 \\ 1 \\ 0 \end{pmatrix}$

Equation of Π is $2x - 2y + z = 0$

Chapter review 6

1 8

2 $\frac{1}{2}\begin{pmatrix} 2 & 0 & 0 \\ -x & 1 & 0 \\ x-6 & -1 & 2 \end{pmatrix}$

3 a An eigenvector corresponding to the eigenvalue 5 is $\begin{pmatrix} 2 \\ 1 \end{pmatrix}$.

An eigenvector corresponding to the eigenvalue -15 is $\begin{pmatrix} 1 \\ -2 \end{pmatrix}$.

b $\begin{pmatrix} \frac{2}{\sqrt{5}} & \frac{1}{\sqrt{5}} \\ \frac{1}{\sqrt{5}} & -\frac{2}{\sqrt{5}} \end{pmatrix}$

4 a $\begin{pmatrix} 2 & -1 \\ 0 & 0 \end{pmatrix}$

b $(AB)^T = \begin{pmatrix} 2 & 0 \\ -1 & 0 \end{pmatrix}$

$B^T A^T = \begin{pmatrix} 2 & -4 \\ -1 & 2 \end{pmatrix}\begin{pmatrix} 5 & 2 \\ 2 & 1 \end{pmatrix} = \begin{pmatrix} 10-8 & 4-4 \\ -5+4 & -2+2 \end{pmatrix} = \begin{pmatrix} 2 & 0 \\ -1 & 0 \end{pmatrix}$
$= (AB)^T$, as required

5 a $-1, -11$.

b $y = \frac{1}{2}x, y = -\frac{3}{4}x$

6 a $\begin{pmatrix} 0 \\ 0 \\ 1 \end{pmatrix}$ b $2, 5$

7 $\frac{x-4}{-1} = \frac{y-2}{6} = \frac{z+1}{3}$

8 a $A - \lambda I = \begin{pmatrix} 3-\lambda & 4 & -4 \\ 4 & 5-\lambda & 0 \\ -4 & 0 & 1-\lambda \end{pmatrix}$

$\begin{vmatrix} 3-\lambda & 4 & -4 \\ 4 & 5-\lambda & 0 \\ -4 & 0 & 1-\lambda \end{vmatrix}$

$= (3-\lambda)\begin{vmatrix} 5-\lambda & 0 \\ 0 & 1-\lambda \end{vmatrix} - 4\begin{vmatrix} 4 & 0 \\ -4 & 1-\lambda \end{vmatrix} + (-4)\begin{vmatrix} 4 & 5-\lambda \\ -4 & 0 \end{vmatrix}$

$= (3-\lambda)(5-\lambda)(1-\lambda) - 16 + 16\lambda - 80 + 16\lambda$
$= (3-\lambda)(5-\lambda)(1-\lambda) - 96 + 32\lambda$

$= (3 - \lambda)(5 - \lambda)(1 - \lambda) - 32(3 - \lambda)$
$= (3 - \lambda)((5 - \lambda)(1 - \lambda) - 32) = (3 - \lambda)(\lambda^2 - 6\lambda - 27)$
$= (3 - \lambda)(\lambda + 3)(\lambda - 9)$

$\det(A - \lambda I) = 0 \Rightarrow (3 - \lambda)(\lambda + 3)(\lambda - 9) = 0 \Rightarrow \lambda = 3, -3, 9$

3 is an eigenvalue of **A** and the other eigenvalues are -3 and 9.

b $\begin{pmatrix} 1 \\ -2 \\ -2 \end{pmatrix}$

c $\begin{pmatrix} \frac{1}{3} & \frac{2}{3} & \frac{2}{3} \\ -\frac{2}{3} & \frac{2}{3} & -\frac{1}{3} \\ -\frac{2}{3} & -\frac{1}{3} & \frac{2}{3} \end{pmatrix}$

9 a $\begin{pmatrix} 2 & -2 & 0 \\ -2 & 1 & 2 \\ 0 & 2 & 5 \end{pmatrix} \begin{pmatrix} 2 \\ 3 \\ -1 \end{pmatrix} = \begin{pmatrix} 4 - 6 + 0 \\ -4 + 3 - 2 \\ 0 + 6 - 5 \end{pmatrix} = \begin{pmatrix} -2 \\ -3 \\ 1 \end{pmatrix} = -1 \begin{pmatrix} 2 \\ 3 \\ -1 \end{pmatrix}$

$\begin{pmatrix} 2 \\ 3 \\ -1 \end{pmatrix}$ is an eigenvalue of A corresponding to the eigenvalue -1.

$\begin{pmatrix} 2 & -2 & 0 \\ -2 & 1 & 2 \\ 0 & 2 & 5 \end{pmatrix} \begin{pmatrix} 2 \\ -1 \\ 1 \end{pmatrix} = \begin{pmatrix} 4 + 2 + 0 \\ -4 - 1 + 2 \\ 0 - 2 + 5 \end{pmatrix} = \begin{pmatrix} 6 \\ -3 \\ 3 \end{pmatrix} = 3 \begin{pmatrix} 2 \\ -1 \\ 1 \end{pmatrix}$

$\begin{pmatrix} 2 \\ -1 \\ 1 \end{pmatrix}$, is an eigenvalue of A corresponding to the eigenvalue 3.

b $\begin{pmatrix} 1 \\ -2 \\ -4 \end{pmatrix}$

c $\begin{pmatrix} \frac{2}{\sqrt{14}} & \frac{2}{\sqrt{6}} & \frac{1}{\sqrt{21}} \\ \frac{3}{\sqrt{14}} & -\frac{1}{\sqrt{6}} & -\frac{2}{\sqrt{21}} \\ -\frac{1}{\sqrt{14}} & \frac{1}{\sqrt{6}} & -\frac{4}{\sqrt{21}} \end{pmatrix}$

10 a $\frac{1}{2x - 5} \begin{pmatrix} -2 & -2 & 2x \\ 2 & 1 & -5 \\ 3 & x - 1 & -3x \end{pmatrix}$

b $a = 19, b = 214, c = 227$

11 b $\alpha = 22, \beta = 21$

c 2

12 a $\frac{1}{9 - u} \begin{pmatrix} 1 - u & 4 & -3 - u \\ -2 & 1 & 6 - u \\ 2 & -1 & 3 \end{pmatrix}, u \neq 9$

b $a = 2.7, b = 3.1, c = 20.8$

13 a $A - \lambda I = \begin{pmatrix} 3 - \lambda & 0 & 0 \\ 1 & 1 - \lambda & 1 \\ 4 & -1 & 3 - \lambda \end{pmatrix}$

$= \begin{vmatrix} 3 - \lambda & 0 & 0 \\ 1 & 1 - \lambda & 1 \\ 4 & -1 & 3 - \lambda \end{vmatrix}$

$= (3 - \lambda) \begin{vmatrix} 1 - \lambda & 1 \\ -1 & 3 - \lambda \end{vmatrix} - 0 \begin{vmatrix} 1 & 1 \\ 4 & 3 - \lambda \end{vmatrix} + 0 \begin{vmatrix} 1 & 1 - \lambda \\ 4 & -1 \end{vmatrix}$

$= (3 - \lambda)((1 - \lambda)(3 - \lambda) + 1)$
$= (3 - \lambda)(\lambda^2 - 4\lambda + 4)$
$= (3 - \lambda)(\lambda - 2)^2$

$\det(A - \lambda I) = 0 \Rightarrow (3 - \lambda)(\lambda - 2)2 = 0 \Rightarrow 3, 2$ repeated.

There are only two distinct eigenvalues of **A**, 2 and 3.

b An eigenvalue corresponding to the eigenvalue 2 is $\begin{pmatrix} 0 \\ 1 \\ 1 \end{pmatrix}$

An eigenvalue corresponding to the eigenvalue 3 is $\begin{pmatrix} 1 \\ 4 \\ 7 \end{pmatrix}$

14 a $PP^T = \begin{pmatrix} \frac{1}{2} & -\frac{1}{2} & \frac{1}{\sqrt{2}} \\ \frac{1}{2} & -\frac{1}{2} & -\frac{1}{\sqrt{2}} \\ \frac{1}{\sqrt{2}} & \frac{1}{\sqrt{2}} & 0 \end{pmatrix} \begin{pmatrix} \frac{1}{2} & \frac{1}{2} & \frac{1}{\sqrt{2}} \\ -\frac{1}{2} & -\frac{1}{2} & \frac{1}{\sqrt{2}} \\ \frac{1}{\sqrt{2}} & -\frac{1}{\sqrt{2}} & 0 \end{pmatrix}$

$= \begin{pmatrix} \frac{1}{4} + \frac{1}{4} + \frac{1}{2} & \frac{1}{4} + \frac{1}{4} - \frac{1}{2} & \frac{1}{2\sqrt{2}} - \frac{1}{2\sqrt{2}} + 0 \\ \frac{1}{4} + \frac{1}{4} - \frac{1}{2} & \frac{1}{4} + \frac{1}{4} + \frac{1}{2} & \frac{1}{2\sqrt{2}} - \frac{1}{2\sqrt{2}} + 0 \\ \frac{1}{2\sqrt{2}} - \frac{1}{2\sqrt{2}} + 0 & \frac{1}{2\sqrt{2}} - \frac{1}{2\sqrt{2}} + 0 & \frac{1}{2} + \frac{1}{2} + 0 \end{pmatrix}$

$= \begin{pmatrix} 1 & 0 & 0 \\ 0 & 1 & 0 \\ 0 & 0 & 1 \end{pmatrix} = I$

Hence **P** is orthogonal.

b $y = 0$

15 a $-2, 2, 3$

b $\begin{pmatrix} 3 & -3 & 6 \\ 0 & 2 & -8 \\ 0 & 0 & -2 \end{pmatrix} \begin{pmatrix} 3 \\ 1 \\ 0 \end{pmatrix} = \begin{pmatrix} 9 - 3 \\ 2 \\ 0 \end{pmatrix} = \begin{pmatrix} 6 \\ 2 \\ 0 \end{pmatrix} = 2 \begin{pmatrix} 3 \\ 1 \\ 0 \end{pmatrix}$

$\begin{pmatrix} 3 \\ 1 \\ 0 \end{pmatrix}$ is an eigenvector of **A** corresponding to the eigenvalue 2.

c $\begin{pmatrix} 7 & -6 & 2 \\ 1 & 2 & 3 \\ 1 & -3 & 2 \end{pmatrix} \begin{pmatrix} 3 \\ 1 \\ 0 \end{pmatrix} = \begin{pmatrix} 21 - 6 \\ 3 + 2 \\ 3 - 3 \end{pmatrix} = \begin{pmatrix} 15 \\ 5 \\ 0 \end{pmatrix} = 5 \begin{pmatrix} 3 \\ 1 \\ 0 \end{pmatrix}$

$\begin{pmatrix} 3 \\ 1 \\ 0 \end{pmatrix}$ is an eigenvector of **B** corresponding to the eigenvalue 5.

d $\begin{pmatrix} 3 \\ 1 \\ 0 \end{pmatrix}$ is an eigenvector of **AB** corresponding to the eigenvalue 10.

16 a $\frac{1}{7} \begin{pmatrix} 5 & 2 & -1 \\ -17 & 3 & 2 \\ 2 & -2 & 1 \end{pmatrix}$
b $\frac{x}{10} = \frac{y}{1} = \frac{z}{-3}$

Challenge

a $AB = \begin{pmatrix} ae + bg & af + bh \\ ce + dg & cf + dh \end{pmatrix}$ $\text{tr}(AB) = ae + bg + cf + dh$

$BA = \begin{pmatrix} ae + cf & be + df \\ ag + ch & bg + dh \end{pmatrix}$ $\text{tr}(BA) = ae + bg + cf + dh$

So $\text{tr}(AB) = \text{tr}(BA)$

b Using result from part **a**
$\text{tr}(P^{-1}MP) = \text{tr}(P^{-1}[MP]) = \text{tr}([MP]P^{-1}) = \text{tr}(M)$
and $\text{tr}(P^{-1}MP) = p + q$ so $\text{tr}(M) = p + q$

Review exercise 2

1 $2\sqrt{2}$

2 $\frac{1}{3}$

4 a $5i - 3j - 4k$
b 100
c 50

5 a $-i + 8j - 4k$
b $3i + j - k$
c $n_1 \times n_2 = \begin{vmatrix} i & j & k \\ -1 & 8 & -4 \\ 3 & 1 & -1 \end{vmatrix}$

$$= \begin{vmatrix} 8 & 4 \\ 1 & -1 \end{vmatrix} \mathbf{i} - \begin{vmatrix} -1 & -4 \\ 3 & -1 \end{vmatrix} \mathbf{j} + \begin{vmatrix} -1 & -18 \\ 3 & 1 \end{vmatrix} \mathbf{k}$$
$$= -4\mathbf{i} - 13\mathbf{j} - 25\mathbf{k} = -1(4\mathbf{i} + 13\mathbf{j} + 25\mathbf{k})$$

d $\mathbf{r} = \mathbf{i} + \mathbf{j} + \mathbf{k} + t(4\mathbf{i} + 13\mathbf{j} + 25\mathbf{k})$

6 a $\mathbf{i} + 4\mathbf{j} + 2\mathbf{k}$
  b $\mathbf{r}.(\mathbf{i} + 4\mathbf{j} - 2\mathbf{k}) = 7$
  c 2

7 a $\overrightarrow{OA}\,\overrightarrow{OB} = (4\mathbf{i} + \mathbf{j} - 7\mathbf{k}), (2\mathbf{i} + 6\mathbf{j} + 2\mathbf{k})$
    $= 4 \times 2 + 1 \times 6 + (-7) \times 2$
    $= 8 + 6 - 14 = 0$
  Hence $\angle AOB = 90°$, as required.
  b $\mathbf{r} = 4\mathbf{i} + \mathbf{j} - 7\mathbf{k} + \lambda(-3\mathbf{i} + 2\mathbf{j} + 8\mathbf{k})$
  c $\mathbf{r}.(2\mathbf{i} - \mathbf{j} + \mathbf{k}) = 0$

8 a $a(4\mathbf{i} + \mathbf{j} + 2\mathbf{k})(\mathbf{i} - 5\mathbf{j} + 3\mathbf{k}) = a(4 \times 1 + 1 \times (-5) + 2 \times 3)$
    $= a(4 - 5 + 6) = 5a$
  Hence A lies in the plane Π, as required.
  b $\overrightarrow{BA} = a(4\mathbf{i} + \mathbf{j} + 2\mathbf{k}) - a(2\mathbf{i} + 11\mathbf{j} - 4\mathbf{k})$
    $= a(2\mathbf{i} - 10\mathbf{j} + 6\mathbf{k})$
    $\overrightarrow{BA} = 2a(\mathbf{i} - 5\mathbf{j} + 3\mathbf{k})$
    $\overrightarrow{BA} =$ is parallel to the vector $\mathbf{i} - 5\mathbf{j} + 3\mathbf{k}$, which is perpendicular to the plane Π.
    Hence BA is perpendicular to the plane Π, as required.
  c 22.3° (nearest one tenth of a degree)

9 a $\begin{pmatrix} -30 \\ -15 \\ 45 \end{pmatrix}$
  b $\mathbf{r} = \begin{pmatrix} 3 \\ 1 \\ 2 \end{pmatrix} + \lambda \begin{pmatrix} -2 \\ -1 \\ 3 \end{pmatrix}$
  d 35

10 a $6\mathbf{i} + \mathbf{j} - 4\mathbf{k}$
  b An equation for Π_1 has the form
    $\mathbf{r}.(6\mathbf{i} + \mathbf{j} - 4\mathbf{k}) = p$
    $p = (\mathbf{i} + 6\mathbf{j} - \mathbf{k}).(6\mathbf{i} + \mathbf{j} - 4\mathbf{k}) = 6 + 6 + 4 = 16$
    A vector equation of Π_1 is
    $\mathbf{r}.(6\mathbf{i} + \mathbf{j} - 4\mathbf{k}) = 16$
    A Cartesian equation of Π_1 is given by
    $(x\mathbf{i} + y\mathbf{j} + z\mathbf{k}).(6\mathbf{i} + \mathbf{j} - 4\mathbf{k}) = 16$
    $6x + y - 4z = 16$, as required
  c -2
  d $(\mathbf{r} - (-3\mathbf{i} + 6\mathbf{j} - 7\mathbf{k})) \times (-9\mathbf{i} + 10\mathbf{j} - 11\mathbf{k}) = 0$

11 a $\begin{pmatrix} 36 \\ 12 \\ 9 \end{pmatrix}$ **b** $\mathbf{r}.\begin{pmatrix} 12 \\ 4 \\ 3 \end{pmatrix} = 3$ **c** $(1, -3, 1)$

12 a $\begin{pmatrix} -5 - 4c \\ -6 - 5c \\ 1 \end{pmatrix}$
  b $c = -2$
  c $\mathbf{r}.\begin{pmatrix} 3 \\ 4 \\ 1 \end{pmatrix} = 7$
  d $\begin{pmatrix} 5 \\ -3 \\ 8 \end{pmatrix}$

13 a $-15\mathbf{i} - 10\mathbf{j} - 10\mathbf{k}$
  b $\mathbf{r}.(3\mathbf{i} + 2\mathbf{j} + 2\mathbf{k}) = 7$
  c $(\mathbf{r} - (3\mathbf{i} - \mathbf{j})) \times (-2\mathbf{i} + \mathbf{j} + 2\mathbf{k}) = 0$
  d $\left(\dfrac{13}{9}, -\dfrac{2}{9}, \dfrac{14}{9}\right)$

14 a $3\mathbf{i} - 6\mathbf{j} + 6\mathbf{k}$
  c 14
  d $\mathbf{r} = 4\mathbf{i} + 3\mathbf{j} + \mathbf{k} + t(\mathbf{j} + \mathbf{k})$
  e $4\mathbf{i} + \mathbf{j} - \mathbf{k}$

15 a $\begin{pmatrix} -6 \\ 2 \\ -4 \end{pmatrix}$
  b A vector equation of Π is $\mathbf{r}.\begin{pmatrix} -6 \\ 2 \\ -4 \end{pmatrix} = \begin{pmatrix} 2 \\ -1 \\ 0 \end{pmatrix}.\begin{pmatrix} -6 \\ 2 \\ -4 \end{pmatrix}$
    $= -12 - 2 = -14$
  Let $\mathbf{r} = \begin{pmatrix} x \\ y \\ z \end{pmatrix}$
  $\mathbf{r}.\begin{pmatrix} -6 \\ 2 \\ -4 \end{pmatrix} = \begin{pmatrix} x \\ y \\ z \end{pmatrix}.\begin{pmatrix} -6 \\ 2 \\ -4 \end{pmatrix} = -6x + 2y - 4z = -14$
  A Cartesian equation of Π is
  $-6x + 2y - 4z = -14$
  Dividing throughout by -2
  $3x - y + 2z = 7$, as required
  c $(-1, 8, 9)$
  d $\overrightarrow{BT} = \overrightarrow{OT} - \overrightarrow{OB} = \begin{pmatrix} -1 \\ 8 \\ 9 \end{pmatrix} - \begin{pmatrix} 1 \\ 2 \\ 3 \end{pmatrix} = \begin{pmatrix} -2 \\ 6 \\ 6 \end{pmatrix}$
  From part **a**
  $\overrightarrow{AB} = \begin{pmatrix} -1 \\ 3 \\ 3 \end{pmatrix}$
  Hence
  $\overrightarrow{AB} = \dfrac{1}{2}\overrightarrow{BT}$ and AB is parallel to BT.
  Hence A, B and T lie in the same straight line.

16 a $\mathbf{r} = \begin{pmatrix} 1 \\ 2 \\ 3 \end{pmatrix} + t\begin{pmatrix} -3 \\ 5 \\ 1 \end{pmatrix}$ **b** $\dfrac{11}{6}$
  c $\mathbf{r}.\begin{pmatrix} -3 \\ 5 \\ 1 \end{pmatrix} = 21$ **d** $\left(\dfrac{68}{35}, \dfrac{15}{35}, \dfrac{94}{35}\right)$
  f $\left(\dfrac{101}{35}, -\dfrac{40}{35}, \dfrac{83}{35}\right)$

17 a $2\mathbf{i} - 3\mathbf{j} - 2\mathbf{k}$ **b** $\dfrac{\sqrt{17}}{2}$
  c $\mathbf{r}.(2\mathbf{i} - 3\mathbf{j} - 2\mathbf{k}) = 27$ **d** $2x - 3y - 2z = 27$
  e $\dfrac{7}{\sqrt{17}}$ **f** 3.2° (1 d.p.)

18 a Equating the x components
    $-1 - 2s = -t$ (1)
    Equating the y components
    $2 + s = -1 + t$ (2)
    (1) + (2) $1 - s = -1 \Rightarrow s = 2$
    Substitute $s = 2$ into (2) $4 = -1 + t \Rightarrow t = 5$
    Checking the z components
    For l_1: $-4 + 3s = -4 + 6 = 2$
    For l_2: $7 - t = 7 - 5 = 2$
    These are the same, so l_1 and l_2 intersect.
    The lines l_1 and l_2 are parallel to
    $-2\mathbf{i} + \mathbf{j} + 3\mathbf{k}$ and $-\mathbf{i} + \mathbf{j} - \mathbf{k}$ respectively.
    $(-2\mathbf{i} + \mathbf{j} + 3\mathbf{k}).(-1 + \mathbf{j} - \mathbf{k}) = 2 + 1 - 3 = 0$
    Hence l_1 is perpendicular to l_2.
  b $\mathbf{r} = -5\mathbf{i} + 4\mathbf{j} + 2\mathbf{k} + u(9\mathbf{i} + (\lambda - 4)\mathbf{j} - 5\mathbf{k})$
  c $\dfrac{5\lambda + 11}{\sqrt{42}\sqrt{(\lambda^2 - 8\lambda + 122)}}$
  d $-\dfrac{11}{5}$

19 a $\sqrt{10}$ **b** $\mathbf{r}.(\mathbf{i} - 2\mathbf{j} - 2\mathbf{k}) = 26$
  c $2\mathbf{i} + \mathbf{j} + 3\mathbf{k}$

20 a Equating the x components
$2t = 1 - 2u$ (1)
Equating the y components
$1 + t = 1 + u \Rightarrow t = u$ (2)
Substituting (2) into (1)
$2u = 1 - 2u \Rightarrow u = \frac{1}{4}$
As $t = u$, $t = \frac{1}{4}$
Checking the z components
For l: $3 - t = 3 - \frac{1}{4} = \frac{11}{4}$
For m: $-1 + u = -1 + \frac{1}{4} = -\frac{3}{4}$
$\frac{11}{4} \neq -\frac{3}{4}$, so the lines do not intersect.

b $(1 - 2t_1 - 2u_1)\mathbf{i} + (-t_1 + u_1)\mathbf{j} + (-4 + t_1 + u_1)\mathbf{k}$

c $u_1 = \frac{3}{5}$
$t_1 = \frac{3}{5}$

21 $\mathbf{A}^n = \begin{pmatrix} 1 & n & \frac{1}{2}(n^2 + 3n) \\ 0 & 1 & n \\ 0 & 0 & 1 \end{pmatrix}$

Let $n = 1$

$\mathbf{A}^1 = \begin{pmatrix} 1 & 1 & \frac{1}{2}(1^2 + 3 \times 1) \\ 0 & 1 & 1 \\ 0 & 0 & 1 \end{pmatrix}$

$= \begin{pmatrix} 1 & 1 & 2 \\ 0 & 1 & 1 \\ 0 & 0 & 1 \end{pmatrix} = \mathbf{A}$

The formula is true for $n = 1$
Assume the formula is true for $n = k$
That is

$\mathbf{A}^k = \begin{pmatrix} 1 & k & \frac{1}{2}(k^2 + 3k) \\ 0 & 1 & k \\ 0 & 0 & 1 \end{pmatrix}$

$\mathbf{A}^{k+1} = \mathbf{A}^k.\mathbf{A}$

$= \begin{pmatrix} 1 & k & \frac{1}{2}(k^2 + 3k) \\ 0 & 1 & k \\ 0 & 0 & 1 \end{pmatrix} \begin{pmatrix} 1 & 1 & 2 \\ 0 & 1 & 1 \\ 0 & 0 & 1 \end{pmatrix}$

$= \begin{pmatrix} 1 & 1+k & 2 + k + \frac{1}{2}(k^2 + 3k) \\ 0 & 1 & 1+k \\ 0 & 0 & 1 \end{pmatrix}$

$2 + k + \frac{1}{2}(k^2 + 3k) = \frac{1}{2}k^2 + \frac{3k}{2} + k + 2$

$= \frac{1}{2}(k^2 + 5k + 4)$

$= \frac{1}{2}(k^2 + 2k + 1 + 3k + 3)$

$= \frac{1}{2}((k+1)^2 + 3(k+1))$

$\mathbf{A}^{k+1} = \begin{pmatrix} 1 & k+1 & \frac{1}{2}((k+1)^2 + 3(k+1)) \\ 0 & 1 & k+1 \\ 0 & 0 & 1 \end{pmatrix}$

This is the formula with $k + 1$ substituted for n.
Hence, the formula is true for $n = 1$, and, if it is true for $n = k$, then it is true for $n = k + 1$.
By mathematical induction the formula is true for all positive integers n.

22 a $k = 3, 6$

b $\frac{1}{-k^2 + 9k - 18} \begin{pmatrix} -k & -2 & k-2 \\ 9k & 18 & 2k^2 \\ 9 & 9-k & 2k \end{pmatrix}$

23 a 3
b 18
c $a = 2\sqrt{2}$
$b = 2\sqrt{2}$
$c = -2\sqrt{2}$
d $54\sqrt{2}$

24 a $\begin{pmatrix} 3 & 3 & 7 \\ 1 & 4 & 4 \\ 3 & 1 & 6 \end{pmatrix}$

b $\mathbf{A}^3 - 5\mathbf{A}^2 + 6\mathbf{A} - \mathbf{I}$
$= \begin{pmatrix} 10 & 9 & 23 \\ 5 & 9 & 14 \\ 9 & 5 & 19 \end{pmatrix} - 5\begin{pmatrix} 3 & 3 & 7 \\ 1 & 4 & 4 \\ 3 & 1 & 6 \end{pmatrix} + 6\begin{pmatrix} 1 & 1 & 2 \\ 0 & 2 & 1 \\ 1 & 0 & 2 \end{pmatrix} - \begin{pmatrix} 1 & 0 & 0 \\ 0 & 1 & 0 \\ 0 & 0 & 1 \end{pmatrix}$

$= \begin{pmatrix} 10 - 15 + 6 - 1 & 9 - 15 + 6 - 0 & 23 - 35 + 12 - 0 \\ 5 - 5 + 0 - 0 & 9 - 20 + 12 - 1 & 14 - 20 + 6 - 0 \\ 9 - 15 + 6 - 0 & 5 - 5 + 0 - 0 & 19 - 30 + 12 - 1 \end{pmatrix}$

$= \begin{pmatrix} 0 & 0 & 0 \\ 0 & 0 & 0 \\ 0 & 0 & 0 \end{pmatrix} = 0$

c $\mathbf{A}^3 - 5\mathbf{A}^2 + 6\mathbf{A} - \mathbf{I} = 0$
$\mathbf{A}^3 - 5\mathbf{A}^2 + 6\mathbf{A} = \mathbf{I}$
$\mathbf{A}(\mathbf{A}^2 - 5\mathbf{A} + 6) = \mathbf{I}$
$\mathbf{A}(\mathbf{A} - 2)(\mathbf{A} - 3) = \mathbf{I}$

d $\begin{pmatrix} 4 & -2 & -3 \\ 1 & 0 & -1 \\ -2 & 1 & 2 \end{pmatrix}$

25 a $\begin{pmatrix} 1 & 0 & 0 \\ 0 & 4 & 3 \\ 0 & 0 & 1 \end{pmatrix}$ **b** $\begin{pmatrix} 1 & 0 & 0 \\ 0 & 8 & 7 \\ 0 & 0 & 1 \end{pmatrix}$

c $\mathbf{A}^n = \begin{pmatrix} 1 & 0 & 0 \\ 0 & 2^n & 2^n - 1 \\ 0 & 0 & 1 \end{pmatrix}$

Let $n = 1$
$\mathbf{A}^1 = \begin{pmatrix} 1 & 0 & 0 \\ 0 & 2^1 & 2^1 - 1 \\ 0 & 0 & 1 \end{pmatrix}$

$= \begin{pmatrix} 1 & 0 & 0 \\ 0 & 2 & 1 \\ 0 & 0 & 1 \end{pmatrix} = \mathbf{A}$

The formula is true for $n = 1$.
Assume the formula is true for $n = k$.
That is

$\mathbf{A}^k = \begin{pmatrix} 1 & 0 & 0 \\ 0 & 2^k & 2^k - 1 \\ 0 & 0 & 1 \end{pmatrix}$

$\mathbf{A}^{k+1} = \mathbf{A}^k.\mathbf{A}$

$= \begin{pmatrix} 1 & 0 & 0 \\ 0 & 2^k & 2^k - 1 \\ 0 & 0 & 1 \end{pmatrix}\begin{pmatrix} 1 & 0 & 0 \\ 0 & 2 & 1 \\ 0 & 0 & 1 \end{pmatrix}$

$= \begin{pmatrix} 1 & 0 & 0 \\ 0 & 2^k \times 2 & 2^k + 2^k - 1 \\ 0 & 0 & 1 \end{pmatrix}$

$2^k \times 2 = 2^k \times 2^1 = 2^{k+1}$
$2^k + 2^k - 1 = 2 \times 2^k - 1 = 2^{k+1} - 1$

This is the formula with $k + 1$ substituted for n.
Hence the formula is true for $n = 1$, and, if it is true for $n = k$, then it is true for $n = k + 1$.
By mathematical induction the formula is true for all positive integers n.

d $\begin{pmatrix} 1 & 0 & 0 \\ 0 & 2^{-n} & 2^{-n}-1 \\ 0 & 0 & 1 \end{pmatrix}$

26 a $\det \mathbf{A} = 3 \begin{vmatrix} 1 & 1 \\ 3 & u \end{vmatrix} - 1 \begin{vmatrix} 1 & 1 \\ 5 & u \end{vmatrix} + (-1)\begin{vmatrix} 1 & 1 \\ 5 & 3 \end{vmatrix}$

$\quad\quad = 3(u-3) - 1(u-5) - 1 \times (-2)$
$\quad\quad = 3u - 9 - u + 5 + 2 = 2u - 2$
$\quad\quad = 2(u-1)$, as required

b $\dfrac{1}{-(u-1)} \begin{pmatrix} u-3 & -u-3 & 2 \\ -u+5 & 3u+5 & -4 \\ -2 & -4 & 2 \end{pmatrix}$

c $a = 1.2, b = 20.4, c = 0.2$

27 a $a = -4$
$\quad\quad b = -3$
$\quad\quad c = 0$
b -1
c $x = 2y$

28 a $\begin{pmatrix} -1 & 3 & 1 \\ 0 & 1 & 5 \\ 2 & 4 & 2 \end{pmatrix}$

b $z = 8$

29 a $\begin{pmatrix} 1 & -1 & 2 \\ 0 & 1 & -1 \\ 0 & 0 & 1 \end{pmatrix}$

b 9

c $\begin{pmatrix} \frac{1}{3} \\ -\frac{1}{3} \\ 1 \end{pmatrix}$

30 a $-6, 1$
b $y = \dfrac{3}{5}x$

31 a The image under T of the line with equation $y = 2x + 1$ is the point with coordinates $(2, -1)$.
b $-2, 3$
c $y = \dfrac{1}{2}x, y = 2x$

32 a $\lambda_1 = 2, \lambda_2 = 3$
b $\dfrac{1}{6}\begin{pmatrix} 1 & 2 \\ -1 & 4 \end{pmatrix}$
d $y = \dfrac{1}{2}x, y = x$.

33 The eigenvalues of the matrix are $-2, 0$ and 1.
An eigenvector corresponding to the eigenvalue
-2 is $\begin{pmatrix} 4 \\ 3 \\ -7 \end{pmatrix}$
An eigenvector corresponding to the eigenvalue
0 is $\begin{pmatrix} 10 \\ 3 \\ -11 \end{pmatrix}$
An eigenvector corresponding to the eigenvalue
1 is $\begin{pmatrix} 1 \\ 0 \\ -1 \end{pmatrix}$

34 a 2
b $p = 4$
$\quad\quad q = -2$
c $l = 2, m = 1, n = 0$

35 a Substitute $\lambda = 3$ into $\begin{vmatrix} 5-\lambda & 1 & -2 \\ -1 & 6-\lambda & 1 \\ 0 & 1 & 3-\lambda \end{vmatrix}$

$\begin{vmatrix} 2 & 1 & -2 \\ -1 & 3 & 1 \\ 0 & 1 & 0 \end{vmatrix} = 2\begin{vmatrix} 3 & 1 \\ 1 & 0 \end{vmatrix} - 1\begin{vmatrix} -1 & 1 \\ 0 & 0 \end{vmatrix} + (-2)\begin{vmatrix} -1 & 3 \\ 0 & 1 \end{vmatrix}$

$\quad\quad = 2 \times (-1) - 1 \times 0 + 2(-2) \times (-1)$
$\quad\quad = -2 + 2 = 0$

Hence 3 is an eigenvalue of $\mathbf{A}$.

b $5, 6$

c $\begin{pmatrix} \frac{1}{\sqrt{2}} \\ 0 \\ \frac{1}{\sqrt{2}} \end{pmatrix}$

36 a $\det \mathbf{A} = 3 \begin{vmatrix} 0 & 2 \\ 2 & k \end{vmatrix} - 2 \begin{vmatrix} 2 & 2 \\ 4 & k \end{vmatrix} + 4 \begin{vmatrix} 2 & 0 \\ 4 & 2 \end{vmatrix}$

$\quad\quad = 3(0 - 4) - 2(2k - 8) + 4(4 - 0)$
$\quad\quad = -12 - 4k + 16 + 16 = 20 - 4k$, as required

b $\dfrac{1}{20 - 4k}\begin{pmatrix} -4 & -2k+8 & 4 \\ -2k+8 & 3k-16 & 2 \\ 4 & 2 & -4 \end{pmatrix}$

c -1

d $\begin{pmatrix} 2 \\ 1 \\ 2 \end{pmatrix}$

37 a $\begin{pmatrix} 1 & 0 & 4 \\ 0 & 5 & 4 \\ 4 & 4 & 3 \end{pmatrix}\begin{pmatrix} 2 \\ -2 \\ 1 \end{pmatrix} = \begin{pmatrix} 2+4 \\ -10+4 \\ 8-8+3 \end{pmatrix} = \begin{pmatrix} 6 \\ -6 \\ 3 \end{pmatrix} = 3\begin{pmatrix} 2 \\ -2 \\ 1 \end{pmatrix}$

Hence $\begin{pmatrix} 2 \\ -2 \\ 1 \end{pmatrix}$ is an eigenvector of $\mathbf{A}$ and the corresponding eigenvalue is 3.

b Substitute $\lambda = 9$ into
$\begin{vmatrix} 1-\lambda & 0 & 4 \\ 0 & 5-\lambda & 4 \\ 4 & 4 & 3-\lambda \end{vmatrix}$

$\begin{vmatrix} 1-9 & 0 & 4 \\ 0 & 5-9 & 4 \\ 4 & 4 & 3-9 \end{vmatrix} = \begin{vmatrix} -8 & 0 & 4 \\ 0 & -4 & 4 \\ 4 & 4 & -6 \end{vmatrix}$

$= (-8)\begin{vmatrix} -4 & 4 \\ 4 & -6 \end{vmatrix} - 0\begin{vmatrix} 0 & 4 \\ 4 & -6 \end{vmatrix} + 4\begin{vmatrix} 0 & -4 \\ 4 & 4 \end{vmatrix}$

$= (-8)(24 - 16) - 0 + 4(0 + 16)$
$= -8 \times 8 + 4 \times 16 = -64 + 64 = 0$

Hence 9 is an eigenvalue of $\mathbf{A}$.
To find an eigenvector corresponding to 9.

$\begin{pmatrix} 1 & 0 & 4 \\ 0 & 5 & 4 \\ 4 & 4 & 3 \end{pmatrix}\begin{pmatrix} x \\ y \\ z \end{pmatrix} = 9\begin{pmatrix} x \\ y \\ z \end{pmatrix}$

$\begin{pmatrix} x + 4z \\ 5y + 4z \\ 4x + 4y + 3z \end{pmatrix} = \begin{pmatrix} 9x \\ 9y \\ 9z \end{pmatrix}$

Equating the top elements
$x + 4z = 9x \Rightarrow -8x + 4z = 0 \Rightarrow z = 2x$
Let $x = 1$, then $z = 2$
Equating the middle elements
$5y + 4z = 9y \Rightarrow 4z = 4y \Rightarrow y = z$
As $z = 2, y = 2$
An eigenvector is corresponding to the eigenvalue

9 is $\begin{pmatrix} 1 \\ 2 \\ 2 \end{pmatrix}$.

242 ANSWERS

  c $P = \begin{pmatrix} \frac{2}{3} & \frac{1}{3} & \frac{2}{3} \\ -\frac{2}{3} & \frac{2}{3} & \frac{1}{3} \\ \frac{1}{3} & \frac{2}{3} & -\frac{2}{3} \end{pmatrix}$ $D = \begin{pmatrix} 3 & 0 & 0 \\ 0 & 9 & 0 \\ 0 & 0 & -3 \end{pmatrix}$ d $P = \begin{pmatrix} \frac{2}{3} & -\frac{1}{3} & \frac{2}{3} \\ \frac{2}{3} & \frac{2}{3} & -\frac{1}{3} \\ -\frac{1}{3} & \frac{2}{3} & \frac{2}{3} \end{pmatrix}$

38 a 1 **b** $\begin{pmatrix} \frac{4}{\sqrt{21}} \\ \frac{1}{\sqrt{21}} \\ \frac{22}{\sqrt{21}} \end{pmatrix}$

$x_1 \cdot x_2 = \begin{pmatrix} \frac{2}{3} \\ \frac{2}{3} \\ -\frac{1}{3} \end{pmatrix} \cdot \begin{pmatrix} -\frac{1}{3} \\ \frac{2}{3} \\ \frac{2}{3} \end{pmatrix} = \frac{2}{3} \times \left(-\frac{1}{3}\right) + \frac{2}{3} \times \frac{2}{3} + \left(-\frac{1}{3}\right) \times \frac{2}{3}$

  c $\begin{pmatrix} \frac{1}{\sqrt{14}} & \frac{1}{\sqrt{6}} & \frac{4}{\sqrt{21}} \\ \frac{2}{\sqrt{14}} & \frac{22}{\sqrt{6}} & \frac{1}{\sqrt{21}} \\ \frac{3}{\sqrt{14}} & \frac{1}{\sqrt{6}} & \frac{22}{\sqrt{21}} \end{pmatrix}$ d $\begin{pmatrix} 1 & 0 & 0 \\ 0 & -1 & 0 \\ 0 & 0 & 8 \end{pmatrix}$

$= -\frac{2}{9} + \frac{4}{9} - \frac{2}{9} = 0$

Hence x_1 is orthogonal (perpendicular) to x_2.
The matrix **P** is an orthogonal matrix.

39 a The eigenvalues of **M** are 21, 2 and 3.
  An eigenvector corresponding to the eigenvalue
  21 is $\begin{pmatrix} 1 \\ 0 \\ 22 \end{pmatrix}$

Challenge

1 a $l = m = \frac{\sqrt{6}}{4}, n = \frac{1}{2}$
  b $l = \cos\theta\sin\varphi, m = \sin\theta\sin\varphi, n = \cos\varphi$

2 a $AB = \begin{pmatrix} ae + bg & af + bh \\ ce + dg & cf + dh \end{pmatrix}$, tr(**AB**) = $ae + bg + cf + dh$

  An eigenvector corresponding to the eigenvalue
  2 is $\begin{pmatrix} 1 \\ 21 \\ 1 \end{pmatrix}$

$BA = \begin{pmatrix} ae + cf & be + df \\ ag + ch & bg + dh \end{pmatrix}$, tr(**BA**) = $ae + cf + bg + dh$

So tr(**AB**) = tr(**BA**).

  An eigenvector corresponding to the eigenvalue
  3 is $\begin{pmatrix} 1 \\ 0 \\ 2 \end{pmatrix}$

  b Using result from part **a**,
  tr($P^{-1}MP$) = tr($P^{-1}(MP)$) = tr($(MP)P^{-1}$) = tr(**M**)
  and tr($P^{-1}MP$) = $p + q$, so tr(**M**) = $p + q$.

  b $x = \frac{y}{2} = \frac{z}{10}$.

Exam practice

40 a $\begin{vmatrix} 6-\lambda & -2 & 2 \\ -2 & 5-\lambda & 0 \\ 2 & 0 & 7-\lambda \end{vmatrix}$

1 a $\pm\frac{p}{e} = 8$ $\pm pe = 3$

$\frac{p}{e} \times pe = p^2 = 18 \Rightarrow p = 3\sqrt{2}$

$= (6-\lambda)\begin{vmatrix} 5-\lambda & 0 \\ 0 & 7-\lambda \end{vmatrix} - (-2)\begin{vmatrix} -2 & 0 \\ 2 & 7-\lambda \end{vmatrix} + 2\begin{vmatrix} -2 & 5-\lambda \\ 2 & 0 \end{vmatrix}$

  b $q = 3$

2 a $1 + 2\sinh^2 x = 1 + 2\left(\frac{e^x - e^{-x}}{2}\right)^2$

$= (6-\lambda)(5-\lambda)(7-\lambda) + 2(-14 + 2\lambda) + 2(-10 + 2\lambda)$
$= (6-\lambda)(5-\lambda)(7-\lambda) - 28 + 4\lambda - 20 + 4\lambda$
$= (6-\lambda)(5-\lambda)(7-\lambda) + 2\lambda - 48$
$= (6-\lambda)(5-\lambda)(7-\lambda) + 8(\lambda - 6)$
$= (6-\lambda)(35 - 12\lambda + \lambda^2 - 8)$
$= (6-\lambda)(27 - 12\lambda + \lambda^2)$
$= (6-\lambda)(3-\lambda)(9-\lambda)$
$(6-\lambda)(3-\lambda)(9-\lambda) = 0$
$\lambda = 3, 6, 9$

$= \frac{2 + e^{2x} - 2e^x e^{-x} + e^{-2x}}{2} = \frac{e^{2x} + e^{-2x}}{2}$

$\cosh 2x = \frac{e^{2x} + e^{-2x}}{2} \Rightarrow$ LHS = RHS

  b $\ln(-3 + \sqrt{10})$, $\ln\left(\frac{5 + \sqrt{29}}{2}\right)$

3 a $-(9 - x^2)^{\frac{1}{2}} + C$

  b $I_n = \int x^{n-1} \frac{x}{\sqrt{9 - x^2}} dx$

$= -x^{n-1}\sqrt{9 - x^2} + \int (n-1)x^{n-2}\sqrt{9 - x^2}\, dx$

$I_n = \left[-x^{n-1}\sqrt{9-x^2}\right]_0^3 + \int_0^3 \frac{(n-1)x^{n-2}(9-x^2)}{\sqrt{9-x^2}}\, dx$

9 is an eigenvalue of the matrix.

  b 3, 6
  c A normalised eigenvector corresponding to
  3 is $\begin{pmatrix} \frac{2}{3} \\ \frac{2}{3} \\ 2\frac{1}{3} \end{pmatrix}$

$I_n = 0 + 9(n-1)I_{n-2} - (n-1)I_n \Rightarrow nI_n$

$= 9(n-1)I_{n-2} \Rightarrow I_n = \frac{9(n-1)I_{n-2}}{n}$

  c $\frac{243\pi}{16}$

4 a $\operatorname{arsinh} 2x + \frac{2x}{\sqrt{1+4x^2}}$

  A normalised eigenvector corresponding to
  6 is $\begin{pmatrix} 2\frac{1}{3} \\ \frac{2}{3} \\ \frac{2}{3} \end{pmatrix}$

  b $\sqrt{2}\ln(3 + 2\sqrt{2} - 1)$

5 $A = \frac{32\pi}{3}(\cosh^3 1 - 1)$

  A normalised eigenvector corresponding to
  9 is $\begin{pmatrix} \frac{2}{3} \\ 2\frac{1}{3} \\ \frac{2}{3} \end{pmatrix}$

6 a $\vec{AC} \times \vec{BC} = 10\mathbf{i} - 15\mathbf{j} + 30\mathbf{k}$
  b 17.5
  c $\mathbf{r}.(2\mathbf{i} - 3\mathbf{j} + 6\mathbf{k}) = -4$ or correct multiple

Online Worked solutions are available in SolutionBank.

7 i $\dfrac{\pi}{3}$

ii a $5\cosh x - 4\sinh x = 5\left(\dfrac{e^x + e^{-x}}{2}\right) - 4\left(\dfrac{e^x - e^{-x}}{2}\right)$
$= \dfrac{e^x + 9e^{-x}}{2} = \dfrac{e^{2x} + 9}{2e^x}$

b $\dfrac{2}{3}\arctan\left(\dfrac{e^x}{3}\right) + C$

8 a $\lambda = 4$

b Uses the third row and $\lambda = 4$ to give $6k + 6 = 24 \Rightarrow k = 3$

c $\begin{vmatrix} 1-\lambda & 0 & 3 \\ 0 & -2-\lambda & 1 \\ 3 & 0 & 1-\lambda \end{vmatrix} = 0$

$\Rightarrow (1-\lambda)[(-2-\lambda)(1-\lambda) - 0] - 0[0(1-\lambda) - 3]$
$\qquad + 3[0 - 3(-2-\lambda)] = 0$
$\Rightarrow \lambda^3 - 12\lambda - 16 = 0$
$\Rightarrow (\lambda + 2)(\lambda^2 - 2\lambda - 8) = 0$
$\Rightarrow (\lambda + 2)(\lambda + 2)(\lambda - 4) = 0 \Rightarrow \lambda = -2, 4$

d $\dfrac{x+1}{13} = \dfrac{y+1}{10} = \dfrac{z-5}{7}$

INDEX

A
alternating signs rule 147–9
angles
    between lines/planes 122–4
    between vectors 101–5, 111–12
answers to questions 204–43
arc length 79–82
areas 82–7, 106–10
asymptotes 5–6, 20–2, 27

C
Cartesian equations
    ellipses 18–19
    hyperbolas 20–2, 39
    lines 115–16, 157–8, 171–2
    parabolas 39
    planes 117, 118–19, 120–1
    surface of revolution 83–4
characteristic equation 165–6, 168, 170, 171, 172
cofactors 147, 148, 149, 160–1
combined transformations 153
completing the square 67–71
cones 83
conic sections 17–45, 94–7
    eccentricity 22–9
    equations of 18–22, 39
    focus-directrix properties 22–9
    loci problems 38–42
    tangents/normals 29–38, 39–40
coordinate systems 17–45, 94–7
    ellipses 18–20, 22–5, 28–33, 38–9
    hyperbolas 20–3, 26–8, 33–8, 39
    loci 38–42
    surfaces of revolution 83–5
cross product 101–10, 115–17

D
derivatives 47, 49, 50
determinant 142–6, 147–8
determinant form vectors 102–3
determinant method 102–4, 107–8
diagonal matrices 175–85
differentiation 46–53, 97
    equations of normals 30
    equations of tangents 29, 33–4
    hyperbolic functions 47–8
    inverse functions 49–52
dimension (of matrix) 138, 139
direction vectors 122
directrix (directrices) 22–9
distance 124–30
    between lines 126–7
    from point to line 128
    from point to plane 124–5
distributive law 102–3, 106
domain, restricted 7–8
dot product 101, 104, 175

E
eccentricity 22–9
eigenvalues/eigenvectors 165–85
ellipses 18–20, 22–5, 28–9
    loci 38–9
    tangents/normals 29–33, 38–9
equation (line) 115–17, 157–8, 171–2
    intersection of planes 122
    normals and tangents 29–40
    vector equation 115–17, 156, 161–2
equation (plane) 117–21, 157
exam practice 199–200
exponential form hyperbolic functions 2, 59–60

F
focus (foci) 22–9

G
geometrical problems 121–30
glossary 201–3
gradient 29, 30, 34, 36
graphs 4–8

H
hyperbolas 20–3, 26–8, 33–8, 39
hyperbolic functions 1–16, 46, 93–4, 97
    differentiation 47–8
    equations 12–14
    exponential form 2, 59–60
    graphs 4–7
    identities 10–14, 59, 62
    integration 55–67, 69, 70, 80, 84
    substitutions 61–7, 70, 80
    see also inverse hyperbolic functions

I
identities 10–14, 31, 59, 61–2, 74–5
identity matrix 138
images 152–3, 155–8
implicit differentiation 29, 30, 33
integration 54–92, 97–9
    by parts 71–2, 73–5, 84
    completing the square 67–71
    finding arc length 79–82
    finding surface area 82–7
    hyperbolic functions 55–67, 69, 70, 80, 84
    inverse functions 71–3
    quadratic surds 67–71
    reduction formulae 73–8
    standard integrals 55–7
    substitutions 61–7, 69–70, 80
    trigonometric functions 73, 75–6, 85
intersections 121–2, 158
invariant lines 171–2
inverse hyperbolic functions 7–10
    differentiation 49–50
    integration 61–3, 65, 69, 71–3
inverse matrices 146–52, 160–5
inverse trigonometric functions
    differentiation 50–2
    integration 55, 61–4, 69, 71–3

L
line of invariant points 172
linear transformations 152–65
lines
    angle with plane 122–3
    distances between/to 126–30
    intersections 121–2, 158
    transformations 156, 157–8, 161–2, 171–2
    see also equation (line)
loci 22, 38–42

M
major axis 24
matrices 137–90, 194–8
    determinant 142–6, 147–8
    eigenvalues/vectors 165–85
    inverse 146–52, 160–5
    linear transformations 152–65
    multiplication 139, 150, 154
    reducing symmetric to diagonal 175–85
    transpose 138–42, 147
minors 146–9, 160–1

N
natural logarithms 8–10, 12

INDEX

non-singular matrices 142, 149, 150
normalised eigenvectors 166, 167, 169–71, 175, 177–83
normalised vectors 166, 175
normals
    ellipses 29–33
    hyperbolas 34–5, 36–7, 39
    parabolas 39–40
    planes 117, 120, 123–4

O

origin, distance to plane 124–5
orthogonal matrices 175–83
Osborn's rule 11

P

parabolas 22–3, 39
    finding arc length 80
    loci problems 38, 39–40
parallel lines 126
parallel planes 125
parallel vectors 102
parallelepiped 111–12, 113–14
parallelograms 107, 108, 109–10
parametric differentiation 29
parametric equations 79, 80–1
    ellipses 18–19
    hyperbolas 21–2, 39
    parabolas 39
    surface of revolution 83, 85
parts, integration by 71–2, 73–5, 84
perpendicular distances 111, 124–5, 127
perpendicular gradient rule 29, 30, 34, 36
perpendicular vectors 101, 103, 117–18
planes 117–25, 157–8
    angles between/with 122–4
    distances to/from 124–5
    equations of 117–21, 157
    intersections of/with 121–2, 158
position vectors 20, 152–6, 160–1
    finding area/volume 106–10, 111–14
    point on a line/plane 115, 117–21
pyramids 112

Q

quadratic equations 3, 8–9, 30–1, 144
quadratic surds 67–71

R

reciprocal hyperbolic functions 2
rectangular hyperbolas 20, 39
reduction formulae 73–8
review questions 93–9, 191–8
    coordinate systems 42–4, 94–7
    differentiation 52–3, 97
    hyperbolic functions 14–15, 93–4, 97
    integration 87–90, 97–9
    matrices 185–8, 194–8
    vectors 130–4, 191–4
right-hand rule 101
roots 168, 170, 172–3

S

scalar (dot) product 101, 104, 175
scalar product form planes 117–19
scalar triple product 110–14
shortest distance 126–8
simultaneous equations 121, 162–3
singular matrices 142, 143–4, 146, 158
skew lines 127
standard integrals 55–7
substitution, integration by 61–7, 69–70, 80
surface of revolution 82–7
symmetric matrices 138–9, 175–85

T

tangents
    ellipses 29–33, 38–9
    hyperbolas 33–4, 35–8, 39
    parabolas 39
tetrahedrons 112–14
transformation matrices 152–65
transpose 138–42, 147, 175
triangles 106, 107–9
trigonometric functions 50–1, 73, 75–6, 85
    angle between lines/planes 122–4
    angle between vectors 101–2, 104, 111–12
    ellipses 18–19, 29–31, 38–9
    hyperbolas 34
    integration 73, 75–6, 85
    substitutions 61–7, 69
trigonometric identities 11, 31, 61–2, 74, 75

U

unit vectors 101–2, 103, 119, 124–5

V

vector equation (line) 115–17, 156, 161–2
vector equation (plane) 117–21, 157
vector product 101–10, 115–17
vectors 100–36, 191–4
    angle between 101–5, 111–12
    finding areas 106–10
    finding volumes 110, 111–14
    geometrical problems 121–30
    scalar triple product 110–14
    transformations 152–6, 160–1
volumes 110, 111–14